Agricultural Sustainability

Agricultural Sustainability

Progress and Prospects in Crop Research

Edited by

Gurbir S. Bhullar and Navreet K. Bhullar

Swiss Federal Institute of Technology
Zurich, Switzerland

AMSTERDAM • BOSTON • HEIDELBERG • LONDON
NEW YORK • OXFORD • PARIS • SAN DIEGO
SAN FRANCISCO • SINGAPORE • SYDNEY • TOKYO
Academic Press is an Imprint of Elsevier

Academic Press is an imprint of Elsevier
32 Jamestown Road, London NW1 7BY, UK
225 Wyman Street, Waltham, MA 02451, USA
525 B Street, Suite 1800, San Diego, CA 92101-4495, USA

Notice
No responsibility is assumed by the publisher for any injury and/or damage to persons or property as a matter of products liability, negligence or otherwise, or from any use or operation of any methods, products, instructions or ideas contained in the material herein. Because of rapid advances in the medical sciences, in particular, independent verification of diagnoses and drug dosages should be made

British Library Cataloguing-in-Publication Data
A catalogue record for this book is available from the British Library

Library of Congress Cataloging-in-Publication Data
A catalog record for this book is available from the Library of Congress

ISBN : 978-0-12-404560-6

For information on all Academic Press publications
visit our website at www.store.elsevier.com

Typeset by MPS Limited, Chennai, India
www.adi-mps.com

Printed and bound in United States of America

12 13 14 15   10 9 8 7 6 5 4 3 2 1

Working together to grow
libraries in developing countries

www.elsevier.com | www.bookaid.org | www.sabre.org

ELSEVIER   BOOK AID International   Sabre Foundation

Contents

Section I
Agricultural Biodiversity, Organic Farming, and New Crops

1. **Functional Agrobiodiversity: The Key to Sustainability?**
    Paolo Bàrberi

2. **Organic Agriculture—Driving Innovations in Crop Research**
    Dionys Forster, Noah Adamtey, Monika M. Messmer, Lukas Pfiffner, Brian Baker, Beate Huber, and Urs Niggli

3. Guar: An Industrial Crop from Marginal Farms

*N. A. Kuravadi, S. Verma, S. Pareek, P. Gahlot, S. Kumari, U. K. Tanwar,
P. Bhatele, M. Choudhary, K. S. Gill, V. Pruthi, S. K. Tripathi, K. S. Dhugga,
and G. S. Randhawa*

Section II
Effective Management of Resources (Nutrients and Water) and Crop Modelling

4. Nitrogen Use as a Component of Sustainable Crop Systems

Amritbir Riar and David Coventry

Section III
Molecular, Biotechnological, and Industrial Approaches for Enhancement of Crop Production and Quality

9. Biofortification of Staple Crops
Vishal Chugh and Harcharan S. Dhaliwal

10. Nutrient-focused Processing of Rice
Nadina Müller-Fischer

Section IV
Expert Advice on Policy and Developmental Aspects

Contributors

Noah Adamtey Research Institute of Organic Agriculture (FiBL), Frick, Switzerland

Mukhtar Ahmed Department of Agronomy, PMAS Arid Agriculture University Rawalpindi, Pakistan

Mustazar N. Akram Department of Agronomy, PMAS Arid Agriculture University Rawalpindi, Pakistan

Muhammad Asif Agricultural Food and Nutritional Science, University of Alberta, Canada

Paolo Bàrberi Institute of Life Sciences, Scuola Superiore Sant'Anna, Pisa, Italy

Brian Baker Research Institute of Organic Agriculture (FiBL), Frick, Switzerland

P. Bhatele Department of Biotechnology, Indian Institute of Technology Roorkee, India

Gurbir S. Bhullar Swiss Federal Institute of Technology, Zurich, Switzerland

D.S. Brar School of Agricultural Biotechnology, Punjab Agricultural University, Ludhiana, Punjab, India

M. Choudhary Department of Biotechnology, Indian Institute of Technology Roorkee, India

Vishal Chugh Akal School of Biotechnology, Eternal University, Sirmour, Himachal Pradesh, India

David Coventry School of Agriculture, Food and Wine, The University of Adelaide, Glen Osmond, Australia

Harcharan S. Dhaliwal Akal School of Biotechnology, Eternal University, Sirmour, Himachal Pradesh, India

K.S. Dhugga Trait Discovery & Technology, DuPont Pioneer, IA, USA

Dionys Forster Research Institute of Organic Agriculture (FiBL), Frick, Switzerland

P. Gahlot Department of Biotechnology, Indian Institute of Technology Roorkee, India

K.S. Gill Department of Crop & Soil Sciences, Washington State University, Pullman, WA, USA

Kabal S. Gill Smoky Applied Research and Demonstration Association, Falher, Alberta, Canada

Aakash Goyal Bayer Crop Science, Saskatoon, Saskatchewan, Canada

Arvind H. Hirani Department of Plant Science, University of Manitoba, Winnipeg, Canada

Beate Huber Research Institute of Organic Agriculture (FiBL), Frick, Switzerland

Elizabeth Humphreys International Rice Research Institute, Los Baños, Philippines

Shalu Jain Department of Plant Sciences, North Dakota State University, Fargo, ND, USA

S.S. Johl Central University of Punjab, Bathinda, Punjab, India

G.S. Khush University of California, Davis, CA, USA

Ajay Kumar Department of Plant Sciences, North Dakota State University, Fargo, ND, USA

S. Kumari Department of Biotechnology, Indian Institute of Technology Roorkee, India

N.A. Kuravadi Department of Biotechnology, Indian Institute of Technology Roorkee, India

Nadina Müller-Fischer Bühler, Uzwil, Switzerland

Sukhdev S. Malhi Agriculture and Agri-Food Canada, Melfort, Saskatchewan, Canada

Kevin McPhee Department of Plant Sciences, North Dakota State University, Fargo, ND, USA

Monika M. Messmer Research Institute of Organic Agriculture (FiBL), Frick, Switzerland

Urs Niggli Research Institute of Organic Agriculture (FiBL), Frick, Switzerland

S. Pareek Department of Biotechnology, Indian Institute of Technology Roorkee, India

Lukas Pfiffner Research Institute of Organic Agriculture (FiBL), Frick, Switzerland

V. Pruthi Department of Biotechnology, Indian Institute of Technology Roorkee, India

G.S. Randhawa Department of Biotechnology, Indian Institute of Technology Roorkee, India

Amritbir Riar School of Agriculture, Food and Wine, The University of Adelaide, Glen Osmond, Australia

Reyazul Rouf Mir Department of Plant Breeding and Genetics, SKAUST, Jammu, India

Tarlok S. Sahota Thunder Bay Agriculture Research Station, Thunder Bay, Ontario, Canada

Surinder Singh Kukal Punjab Agricultural University, Ludhiana, India

Balwinder-Singh International Rice Research Institute, Los Baños, Philippines

Ravinder Singh Department of Biotechnology, SKAUST, Jammu, India

U.K. Tanwar Department of Biotechnology, Indian Institute of Technology Roorkee, India

S.K. Tripathi Department of Water Resource Development and Management, Indian Institute of Technology Roorkee, India

S. Verma Department of Biotechnology, Indian Institute of Technology Roorkee, India

Sant S. Virmani Plant Breeding, Genetics and Biotechnology Division, International Rice Research Institute, Manila, Philippines

Sudhir-Yadav International Rice Research Institute, Los Baños, Philippines

With an increase in the frequency of occurrence of extreme weather events such as drought, flood, tsunami, and sea level rise, there is also increased volatility in the price of major staple grains in the international market. There was a big rise in the price of rice, wheat, and other cereals in 2008, as a result of which nearly a billion additional children, women, and men went to bed hungry. In 2012 again there is increased price volatility caused partly by the drought in North America and also by the use of corn for the production of ethanol. Hence, for sustainable food security, it is important that we have sustained production of adequate quantities of food grains. This will call for an "evergreen" revolution in agriculture leading to the improvement of productivity in perpetuity without associated ecological harm.

For a long time, sustainability was measured only in economic terms. After the 1972 Stockholm conference on the human environment, environmental parameters were also added to measure sustainability. Fifty years ago Rachel Carson, in her book *Silent Spring*, drew attention to the harmful effects of excessive use of pesticides. Also the loss of biodiversity resulted in some cases in genetic homogeneity in crops, thereby increasing genetic vulnerability to pests and diseases. This was clear during the potato famine of the 1840s in Ireland.

In addition to economic and environmental sustainability, social sustainability has also become important. With increasing emphasis on research for private profit rather than for public good, there will be social exclusion in access to technology depending on the purchasing power of the small farmer. The year 2014 has been declared by the UN as "International Year of Family Farming." The aim is to rekindle and sustain family farming around the world. In developing countries, farming is not only a way of life but a means to livelihood. Agriculture therefore will have to help in generating more income and more jobs, in addition to more food.

In the context outlined above, this book on agricultural sustainability, edited by Gurbir S. Bhullar and Navreet K. Bhullar, is a timely contribution. The book covers different aspects of sustainability in a holistic manner. It also shows how to improve the efficiency of the use of market-purchased inputs such as mineral fertilizers. Sustainable agriculture is the pathway to avoid price volatility and human suffering. I therefore hope that this book will be widely read and used by professionals and policy makers as well as farmers and farmdwellers. We owe a deep sense of gratitude to Gurbir and Navreet, as well as to the authors of the chapters, for their labor of love toward sustainable advances in agricultural productivity.

Prof. M. S. Swaminathan
Member of Parliament of India (*Rajya Sabha*)
Emeritus Chairman, M S Swaminathan Research Foundation

Provision of sufficient amounts of nutritious food for the ever-increasing global population is probably the largest challenge facing mankind. Despite a number of hunger eradication programs a large portion of the human population still remains undernourished. Land degradation and changes in land use patterns limit the area that could be brought under crop cultivation. Diminishing stocks of natural resources (fossil fuels and nutrients such as phosphorus) question the continuation of current agricultural practices, which depend heavily on high-energy inputs. The ongoing environmental changes are projected to seriously hamper agricultural production by increased frequency and intensity of extreme events such as drought and floods, more so in underprivileged parts of the world. Anthropogenic activities have not only contributed towards the climatic changes but have also resulted in degradation of natural resources (e.g., water and air pollution) and loss of biodiversity. Biodiversity losses—that affect a number of ecosystem services—are not only limited to natural habitats; with intensive monoculture farming on a large scale and use/misuse of cultivation and pest control practices, the agricultural landscape has also been deprived of a lot of diversity at species, varietal, and microbial scales. It is also noteworthy that, with changing food habits, we are increasingly shrinking the number of species from which we source a major portion of our food. For example, only 12 plants and five animal species currently contribute 75% of the world's food production; and 60% of plant-based calories and proteins are obtained from only three crops: namely, rice, maize, and wheat.

Agriculture being the primary anthropogenic activity for provision of basic needs for human beings, it is no surprise that agricultural sustainability is one of the most discussed subjects of our times. This book, *Agricultural Sustainability: Progress and Prospects in Crop Research*, presents the views of agricultural experts from across disciplinary and geographical boundaries. The 15 chapters—contributed by internationally recognized scientists from Europe, North America, Australia, and Asia—have been grouped into four distinct sections, each representing a crucial thematic area. The vast array of subject areas discussed in the book range from agrobiodiversity to biotechnology, from marginal crops to industrial approaches, from resource conservation to nutritional enhancement of crops and crop products, and from strengthening of human resources for agricultural research and development to economic and political priorities for effective production, marketing, and distribution of agricultural commodities. The authors of most of the chapters have experienced agricultural research and/or development both in developed and developing worlds and hence benefit from a wider vision in presenting a balanced view. As far as possible, the language of the chapters has been kept simple so that educated non-expert readers may enjoy reading and may benefit from the information provided herein. This book will serve as an educational tool for budding scientists, will provide a comprehensive overview for advanced researchers, and will lay guidelines for important policy decisions.

The Editors

Agricultural Biodiversity, Organic Farming, and New Crops

Functional Agrobiodiversity: The Key to Sustainability?

Paolo Bàrberi

Institute of Life Sciences, Scuola Superiore Sant'Anna, Pisa, Italy

1. INTRODUCTION

It is a common assumption that biodiversity has an important role to play in efforts to achieve agricultural sustainability (Altieri, 1995). However, the scientific literature is often unclear as to what type of "biodiversity" is necessary for agriculture. The problem results in part from lack of agreement in defining agrobiodiversity and related priorities.

Some scientists tend to identify biodiversity (species and habitats) conservation as the main priority for agriculture (e.g., Jeanneret et al., 2003). This is a dominant view in Europe, where it has resulted in considerable public financial support to farmers engaging in "agri-environmental schemes" (AESs) aimed at biodiversity conservation (European Commission, 2005). However, AESs have often failed, due to unclear definition of goals and management tools expected to meet them (Kleijn and Sutherland, 2003).

Other scientists tend to prioritize (agro)ecological functions aimed to optimize and/or stabilize agricultural production through increased agroecosystem biodiversity, e.g., crop rotation, intercropping, and cultural practices diversification (e.g., Malézieux et al., 2009). Priority is then given to species and habitats linked to production and related functions, e.g., soil fertility, biological pest control, and pollination (Altieri and Nicholls, 2004).

Indeed the concept of "agricultural sustainability" is also fuzzy, because priorities (and indicators) of environmental, economic, and social sustainability are continuously evolving (Pretty, 2008). In order to fully evaluate the potential of biodiversity to meet global agricultural challenges, we first need to define what is agricultural sustainability nowadays and then identify the elements of agricultural biodiversity that are more likely to help humans tackle those challenges. In the context of this chapter, the term "function" is considered to be a synonym of "service" and refers to processes that—either directly or indirectly—contribute to the provision of agricultural goods. It is worthwhile noticing that ecologists instead tend to distinguish between "functions" (self-regulating ecosystem processes) and "services" (processes providing material or immaterial outputs that are valued by humans). A thorough discussion on usage of these terms can be found in Jax (2005) and Violle et al. (2007).

2. AGRICULTURAL SUSTAINABILITY AT THE ONSET OF THE THIRD MILLENNIUM

Nowadays we can identify three major challenges to agricultural sustainability: (a) climate change, (b) energy availability, and (c) global economic insecurity.

There seems to be a general consensus on the likelihood of human (co) cause of global climate change (IPCC, 2007). The most common manifestation of climate change is the intensification of extreme climate events such as floods, droughts, and heat/cold waves that are jeopardizing agriculture in many areas around the world. Agricultural science and practice are asked to provide solutions to both mitigate the effects of climate change and increase adaptation of cropping/farming systems (Fleming and Vanclay, 2010).

Fossil fuels are becoming short in supply and increasingly inaccessible (Giampietro et al., 2012), with resulting increased prices. Biofuels have long been suggested as a possible solution to fossil fuel shortage, but their cost/benefit analyses and energy budgets do not seem to play for long-term sustainability, at least for some annual crops, e.g., corn (de Vries et al., 2010). Consequently, it seems wiser to reduce external energy use and increase energy efficiency in agriculture. This can be achieved through substitution of external inputs (seeds, fertilizers, and pesticides) with renewable and local resources (Altieri et al., 1983).

Like extreme climate events, price fluctuation of agricultural food commodities (e.g., wheat, corn, or rice) has been intensifying in the latest decade (Akram, 2009). Coupled with the current long-lasting global economic crisis, this is causing dire problems for many farmers and farmdwellers worldwide. Farmers producing for the global market are particularly vulnerable, because they are facing increasingly unpredictable market trends whilst cost of agricultural inputs is rising following the rise in oil price (Mitchell, 2008). Concurrently, recent phenomena like urban sprawl (Couch et al., 2007) and land grabbing (De Schutter, 2011) are eating up agricultural land. In the light of a continuously increasing world population, approaches and solutions to overcome these crises are urgently needed.

The emerging question is then: is increased biodiversity in agricultural practices and systems a possible solution to global challenges that jeopardize agricultural sustainability?

3. AGROBIODIVERSITY: A CONCEPTUAL FRAMEWORK

The United Nations Convention on Biological Diversity (CBD) and The Organisation for Economic Co-operation and Development (OECD) (Parris, 2001)

TABLE 1.1 Elements of Planned and Associated Agrobiodiversity Included (+) and Missing (−) in the OECD Definition (Parris, 2001), with Relevant Examples

Type of Agrobiodiversity	Level of Agrobiodiversity		
	Genetic	Species	Ecosystem
Planned	Crop cultivars and livestock breeds (+)	Crops included in the rotation, beneficial arthropods released for biological pest control (−)	Crop management practices, planted hedgerows, ratio of UAA[a] in the territory (+)
Associated	Genetic variation within wild plants and insects (−)	Wild plants and animals, endemic (non-released) beneficial arthropods (+)	Natural (non-planted) hedgerows and other non-cropped areas, e.g., natural woodland (+)

[a] UAA, utilized agricultural area.

define three levels of agricultural biodiversity (hereafter "agrobiodiversity"): genetic, species, and ecosystem. The elements of each level that are included in the OECD definition, as well as the missing ones, are summarized in Table 1.1.

3.1   Genetic Agrobiodiversity

Genetic agrobiodiversity refers to any variation in the nucleotides, genes, chromosomes, or whole genomes of organisms, i.e., it deals with within-species diversity. In the OECD definition, genetic agrobiodiversity includes variation within species of crops, livestock, and their wild relatives. Conservation and use of locally adapted cultivars of major crops, of neglected and underutilized crop species, and of livestock breeds has a value in itself because it contributes to save overall biodiversity. However, its main value is that a wider genetic pool in crops and livestock ensures adaptation to a changing environment and provides useful traits to be used in genetic breeding programs (Hajjar et al., 2008).

Genetic agrobiodiversity can be conserved *ex situ* or *in situ*. For plants, *ex situ* conservation is commonly done in seed banks under sterile and strictly controlled environmental conditions and protocols (Li and Pritchard, 2009). Considerable financial investments have recently been done in facilities hosting wide germplasm collections, e.g., the Kew Gardens' Millennium Seed Bank Partnership, UK (http://www.kew.org/science-conservation/save-seed-prosper/millennium-seed-bank/index.htm) or the Svalbard Global Seed Vault, Norway (http://www.regjeringen.no/en/dep/lmd/campain/svalbard-global-seed-vault. html?id=462220).

Despite the importance of global seed collections, it is being recognized that *in situ* methods offer a higher potential to ensure conservation and use of genotypes, because of direct involvement of farmers in the selection and maintenance processes (Altieri and Merrick, 1987). Furthermore, *in situ* methods make conservation of genetic agrobiodiversity also accessible to developing countries, due to reduced facilities costs (Jarvis et al., 2000).

3.2   Species Agrobiodiversity

For the OECD, species agrobiodiversity includes the variation between *wild* species that are directly or indirectly relevant to agriculture (Table 2.1). These are grouped in three categories: (a) species supporting agricultural production; (b) wild species depending directly or indirectly on agriculture and its effects, and (c) non-native species threatening agroecosystems.

Category (a) includes species guilds sustaining agricultural production through their effects on agroecosystem functions like soil fertility, pollination, and biological pest control. Examples are soil microorganisms involved in the organic matter cycle, arbuscular-mycorrhyzal fungi, earthworms, natural enemies of crop pests, and pollinators (Parris, 2001). This category only includes species exerting a positive function in the agroecosystem.

Category (b) includes wild species like bats, birds, or rodents that are directly dependent on agroecosystems for their survival and reproduction. Two thirds of European endangered or vulnerable bird species live exclusively in agroecosystems (Tucker and Heath, 1994), therefore adequate farming practices are essential for their conservation (European Environment Agency, 2004). Category (b) also includes marine or fluvial species that indirectly depend on agricultural activities, e.g., because they suffer from agricultural pollution due to fertilizers or pesticides runoff. Regarding their effects on agroecosystem functions, category (b) species can be considered as neutral.

Category (c) includes exotic species that can directly threaten agricultural production, e.g., newly introduced pests, diseases, and weeds. This is a worldwide problem, exacerbated by the development of global trade (Bright, 1999). Recent introductions of highly noxious organisms are, for example: *Tuta absoluta* Meyrick, a Lepidoptera pest of Mediterranean tomato crops native to South America (Desneux et al., 2010); wheat stem rust (*Puccinia graminis* f. sp. *tritici* Eriks. E. Henn.) race Ug99, spreading from Uganda to East Africa, the Middle East, and Asia (Singh et al., 2008); and *Commelina benghalensis* L., a creeping herb native to Africa and Asia that is invading vast pasture and cropland areas in southern USA (Webster et al., 2005). Clearly, all species in category (c) exert a negative function in the agroecosystem.

3.3   Ecosystem Agrobiodiversity

The OECD definition of ecosystem diversity embraces three components: (a) the diversity in farming systems and cultural practices and their change in time and space, (b) the ratio between land utilized for agriculture and for other uses (e.g., natural or urban areas), and (c) the interactions between agroecosystems and nearby ecosystems (Parris, 2001).

It must be noticed that this definition extends well beyond biodiversity *per se*, including elements of agroecosystem structure and management. This is very much in line with a functional approach to agrobiodiversity, as envisaged in agroecology (Altieri and Nicholls, 2004).

3.4   Limitations of the OECD Definition of Agrobiodiversity

Many authors distinguish between *planned* agrobiodiversity (the biodiversity elements—at gene, species, or ecosystem level—deliberately introduced in the agroecosystem) and *associated* agrobiodiversity (the biodiversity elements—at any level—inhabiting the agroecosystem without being introduced) (Jackson et al., 2007). If we examine the OECD agrobiodiversity definition in the light of planned and associated biodiversity at any of the three levels (Table 2.1), we notice that all combinations "planned/associated biodiversity × level" are taken into account except two: associated agrobiodiversity at gene level and planned agrobiodiversity at species level.

Associated agrobiodiversity at gene level includes, e.g., the genetic variation within populations of weeds, crop pests, and diseases and of natural (endemic) enemies of crop pests, which is important to determine the level and extent of their interactions (Crutsinger et al., 2008). Planned agrobiodiversity at the species level includes, e.g., the diversity of crops grown in rotation and cover crops introduced for various purposes (soil fertility building, weed suppression, attraction of beneficial arthropods and/or repulsion of crop pests, etc.). These are very important *functional* tools to increase crop performance in the framework of external input reduction, and hence they are likely to contribute to agroecosystem sustainability (Tilman et al., 2002). The reasons why these two categories have not been included in the OECD definition is unknown; the latter, especially, is an important component of the arsenal of biodiversity tools available to farmers and land managers.

Additionally, there is no reason to focus only on the diversity of non-native (exotic) pests, diseases, and weeds—category (c) of species agrobiodiversity—since native biotic stressors can be more detrimental than newly introduced ones (Gressel, 2006).

Lastly, the OECD definition does not mention elements that can exert multiple functions, positive or negative depending on the context. For example, *Rubus fruticosus* L. (blackberry) can be a crop, an invasive weed, a windbreak, a hedgerow supporting both pests and beneficial arthropods, and a pleasant or unpleasant landscape element depending on varying human perceptions in different environments (Moonen and Bàrberi, 2008). It is then clear that functions associated with agrobiodiversity components are strictly context-dependent, and thus definition of the context (agroecosystem) and its priorities is a fundamental step in agrobiodiversity evaluation.

4. FROM AGROBIODIVERSITY TO FUNCTIONAL AGROBIODIVERSITY

The OECD definition of agrobiodiversity already refers to biodiversity components (e.g., species supporting agricultural production) that are particularly relevant to the functionality of agroecosystems. However, there is neither a clear definition of functional agrobiodiversity nor an indication of which functions should have priority in agroecosystems. This is an issue to be clarified if agrobiodiversity is to become a key component of sustainable cropping/farming systems.

4.1 Functional Biodiversity: A Plethora of Definitions

From the analysis of the scientific literature it emerges that there is no commonly accepted definition of functional biodiversity. As already said, this is due partly to lack of clear objectives—see Moonen and Bàrberi (2008) for in-depth discussion—and partly to the different views and priorities set forth by ecologists, agroecologists, and agronomists. A sample of the different definitions is reported hereafter.

A classical ecological definition is, e.g., that of Pearce and Moran (1994), who define "functional diversity" simply as the relative abundance of organisms expressing different functions. There is reference neither to the role that these organisms exert in an ecosystem nor to the positive or negative functions they are likely to influence. Although formally correct, this definition is not useful for agroecosystems, whose objectives and priorities (production, to start with) are usually clear.

The term "multi-function agricultural biodiversity" has been used, after Gurr et al. (2003), to indicate the positive "domino effect" observed between plant diversity and biological pest control at multiple spatial scales. Plant diversity has been seen to improve biological pest control at field scale, which in turn improves the pest control function at landscape scale. The magnitude of this effect depended on agroecosystem diversity at either scales. The value of this definition is that it recognizes the importance of biodiversity-driven interactions across different trophic levels and spatial scales, but it only focuses on one function and on positive plant–insect interactions. Indeed, Conservation Biological Control—i.e., the maintenance or (re)introduction of (semi)natural habitats in agroecosystems to support native populations of biological control agents (Barbosa, 1998)—is a synonym of functional biodiversity for many entomologists (see, e.g., the IOBC Working Group "Landscape management for functional biodiversity": http://www.iobc-wprs.org/expert_groups/19_wg_landscape_management.html).

Peeters et al. (cited in Clergue et al., 2005) distinguished three types of agrobiodiversity: (1) agrobiodiversity *sensu stricto*, i.e., the diversity of organisms *directly* useful for production (crops, varieties, and livestock species and breeds); (2) para-agrobiodiversity (also called "functional biodiversity"), i.e., the diversity of organisms *indirectly* useful for production (e.g., soil microorganisms, beneficial arthropods, unsown grassland plants), corresponding to category (a) of species agrobiodiversity in the OECD definition; (3) extra-agricultural biodiversity, i.e., all biodiversity present in an agroecosystem which is unrelated to production (e.g., wild species of plants and animals that are not providing a function). Peeters et al.'s types are interesting because they clearly relate agrobiodiversity to the main agroecosystem function (production), but they do not clearly address the three OECD/CBD levels and do not incorporate species exerting a negative function.

Instead, a comprehensive and objective evaluation of the effects of biodiversity should include the positive as well as the neutral and negative functions exerted by agroecosystem components at any of the three levels (gene, species, and ecosystem). A fine-tuned definition of functional (agro) biodiversity could then be "that part of the total biodiversity composed of clusters of elements (at the gene, species, or habitat level) providing the same (agro)ecosystem service, that is driven by within-cluster diversity" (Moonen and Bàrberi, 2008). Its aims and consequences for agrobiodiversity evaluation are illustrated below.

4.2   Functional Agrobiodiversity: A Methodological Approach

If the goal of functional biodiversity study is to understand which components can help to improve crop production and thereby agricultural sustainability, the analysis should encompass four subsequent steps, summarized in Table 1.2.

First, the operational context and the related objectives must be clearly defined: this includes the description of the agroecosystem and its goals, which may differ between, e.g., conventional and organic management of the same crop.

Second, one should list the agroecosystem functions that are deemed a priority in a given context. For example, in olive (*Olea europaea* L.) groves, one of the major problems is olive fly (*Bactrocera oleae* Rossi) control. As such, biological pest control should be a priority function for the study and application of functional agrobiodiversity in olive.

Third, the "agroecosystem functional group" (the "cluster" in Moonen and Bàrberi's definition) comprising all elements (at gene, species, and ecosystem level) that are relevant for the target function in the given context should be defined. This group will be the subject of the functional agrobiodiversity

TABLE 1.2 Steps to be Included in a Functional Agrobiodiversity Analysis, with Relevant Examples

Step No.	Description	Example
1	Definition of the context (target agroecosystem) and related objectives	Olive grove with organic management
2	Definition of priority agroecosystem functions	Control of the olive fly
3	Definition of the agroecosystem functional group	Species of parasitoids and hyper-parasitoids of the olive fly, wild plant species and structures (e.g., hedgerows, woodland) attracting natural enemies of the olive fly, olive cultural practices known to affect olive fly (e.g., cultivar type, pruning, types and amount of natural pesticides sprayed)
4	Definition of space and time boundaries for the study of the agroecosystem functional group and of pertinent indicators	Field and landscape scale, whole year, number of fruits with fly punctures (sample), number of parasitized fruits and natural enemies species (sample), presence and abundance (e.g., percent area cover) of wild plants and structures supporting natural enemies of the olive fly, details (e.g., type and rates of active ingredients used) of cultural practices known to affect the olive fly (see step 3)

Modified from Moonen and Bàrberi (2008).

analysis. In the olive fly case, the agroecosystem functional group would, e.g., include the guilds of parasitoids and hyper-parasitoids potentially able to keep the pest under control, plant species (other than olive) supporting the complex of beneficial arthropods, and (semi)natural areas (e.g., woodland and hedgerows) ecologically important for them. It must be stressed that, in agreement with the OECD definition of ecosystem agrobiodiversity, agricultural management is very much part of an agroecosystem functional group. If one is able to identify the management elements (e.g., mowing, pruning, fertilizer and pesticides application, as well as their details—timing, rates, etc.) that are likely to enhance the function by favoring the components of the agroecosystem functional group, these would form a "management functional group" (Moonen and Bàrberi, 2008).

Fourth, the agroecosystem functional group should be studied by selecting the most pertinent indicators, level(s) (*sensu* CBD/OECD) and spatio-temporal scale(s). The methodological details are determined by the type and extent of the ecological interactions occurring among the agroecosystem functional groups components (Bàrberi et al., 2010). Here, it must be pointed out that the ultimate goal of this study is to determine whether or not *diversity* within the functional group is important for the fulfillment of the function. For example, would the presence of three species of aphid predators instead of just one increase the biological pest control function (i.e., aphid predation)? If the answer is yes, the conclusion is that diversity in the agroecosystem functional group matters, thus functional biodiversity helps. If the answer is no, the conclusion is that the *identity* ("biofunctionality") of the functional group components (in this case the only predator species present) is more important than the diversity within the functional group, thus functional biodiversity does not help. However, it should not be neglected that, according to the "insurance hypothesis" (Yachi and Loreau, 1999), a higher level of biodiversity insures (agro)ecosystems against declines in their functioning because the presence of many species guarantee that some will maintain the function if others disappear. This is particularly relevant in the presence of the major challenges to agricultural sustainability outlined in this chapter.

5.   FUNCTIONAL AGROBIODIVERSITY IN PRACTICE

In a cropping system, once the priority functions have been selected, one important question to be answered is: what are the traits that crops should possess to express these functions at best? Similarly to the use of the term "function", use of the term "functional trait" is somewhat different between ecologists (see, e.g., Garnier and Navas, 2012) and agroecologists. For consistency, the term "functional trait" is used here to indicate any characteristics of a crop that might favor the expression of an agronomically important function: e.g., crop production, crop nutrition, or weed suppression. Similar to what was said before, these functions can be accomplished either by a single, very effective functional trait or by a suite of functional traits, thus highlighting the importance of functional trait diversity. The second option is more promising because it is in accordance with the aforementioned insurance hypothesis.

Some case studies illustrating how the concept of functional agrobiodiversity can be turned into practice through selection of suitable functional traits are reported hereafter.

5.1   Genetic Agrobiodiversity

Modern cultivars and hybrids are characterized by a narrow genetic base, carrying a combination of traits which are primarily aimed at increasing yield. However, their yield potential can only be expressed in an optimized environment created by continuous assistance from the farmer, with large supply of external inputs (water, fertilizers, and pesticides). As such, these cultivars do not seem well equipped to withstand the challenge implied by the necessity to reduce energy (input) consumption and to adapt to harsher environments consequent to global climate change (Mendelsohn and Dinar, 1999).

Crop genetic diversity can be increased by sowing cultivar mixtures instead of a single, high-yielding cultivar. The cultivars in the mixture should differ in their functional traits: e.g., resistance or tolerance to diseases or the ability to exploit unlimited (light) and limited (soil nutrients and water) resources. The assumption is that a more diverse crop stand should better cope with environmental variation consequent to climate and/or management conditions. Cultivar mixtures have been shown, e.g., to reduce incidence of rusts and powdery mildews in small grain crops (Mundt, 2002) and of late blight in potato (Garrett et al., 2001).

One step beyond cultivar mixtures is the use of Composite Cross Populations (CCPs) in small grain cereals like wheat (*Triticum* spp.) or barley (*Hordeum vulgare* L.). CCPs originate from a high number of initial crosses between several cultivars, grown in the target environment. Every year part of the grain is saved and sown in the next season, giving rise to a composite population whose traits co-evolve with local climate and management conditions, upon the concept of "evolutionary breeding" (Wolfe et al., 2008). Seven or eight years are usually enough to segregate the functional traits that are best "collectively" adapted to the target environment. Due to their very high genetic diversity, composite populations will be able to withstand high fluctuations in climate or suboptimal management conditions, which make them a well suited genotype for adaptation to climate change or to low input and organic systems. CCPs are often the outcome of participatory plant breeding schemes, an example of *in situ* conservation of genetic diversity seeking the highest involvement of local farmers and other stakeholders (Pimbert, 2011).

5.2   Species Agrobiodiversity

Optimum design of crop rotation is the most common way to exploit species agrobiodiversity to increase agroecosystem sustainability. The benefits of crop rotation are manifold and well known (Karlen et al., 1994). In the context of low-input and organic agriculture, functional traits of component crops in the rotation as well as their functional interactions are particularly important because they are expected to surrogate (part of) external inputs while maintaining adequate levels of crop production and related agroecosystem functions. In temperate organic arable cropping systems, increasing species agrobiodiversity through the introduction of winter cover crops (green manures) is an important tool to increase agroecosystem functioning and hence sustainability. In a long-term experiment, Mazzoncini et al. (2010) compared organic and conventional management of a five-year arable crop rotation where the organic cropping system included a red clover (*Trifolium pratense* L.) green manure crop not included in the conventional system. Five years after the onset of the experiment the organic system already showed higher soil carbon sequestration

(+22%) and potentially mineralizable C (+9%) than the conventional system. In contrast, mites/collembolans ratio was nearly double in the conventional system, likely a consequence of more frequent soil tillage disturbance (mechanical weed control) in the organic system. These findings suggest that although higher species agrobiodiversity in crop rotations has the potential to improve the soil fertility function in a relatively short period of time, some of the accompanying management practices can partially overcome the gained benefits. A current major challenge in organic farming research is to try to optimize cropping systems functionality and sustainability by combining cover crops with conservation tillage techniques (including no-till) without jeopardizing crop yield (Peigné et al., 2007).

Intercropping and living mulches (i.e., systems where a non-cash crop is grown alongside a cash crop) are alternative management tools to improve agroecosystem sustainability through increased species agrobiodiversity. An analysis of the various positive effects ascribed to intercropping can be found, e.g., in Coolman and Hoyt (1993). In both intercropping and living mulch systems the key point is to combine crops that are complementary in the use of environmental resources (light, soil nutrients, and water), to optimize it and to minimize competitive effects between companion crops. As such, they should clearly differ in, e.g., height, growth habit, and root architecture, i.e., in those functional traits ultimately resulting in, e.g., increased (overall) yield, soil fertility, crop nutrition, and weed suppression (Vandermeer, 1989). Intercropping and living mulches are clear examples of cultural practices that should be based on the concept of functional agrobiodiversity to be successful.

One example of intercropping which is common in part of the Horn of Africa is 'hanfets', i.e., a mixture of wheat and barley. Woldeamlak et al. (2008) tested 16 hanfets constituted by all combinations of four barley landraces and four wheat (two landraces and two varieties) at three locations in Eritrea, where farmers (both men and women) were involved in plant selection. Hanfets grain yield was on average similar to that of pure barley but significantly higher than that of wheat. The Land Equivalent Ratio (LER) did not differ between hanfets types but was on average over 50% greater than in the pure crops, showing a clear advantage of the diversified system. Stability analysis showed that the most stable entries always included some hanfets, although not all of them were more stable than pure crops. Both men and women selected for traits like high grain yield, earliness, short heads, low kernel weight, and short-statured plants, and seemed to prefer hanfets in which both components were early heading and maturing. It has been hypothesized that differences in root architecture among component genotypes might make hanfets more efficient in exploiting soil water resources than pure crops, a functional trait which is of utmost importance in a geographical area with chronic water shortage and famine. It is worthwhile noticing that hanfets is a cropping system which is based on the concurrent application of genetic and species agrobiodiversity.

5.3   Ecosystem Agrobiodiversity

By their own nature, the management tools ascribable to ecosystem agrobio-diversity often have an effect which goes beyond the farm gate (Bàrberi et al., 2010; Gabriel et al., 2010), and therefore should be tackled by consortia of neighboring farmers rather than by individual farmers, except where the farm size is wide enough to encircle any possible ecological interactions occurring among the elements in the agroecosystem functional group.

Besides Conservation Biological Control, another way to exploit ecosys-tem agrobiodiversity to improve the pest control function is via the use of "push-pull" strategies. These involve the behavioral manipulation of insect pests and their natural enemies via the integration of stimuli that make crops unattractive or unsuitable to the pests ("push") while luring them toward an attractive source ("pull") from where they are subsequently removed (Cook et al., 2007). Naturally generated plant stimuli can be exploited using vegeta-tion diversification. This includes tools pertaining to both species agrobio-diversity (e.g., intercropping and trap cropping; see, e.g., Finch and Collier, 2000) and ecosystem agrobiodiversity (e.g., flower strips, grass strips, or other measures targeting the field margin rather than the field itself; see, e.g., Olson and Wäckers, 2007). The "push" tools act through visual distraction and production of semiochemicals like non-host volatiles, anti-aggregation or alarm pheromones, oviposition deterrents, and antifeedants. Instead, the "pull" tools provide visual stimulants, host volatiles, aggregation or sex pheromones, and oviposition or gustatory stimulants (Cook et al., 2007).

Field Margin Complexes (FMCs) comprise all structural elements in the space between crop outer rows and the adjacent field (Moonen et al., 2006). FMCs can be simple (e.g., a barbed wire fence or a ditch) or complex, includ-ing different layers/types of structure (e.g., ditches, grass or flower strips, headlands, hedgerows, different width/height), vegetation (e.g., grasses, forbs, shrubs, trees), and management (e.g., pruning, mowing, spraying) (Greaves and Marshall, 1987). Moonen et al. (2006) tested the hypothesis that more structurally complex FMCs (i.e., higher ecosystem agrobiodiversity) would improve the weed and pest control functions, by reducing the number of weed propagules invading the field from the margin and creating habitats more suit-able for natural enemies of insect pests, respectively. Structure, vegetation, and management diversity was analyzed in 62 FMCs in an organic arable farm and related to the presence of beneficial arthropods (Coccinellidae, Syrphidae, and Chrysopidae, representing the agroecosystem functional group) known to potentially control aphids in the study area. FMC diversity was positively cor-related with plant species richness in the margins ($r = 0.35$, $P = 0.005**$), which in turn was negatively correlated with the percentage of weeds ("weed-iness") present in the margins ($r = -0.47$, $P = 0.0001***$). However, FMC weediness was strongly positively correlated with the presence of aphid natu-ral enemies ($r = 0.93$, $P = 0.002**$), likely because they need weeds for part of their life cycle, e.g., as overwintering or oviposition habitat. These findings show that the same element of ecosystem agrobiodiversity may cause conflicts between two functions. In this situation, one must define which of the two functions is more important. In the previous case study, if biological pest control is considered high priority, for example, it is advisable to reduce FMC structural complexity, since this would promote higher field margin weediness which would favor natural enemies' populations. In that case, however, farmers should accept a higher risk of weed invasion from the margin to the field.

6. FUNCTIONAL AGROBIODIVERSITY: OPPORTUNITIES AND BOTTLENECKS

The previous case studies show that functional agrobiodiversity has potential to improve agroecosystem sustainability and that this potential is higher when genetic, species, and ecosystem agrobiodiversity are combined in novel cropping systems. However, despite scientific evidence, actual large-scale adoption of functional agrobiodiversity will only be possible if global policy, science, and public opinion objectives converge. Opportunities and threats for functional agrobiodiversity in the foreseeable future are discussed below.

6.1 What Could Favor Functional Agrobiodiversity?

In the past 20 years, several international treaties and policies developed at continental or regional scales have set the ground for the conservation and valorization of (agro)biodiversity, a cause and consequence of increased public opinion awareness. The starting point has been the 1992 Convention on Biological Diversity, which includes a Thematic Programme on Agricultural Biodiversity (www.cbd.int/agro). Regarding genetic agrobiodiversity, one important milestone has been the promulgation (2004) of the FAO-supported International Treaty on Plant Genetic Resources for Food and Agriculture (www.planttreaty.org), which aims to favor plant germplasm exchange and equitable sharing of benefits deriving from its use among signatory states, to reach worldwide food security. A synopsis of policy developments to support genetic agrobiodiversity can be found in Bragdon et al. (2009). Regarding common agricultural policies, in October 2009 the European Union released a Framework Directive on the Sustainable Use of Pesticides (EU, 2009), binding all Member States to use Integrated Pest Management (IPM) as the reference agricultural management approach for all farmers. In Annex III, the Directive indicates eight IPM principles which must be taken up in EU agriculture from January 1, 2014; these principles clearly state that diversity in crop rotations and associated cultural practices should be a common goal. This calls upon the importance of (functional) agrobiodiversity for the design of future EU cropping systems, even though the Directive does not mention it explicitly.

In the scientific arena there is increasing awareness of the importance of agrobiodiversity for sustainable agriculture. International networks bringing together scientists from various parts of the world are expanding—see, e.g., the Agrobiodiversity cross-cutting network of the Diversitas platform (www.agrobiodiversity-diversitas.org), and the Platform for Agrobiodiversity Research (PAR, http://agrobiodiversityplatform.org). It is worth noticing that the concept of functional agrobiodiversity (as illustrated in this chapter) is gaining ground—see, e.g., the recent establishment of the European Learning Network on Functional Agro Biodiversity (ELN-FAB, www.eln-fab.eu). Basic and applied research on the functional value of genetic, species, and habitat agrobiodiversity is in progress, often supported by substantial public funding (see, e.g., http://cordis.europa.eu).

6.2 What Could Hinder Functional Agrobiodiversity?

Despite support from policy and public opinion, current global agreements on international trade and intellectual property rights (Correa, 2007) play against widespread adoption of functional agrobiodiversity because they are tailored to the needs of standardized agricultural and food systems (Gonzalez, 2002). In particular, there are strong arguments against farmers' reuse of seeds saved from the previous harvest because this would violate international regulations on seed royalties (Howard, 2009). As such, adoption of innovative solutions based on increased genetic agrobiodiversity (e.g. CCPs) might be hindered by lack of regulatory support. Another issue possibly limiting adoption of cultivar mixtures and CCPs is that they provide produce that is different every year. Does this comply with the requirements and expectations of the food processing industry and of consumers? To date, due to the structure of agricultural trade systems, diversified solutions seem better suited to small-scale farming and local markets attended by consumers well aware of what they buy and why they buy it (Wolfe et al., 2008). From the viewpoint of agricultural management, a major question is whether or not large-scale farms can afford the higher degree of attention, time, and consequent labor costs implied by agroecological solutions. In fact, larger-scale organic farms tend to be based on a substitution approach (use of non-synthetic inputs instead of synthetic ones) without really implementing agroecological solutions, likely because they seek scale economies and standardization of produce and consequently of cultural practices (Best, 2008). The same reasons hinder large-scale introduction of ecosystem agrobiodiversity such as hedgerows or other non-cropped areas, unless they are supported by public funding, or research outcome can clearly be translated into operational guidelines (Kleijn and Sutherland, 2003).

Although research on the relationships between agriculture and biodiversity is progressing, lack of a commonly accepted definition of agrobiodiversity and, to some extent, of agroecology (Wezel et al., 2009) slows down the pace of research findings and their implementation in actual cropping/farming/agricultural systems. The very nature of agroecology requires long-term experiments, multidisciplinary collaboration (hence open-minded scientists), and studies across multiple time and spatial scales (Francis et al., 2008). This extends time to publication of results, which is against present expectations from scientists. In perspective, this might jeopardize public funding to agroecological research (Vanloqueren and Baret, 2009).

7. CONCLUSIONS

Functional agrobiodiversity can provide practical solutions to help shape agro-ecosystems of the future, but needs time to provide evidence through complex studies carried out beyond the disciplinary boundaries that still characterize science. As well as agroecology, functional agrobiodiversity does not provide standardized solutions but rather an approach to the development of these solutions locally. Society needs to understand that the pace of economic systems would need to be reconciled with that of biological systems, otherwise biodiversity-driven solutions, although they might work at a local scale, would never be instrumental in solving global (agri)environmental problems.

ACKNOWLEDGEMENTS

Part of this work has been carried out with the support of the EU-funded Projects SOLIBAM (Grant agreement no. FP7-KBBE 245058) and OSCAR (Grant agreement no. FP7-KBBE 289277) and of the ERA-NET Core Organic II Project TILMAN-ORG.

REFERENCES

Akram, Q.F., 2009. Commodity prices, interest rates and the dollar. Energy Econ. 31, 838–851.

Altieri, M.A. (Ed.), 1995. Agroecology: The Science of Sustainable Agriculture. Springer, Netherlands, Dordrecht.

Altieri, M.A., Letourneau, D.K., Davis, J.R., 1983. Developing sustainable agroecosystems. Biosci. 33, 45–49.

Altieri, M.A., Merrick, L.C., 1987. *In situ* conservation of crop genetic resources through maintenance of traditional farming systems. Econ. Bot. 41, 86–96.

Altieri, M.A., Nicholls, C.I., 2004. Biodiversity and Pest Management in Agroecosystems, Second ed. The Haworth Press, New York.

Bàrberi, P., Burgio, G., Dinelli, G., Moonen, A.C., Otto, S., Vazzana, C., et al., 2010. Functional biodiversity in the agricultural landscape: relationships between weeds and arthropod fauna. Weed Res. 50, 388–401.

Barbosa, P. (Ed.), 1998. Conservation Biological Control Academic Press, San Diego.

Best, H., 2008. Organic agriculture and the conventionalization hypothesis: a case study from West Germany. Agric. Human Values. 25, 95–106.

Bragdon, S., Jarvis, D.I., Gauchan, D., Mar, I., Hue, N.N., Balma, D., et al., 2009. The agricultural biodiversity policy development process: exploring means of policy development to support the on-farm management of crop genetic diversity. Int. J. Biodivers. Sci & Manage. 5, 10–20.

Bright, C., 1999. Invasive species: pathogens of globalization. Foreign Policy. 116 (50–60 + 62–64).

Clergue, B., Amiaud, B., Pervanchon, F., Lasserre-Joulin, F., Plantureux, S., 2005. Biodiversity: function and assessment in agricultural areas – a review. Agron. Sustainable Dev. 25, 1–15.

Cook, S.M., Khan, Z.R., Pickett, J.A., 2007. The use of push-pull strategies in integrated pest management. Annu. Rev. Entomol. 52, 375–400.

Coolman, R.M., Hoyt, G.D., 1993. Increasing sustainability by intercropping. HortTechnology. 3, 309–312.

Correa, C., 2007. Trade Related Aspects of Intellectual Property Rights. A Commentary on the TRIPS Agreement. Oxford University Press, Oxford.

Couch, C., Petschel-Held, G., Leontidou, L. (Eds.), 2007. Urban Sprawl in Europe: Landscape, Land-Use Change and Policy. Wiley-Blackwell, Oxford.

Crutsinger, G.M., Collins, M.D., Fordyce, J.A., Sanders, N.J., 2008. Temporal dynamics in non-additive responses of arthropods to host-plant genotypic diversity. Oikos. 117, 255–264.

De Schutter, O., 2011. How not to think of land-grabbing: three critiques of large-scale investments in farmland. J. Peasant Stud. 38, 249–279.

Desneux, N., Wajnberg, E., Wyckhuys, K.A.G., Burgio, G., Arpaia, S., Narváez-Vasquez, C.A., et al., 2010. Biological invasion of European tomato crops by *Tuta absoluta*: ecology, geographic expansion and prospects for biological control. J. Pest. Sci. 83, 197–215.

de Vries, S.C., van de Ven, G.W.J., van Ittersum, M.K., Giller, K.E., 2010. Resource use efficiency and environmental performance of nine major biofuel crops, processed by first-generation conversion techniques. Biomass Bioenergy. 34, 588–601.

EU, 2009. Directive 2009/128/Ec of the European Parliament and of the Council of 21 October 2009 establishing a framework for community action to achieve the sustainable use of pesticides. Official Journal of the European Union, 16. (24.11.2009).

European Commission (2005). Agri-Environment Measures – Overview on General Principles, Types of Measures, and Application. European Commission, DG for Agriculture and Rural Development, p. 24.

European Environment Agency, 2004. High Nature Value Farmland – Characteristics, Trends, and Policy Challenges. European Environment Agency, Copenhagen.

Finch, S., Collier, R.H., 2000. Host-plant selection by insects – a theory based on 'appropriate/inappropriate landings' by pest insects of cruciferous plants. Entomol. Exp. Appl. 96, 91–102.

Fleming, A., Vanclay, F., 2010. Farmer responses to climate change and sustainable agriculture. A review. Agron. Sustainable Dev. 30, 11–20.

Francis, C.A., Lieblein, G., Breland, T.A., Salomonsson, L., Geber, U., Sriskandarajah, N., et al., 2008. Transdisciplinary research for a sustainable agriculture and food sector. Agron. J. 100, 771–776.

Gabriel, D., Sait, S.M., Hodgson, J.A., Schmutz, U., Kunin, W.E., Benton, T.G., 2010. Scale matters: the impact of organic farming on biodiversity at different spatial scales. Ecol. Lett. 13, 858–869.

Garnier, E., Navas, M.L., 2012. A trait-based approach to comparative functional plant ecology: concepts, methods and applications for agroecology. A review. Agron. Sustainable Dev. 32, 365–399.

Garrett, K.A., Nelson, R.J., Mundt, C.C., Chacón, G., Jaramillo, R.E., Forbes, G.A., 2001. The effects of host diversity and other management components on epidemics of potato late blight in the humid highland Tropics. Phytopathology 91, 993–1000.

Giampietro, M., Ramos-Martin, J., Ulgiati, S., 2012. Can we break the addiction to fossil energy? Energy 37, 2–4.

Gonzalez, C.G., 2002. Institutionalizing inequality: the WTO Agreement on Agriculture, food security, and developing countries. Columbia. J. Environ. Law. 27, 433–489.

Greaves, M.P., Marshall, E.J.P., 1987. Field margins: definitions and statistics. In: Way, J.M., Greig-Smith, P.J. (Eds.), Field Margins British Crop Protection Council, Thornton Heath, pp. 3–10. (Monograph No. 35).

Gressel, J. (2006). Native species are the most intractable evolving agricultural weeds. In "Proceedings International Symposium on Intractable Weeds and Plant Invaders", Ponta Delgada, Azores, Portugal, 17–21 July, 10–11.

Gurr, G.M., Wratten, S.D., Luna, J.M., 2003. Multi-function agricultural biodiversity: pest management and other benefits. Basic. Appl. Ecol. 4, 107–116.

Hajjar, R., Jarvis, D.I., Gemmill-Herren, B., 2008. The utility of crop genetic diversity in maintaining ecosystem services. Agric., Ecosyst. Environ. 123, 261–270.

Howard, P.H., 2009. Visualizing consolidation in the global seed industry: 1996–2008. Sustainability 1, 1266–1287.

IPCC, 2007. In: Allali, A., Bojariu, R., Diaz, S., Elgizouli, I., Griggs, D., Hawkins, D. (Eds.), Climate Change 2007: Synthesis Report Intergovernmental Panel on Climate Change, Geneva.

Jackson, L.E., Pascual, U., Hodgkin, T., 2007. Utilizing and conserving agrobiodiversity in agricultural landscapes. Agric., Ecosyst. Environ. 121, 196–210.

Jarvis, D.I., Myer, L., Klemick, H., Guarino, L., Smale, M., Brown, A.H.D., et al., 2000. A Training Guide for In Situ Conservation On-farm, Version I. International Plant Genetic Resources Institute, Rome.

Jax, K., 2005. Function and "functioning" in ecology: what does it mean? Oikos 111, 641–648.

Jeanneret Ph., Schüpbach, B., Luka, H, 2003. Quantifying the impact of landscape and habitat features on biodiversity in cultivated landscapes. Agric., Ecosyst. Environ. 98, 311–320.

Karlen, D.L., Varvel, G.E., Bullock, D.G., Cruse, R.M., 1994. Crop rotations for the 21st century. Adv. Agron. 53, 1–45.

Kleijn, D., Sutherland, W.J., 2003. How effective are European agri-environment schemes in conserving and promoting biodiversity? J. Appl. Ecol. 40, 947–969.

Li, D.Z., Pritchard, H.W., 2009. The science and economics of *ex situ* plant conservation. Trends. Plant. Sci. 14, 614–621.

Malézieux, E., Crozat, Y., Dupraz, C., Laurans, M., Makowski, D., Ozier-Lafontaine, H., et al., 2009. Mixing plant species in cropping systems: concepts, tools and models. A review. Agron. Sustainable Dev. 29, 43–62.

Mazzoncini, M., Canali, S., Giovannetti, M., Castagnoli, M., Tittarelli, F., Antichi, D., et al., 2010. Organic *vs* conventional stockless arable systems: a multidisciplinary approach to soil quality evaluation. Applied Soil Ecology. 44, 124–132.

Mendelsohn, R., Dinar, A., 1999. Climate change, agriculture, and developing countries: does adaptation matter? World Bank Res. Obs. 14, 277–293.

Mitchell, D. (2008). A note on rising food prices. World Bank Policy Research Working Paper Series, Available at SSRN: <http://ssrn.com/abstract=1233058>.

Moonen, A.C., Bàrberi, P., 2008. Functional biodiversity: an agroecosystem approach. Agric., Ecosyst. Environ. 127, 7–21.

Moonen, A.C., Castro Rodas, N., Bàrberi, P., Petacchi, R., 2006. Field margin structure and vegetation composition effects on beneficial insect diversity at farm scale: a case study on an organic farm near Pisa in: Rossing, W.A.H. Eggenschwiler, L. Poehling, H.M. (Eds.), Landscape Management for Functional Biodiversity, 29 *IOBC wprs Bulletin*, pp. 77–80.

Mundt, C.C., 2002. Use of multiline cultivars and cultivar mixtures for disease management. Annu. Rev. Phytopathol. 40, 381–410.

Olson, D.M., Wäckers, F.L., 2007. Management of field margins to maximize multiple ecological services. J. Appl. Ecol. 44, 13–21.

Parris, K, 2001. OECD Agri-biodiversity indicators: background paper. Proceedings OECD expert meeting on agri-biodiversity indicators. Zurich, Switzerland, p. 42.

Pearce, D., Moran, D., 1994. The Economic Value of Biodiversity. Earthscan, London.

Peigné, J., Ball, B.C., Roger-Estrade, J., David, C., 2007. Is conservation tillage suitable for organic farming? A review. Soil Use Manage. 23, 129–144.

Pimbert, M., 2011. Participatory research and on-farm management of agricultural biodiversity in Europe. IIED, London.

Pretty, J., 2008. Agricultural sustainability: concepts, principles and evidence. Phil. Trans. R. Soc. B. 363, 447–465.

Singh, R.P., Hodson, D.P., Huerta-Espino, J., Jin, Y., Njau, P., Wanyera, R., et al., 2008. Will stem rust destroy the world's wheat crop? Adv. Agron. 98, 271–309.

Tilman, D., Cassman, K.G., Matson, P.A., Naylor, R., Polasky, S., 2002. Agricultural sustainability and intensive production practices. Nature 418, 671–677.

Tucker, G.M., Heath, M.F., 1994. Birds in Europe. Their Conservation Status. Birdlife Conservation Series No 3. Birdlife International, Cambridge.

Vandermeer, J., 1989. The Ecology of Intercropping. Cambridge University Press, Cambridge.

Vanloqueren, G., Baret, P.V., 2009. How agricultural research systems shape a technological regime that develops genetic engineering but locks out agroecological innovations. Res. Policy 38, 971–983.

Violle, C., Navas, M.L., Vile, D., Kazakou, E., Fortunel, C., Hummel, I., et al., 2007. Let the concept of trait be functional! Oikos 116, 882–892.

Webster, T.M., Burton, M.G., Culpepper, A.S., York, A.C., Prostko, E.P., 2005. Tropical spiderwort (*Commelina benghalensis*): a tropical invader threatens agroecosystems of the southern United States. Weed Technol. 19, 501–508.

Wezel, A., Bellon, S., Doré, T., Francis, C., Vallod, D., David, C., 2009. Agroecology as a science, a movement and a practice. A review. Agron. Sustainable Dev. 29, 503–515.

Woldeamlak, A., Grando, S., Maatougui, M., Ceccarelli, S., 2008. Hanfets, a barley and wheat mixture in Eritrea: yield, stability and farmer preferences. Field Crops Res. 109, 50–56.

Wolfe, M.S., Baresel, J.P., Desclaux, D., Goldringer, I., Hoad, S., Kovacs, G., et al., 2008. Euphytica 163, 323–346.

Yachi, S., Loreau, M., 1999. Biodiversity and ecosystem productivity in a fluctuating environment: The insurance hypothesis. Proc. Natl. Acad. Sci. USA. 96, 1463–1468.

Chapter 2

Organic Agriculture—Driving Innovations in Crop Research

Dionys Forster, Noah Adamtey, Monika M. Messmer, Lukas Pfiffner, Brian Baker, Beate Huber, and Urs Niggli
Research Institute of Organic Agriculture (FiBL), Frick, Switzerland

1. INTRODUCTION

At present, agriculture faces the unprecedented challenge of securing food supplies for a rapidly growing human population, while seeking to minimize adverse impacts on the environment and to reduce the use of non-renewable resources and energy. A shift towards sustainable agricultural production entails the adoption of more system-oriented strategies that include farm-derived inputs and productivity based on ecological processes and functions (Garnett and Godfray, 2012). Sustainable agricultural systems involve the traditional knowledge and entrepreneurial skills of farmers (IAASTD, 2008). System-oriented sustainable practices include organic farming and low external input sustainable agriculture (LEISA). Elements of agroecology—such as integrated pest management, integrated production (IP), and conservation tillage—have been successfully adopted by conventional farms.

Organic farming offers the most consistent approach to agroecological progress. Because of the ban or restricted use of many direct control techniques such as pesticides, herbicides, synthetic soluble fertilizers, and veterinary medicines, organic farmers rely heavily on preventive and system-oriented practices. Organic farm management aims to maximize the stability and homeostasis of agroecosystems. It improves soil fertility through the incorporation of legumes and compost and by recycling local nutrients and organic matter. Organic practices rely on preventive measures found in nature to regulate pests and diseases in crops and livestock. Because organic farming systems are relatively free from the use of synthetic pesticides, and organic processors use only a few additives, organic agriculture offers consumers high-quality and healthy food. Organic farmers may profit from ready access to local markets as part of a participatory guarantee system (PGS) or from high-value export markets when certified by accredited third-parties.

Agricultural Sustainability. DOI: http://dx.doi.org/10.1016/B978-0-12-404560-6.00002-2

Organically farmed, third-party certified land—including in-conversion areas—comprises 37 million hectares or 0.9% of total global agricultural land area. Organic agriculture is the most advanced and widely practiced in Europe. In some countries, organic agriculture is becoming mainstream: e.g., in Austria where 20% of the agricultural land area is organically managed. In developing countries, permanent crops such as coffee, tea, cocoa, coco nuts, and olives are increasingly produced according to organic standards to satisfy fast-changing consumer habits. The global market for certified organic products has grown to 44.6 billion Euros (Willer and Kilcher, 2012).

In the past, the unsustainable production of food, feed, fiber, and fuel strongly degraded global ecosystems and the services those systems provided for human survival (Millennium Ecosystem Assessment, 2005). Such ecosystem services include the provision of pure water, the recycling of organic matter and nutrients, the adaptation to climate and weather events by fertile soils, the regulation of crop pests and diseases through biodiversity and natural enemies, and the pollination of crops by wild animals, to name a few. Such degradations have not been halted or reversed yet, despite the fact that sustainability has become the axiom of agricultural policy. Global loss of fertile soils caused by wind and water erosion is continuing at an annual rate of 10 million hectares because of unsustainable farming techniques (Pimentel et al., 1995).

While the benefits to consumers and the bounty of public or common goods of organic farming and food systems are well documented (Scialabba El Hage and Hattam, 2002; UNCTAD, 2006; Niggli, 2010; Schader et al., 2012), scientists and global food security experts linger on the insufficient productivity of organic farming systems. Two most recently published scientific meta-analyses shed light on this important aspect. The overall yield gaps of organic for all crops are estimated to be 25% (Seufert et al., 2012) based on 316 comparisons and at 20% based on 362 comparisons (De Ponti et al., 2012).

Nitrogen availability was identified as a major yield-limiting factor in organic systems (Seufert et al., 2012). Nitrogen fertilizers exemplify the trade-offs between productivity and soil fertility. Mulvaney et al. (2009) pointed out that synthetic nitrogen fertilizers deplete organically bound soil nitrogen as well as soil organic carbon, leading to stagnating yields in a wide variety of soils, geographic regions, and tillage practices. In Asia, yields of cereal production have even decreased as a consequence of these depletion phenomena (Mulvaney et al., 2009).

Mulvaney et al. (2009) suggested a "gradual transition from intensive synthetic N inputs to legume-based crop rotations" for sustainable cropping systems. This very recommendation qualifies the yield gap between organic and conventional systems because, in the latter, yields are based upon an unsustainably excessive use of synthetic nitrogen fertilizer.

Some of the data of the two yield meta-analyses published in 2012 are derived from long-term field experiments. De Ponti et al. (2012) reported a decrease of the yield gap of organic farming to 16% on farms and fields managed organically for over 5 years. For countries where organic agriculture has a long tradition and is technologically advanced, the yield gap is further reduced. In Switzerland, with 11% of all agricultural land transitioned to organic, and Austria, where 20% of all agricultural land has transitioned to organic, the yield gap is reduced further to about 12% (De Ponti et al., 2012). More importantly, 25% of all data from the 362 comparisons shows relative yields of organic farms to be 90–180% of conventional cropping. These cases are relevant when estimating the potentials of organic farming and also those of further research on eco-functional intensification (Niggli et al., 2008), as described in the last part of this chapter.

Individual long-term system comparison field trials further reveal that organic systems have relatively high yields with an excellent input–output efficiency. The DOK trial (referring to bio-**d**ynamic, bio-**o**rganic and conventional) in Switzerland compared the same 7-year crop sequence for organic and conventional management since 1978. The organic plots produced 80% of the conventional yields, but fertilizer use and energy use were reduced by 34% and 53%, respectively, and pesticide applications were reduced by 97% (Mäder et al., 2002). "Enhanced soil fertility and higher biodiversity found in organic plots may render these systems less dependent on external inputs" was the conclusion of the authors. The field experiment has continued for 34 years, and the efficient input–output ratio of the organic systems has remained stable.

2. SOIL FERTILITY AND ORGANIC FARMING IN THE TROPICS—CHALLENGES AND THE WAY FORWARD

One of the most important issues faced by organic farming and sustainable production systems is how to maintain and improve soil fertility. Soil fertility is a measure of the ability of soil to sustain crop growth in the long-term, and can be determined by physical, chemical, and biological processes intrinsically linked to soil organic matter content and quality (Bhupinderpal-Singh and Rengel, 2007; Diacono and Montemurro, 2010). Agricultural lands are being degraded through depletion of soil organic matter, nutrient loss and imbalance, accelerated soil erosion, waterlogging and salinity in irrigated areas, degradation of soil structure leading to crusting and compaction of the surface soils, and decline in soil water and nutrient retention capacities (Lal, 2009). Degraded lands account for about 65% of the arable land in Africa, 74% of the arable land in Central America, and 45% of the arable land in South America (Oldeman et al., 1991; Scherr, 1999). While nutrient loss is the main form of land degradation in Central and South America, salinization and nutrient loss accounted for about 36% of arable land degradation in Asia (Oldeman et al., 1991).

African soils are inherently low in fertility because they developed from poor parent material, are old, and lack volcanic rejuvenation. African soil types are highly weathered, with low organic matter, low capacity to retain and supply nutrients to plants, high nitrogen leaching and phosphate fixation potential, low to medium water-holding capacity, weak soil structure, and deficiency in minor nutrients (Deckers, 1993). Soil fertility depletion is a major constraint to food security and income generation for African small-holder farmers as tropical soils lose nutrients every year (Pinstrup-Andersen et al., 1999; Sanchez and Swaminathan, 2005; AFS, 2006; Henao and Baanante, 2006).

By contrast, South Asian countries have progressed from food-deficit status in the early 1970s. Almost all countries in the region have increased per capita food production. The Green Revolution is credited with solving South Asia's food crises from the 1970s by promoting widespread diffusion of fertilizer-responsive wheat and rice varieties in areas with access to inputs, credits, and irrigation. In 1998, the respective consumption of N, P, and K corresponded to 71, 23, and 8 kg ha^{-1} (Katyal and Reddy, 2012). Monoculture crop production, cereal dominated crops such as rice–wheat and rice–rice–wheat rotations, mechanization, and increased intensity of land and water use occurred simultaneously with reduced return of organic matter to the soil, to cause deterioration and compaction in most of the soil structure in South Asia (Gill, 1995). Long-term problems resulting from irrigation have caused lands to retire from use in places where Green Revolution technologies have been widely adopted (Paarlberg, 1993). Soil acidity and aluminum toxicity are growing problems in hilly areas, and as this builds up, farmers find it increasingly difficult to realize an acceptable return to their labor and inputs (Gill, 1995).

Attempts to improve African crop production during the Green Revolution often neglected impacts on soil fertility. Traditional agricultural systems maintained soil fertility through long-term bush fallows of 10 or more years (Schlecht et al., 2006), but this has given way to continuous cropping and intensive use of land under an external inputs system. External inputs—mineral fertilizers, lime, irrigation water, and improved cereal germplasm—were promoted in the 1960s to overcome constraints on crop production. Unlike in Asia and Latin America, where these technologies boosted agriculture production, African production did not respond favourably, because of the diversity of the agroecologies and cropping systems, variability in soil fertility, weak institutional arrangements, and unfavorable policies (Bationo et al., 2012). Instead, continuous cultivation of the land caused a significant decline in soil organic matter content and nutrients (Abdullah and Lombin, 1978; FAO-RAF, 2000). Long-term field experiments in west African agroecosystems showed that the use of mineral fertilizers without recycling of organic materials resulted in high yields, but severe loss of soil organic matter: 5% and 2% per annum on sandy soils and more textured soils, respectively (Bationo et al., 2012). The use of synthetic fertilizer in Africa is also limited by inherent low conversion efficiency, high cost, lack of capital, inefficient distribution systems, unfavorable policies, and other socio-economic factors (Kherallah et al., 2002; Omotayo and Chukwuka, 2009). The global phosphorus scarcity is likely to threaten global food security. Nitrogen and phosphorus fertilizers will become more expensive in the future, putting upward pressure on fertilizer and food prices, and geopolitical source concentrations may lead to high price fluctuations and supply and accessibility risks. Making the most efficient use of limited nitrogen and phosphorus inputs will become a key driver for agricultural systems in the future.

Organic agriculture is one of the valid alternative approaches having the potential to meet the above-mentioned challenges. Organic agriculture is a holistic production system aiming to sustain the health of soils, ecosystems, and people (FAO/WHO, 1999; IFOAM, 2007; Ramesh et al., 2010). Organic agriculture fuels nutrient cycling in food production, based on organic fertilizers such as green manure, compost or animal manure.

Four fertility management practices typically used in organic farming systems are: (i) application of organic residues as soil amendments or nutrient source; (ii) use of biological nitrogen (N)-fixation; (iii) rotations that include cover crops, intercrops, and alley cropping; and (iv) diversification of plant species in space and time to fulfill a variety of ecosystem services. Integration of cover crops such as Mucuna in cereal–Mucuna relay intercropping improved physical, chemical, and biological soil properties as compared with sole maize monocropping or legumes—e.g., beans—intercropped with maize (Azontonde, 1993; Tian et al., 2001; Bationo et al., 2011). Soil erosion was reduced to about 10 times in maize–Mucuna relay intercropped as compared with sole maize cropping in southern Benin (Azontonde, 1993). Mucuna and Pueraria increased soil organic matter, available phosphorus, and total potassium on degraded land (Wiafe, 2010). Pueraria cover crops grown in association with food crops stimulated biological activity, improved soil nutrient availability, and N accumulation in P- and K-poor soils (Tian et al., 2001). Combined use of crop residues, manure, and inorganic P supplied increased yields of cowpea and associated cereals (Bationo et al., 2011).

Timely applications of organic materials with low carbon-to-nitrogen (C:N) ratios, such as green manure and compost, can synchronize nutrient release with plant demand (Omotayo and Chukwuka, 2009). Organically managed farms in India recorded lower productivity and yield losses but gave higher net profit to farmers and improvement in soil quality parameters (Ramesh et al., 2010), indicating better soil health. According to Ramesh et al. (2010), there is less bulk density, a slight increase in soil pH and electrical conductivity, increase in organic carbon, increase in availability of both macronutrients (N, P, K) and micronutrients (Zn, Cu, Fe, Mn) under organic farms as compared with conventional farms. When compared with conventional farms, organically managed soils had higher levels of dehydrogenase (by 52.35%), alkaline phosphatase (28.4%), and microbial biomass carbon (33.4%). This indicates higher microbial activity in organically amended soils, which is essential for nutrient transformations and increased availability of nutrients to plants.

Some constraints on organic-based soil management systems include: requirement for a large labor force to collect, process, and transport bulky organic materials; need for large quantities of organic materials to supply adequate nutrients to the soil to meet crops' nutritional demands; and need for nutrients from low-quality and slowly decomposing organic materials to be managed to fulfill crop requirements (Meertens, 2003; Danso et al., 2006; Omotayo and Chukwuka, 2009; Bationo et al., 2011). Other major challenges are competing use of organic matter in local farmland systems other than for soil fertility improvement, such as the demand for fodder crops, construction material and energy use, and lack of supportive institutions (Lele, 1994; Bumb and Baanante, 1996; Meertens, 2003; Chianu and Tsujii, 2005; Thierfelder and Wall, 2011).

While the improvement of soil fertility under organic farming is well documented in temperate systems (Mäder et al., 2002, 2006; Hepperly et al., 2006; Fliessbach et al., 2007; Teasdale et al., 2007; Birkhofer et al., 2008; Niggli et al., 2009; Gattinger et al., 2011), few scientifically sound studies are available demonstrating the long-term effect of organic farming in the tropical and subtropical regions on soil fertility. Therefore, the Research Institute of Organic Agriculture (FiBL) has implemented long-term farming system comparison trials in Kenya, India, and Bolivia in 2007. The experiments are expected to provide solid agronomic soil fertility, and socio-economic data of the predominant organic and conventional agriculture production systems representative for smallholder farms in the respective geographic regions.

Organic farming is an alternative and holistic approach that uses practices to improve soil physical, biological, and chemical properties, and can deliver agronomic and environmental benefits, particularly long-term improvement of soil fertility and quality. Organic agriculture aims for resilience, or the capacity to resist shocks and stresses, and persistence, or the capacity to continue over long periods. Soil fertility is knowledge and management intensive. Farmers' knowledge of soil fertility evaluation needs to be understood (Mowo et al., 2006) in addition to their social and economic realities to improve their sustainable organic farming practices.

3. PLANT BREEDING STRATEGIES FOR ORGANIC AND LOW EXTERNAL INPUT FARMING

Domestication and selection of plants is linked to human settlement and our cultural heritage. Nearly 10,000 years of domestication has resulted in more than 7,000 different crop plants selected for different purposes and ecological niches (FAO, 1996). However, only 30 crop species—comprising less than 0.1% of all edible plants—account for 95% of the global calorie intake, with rice, maize, and wheat accounting for more than 50% (FAO, 1996). Enormous efforts have been made to increase food production to feed a growing world population. Production of the main staple crops has increased by 145% since 1960.

Breeding and agronomic improvements are partly responsible, but so are increased inputs. Over the same period, agricultural land increased by 11%, the amount of irrigated land doubled, and fertilizer and pesticide use increased four-fold, all causing negative side effects (Pretty, 2008).

Sustainable plant breeding and seed networks are essential for food sovereignty and adaptation to climate change (Sthapit et al., 2008; FAO, 2010). Plant breeding has selected for increase of gross yield of crops that rely on unsustainably high inputs grown as monocultures of homogeneous cultivars (Vanloqueren and Baret, 2008). In recent decades plant breeding has moved from family owned or public breeding programs to large-scale multinational breeding companies, and seed propagation has shifted from mainly farm-saved seed to F1 hybrid seed that need to be purchased each year (Vernooy, 2003; da Silva Dias, 2010). Genetic diversity has been lost as companies concentrated on breeding the most profitable crop species (Haussmann and Parzies, 2009; da Silva Dias, 2010; FAO, 2010). Numerous local varieties and landraces have disappeared, causing genetic erosion within species (Haussmann and Parzies, 2009).

The transformation of the seed sector was accompanied over the same period by a rapid consolidation of seed companies closely interlinked with the agrochemical sector (Howard, 2009). The three largest seed companies—Monsanto, DuPont, and Syngenta—control over 50% of the global seed market (ETC Group, 2011). These companies seek patents that restrict breeders' rights to use a released cultivar for further breeding purposes. Access to plant genetic resources is also limited by laws (Engels et al., 2011; Chable et al., 2012).

Loss of species and varieties cultivated, and narrow control of the seed sector, limit options for smallholder producers. Most of the underutilized crops are strongly linked to tradition and cultural knowledge and are adapted to specific niches, with short supply and sparse documentation (Hoeschle-Zeledon and Jaenicke, 2007). Diversification of crop species—especially with legumes and vegetables—reduces risk and prevents malnutrition (Hoeschle-Zeledon and Jaenicke, 2007; Ogoke et al., 2009; Keatinge et al., 2011, Mayes et al., 2012).

Farmers' crop choices depend on several criteria, including available inputs, relative prices, government policy, and various environmental factors (FAO, 2010). Poor supply of reasonably priced high-quality seed is a major constraint on production of vegetable crops in Africa (Keatinge et al., 2011). To counteract this trend, nutritious local crop species were successfully re-introduced in the scope of the project "Indigenous African leafy vegetables for enhancing livelihood security of smallholder farmers in Kenya" (Mwangi and Kimathi, 2006; Gotor and Irungu, 2010).

Resource-poor farmers in marginal, complex agricultural environments have been slow to adopt modern varieties. Such farmers lack access to external inputs, and the new cultivars seldom meet farmers' socio-economic and cultural priorities and needs (Keneni and Imtiaz, 2010; Temudo, 2011). High-input varieties are generally selected under uniform and highly controlled conditions where soluble fertilizer, seed treatment, herbicides, and other pesticides are applied. Such conditions do not reflect the situation on organic and low external input farms that operate primarily on closed nutrient cycles with minimal external inputs (Ceccarelli et al., 2007). Nutrient release from soil organic matter depends on complex biological, chemical, and physical factors (Messmer et al., 2012). High-yielding varieties selected for high input may suffer from temporarily insufficient nutrient supply, causing considerable yield reduction under organic and low external input systems (Dawson et al., 2008a). Cultivars are needed that show high nutrient use efficiency, optimized root morphology, and the capacity to establish beneficial plant microbial interactions that play important roles in nutrient uptake efficiency and disease suppression (Lynch., 2007; Fageria et al., 2008; Hartmann et al., 2009; Wissuwa et al., 2009). Such traits might become more important in future as, for example, soil-borne diseases may increase with climate change (Jaggard et al., 2010).

Breeders need to develop cultivars that perform well in very different environments. Genotype × environment × management interactions need to be considered for the most-promising breeding strategies. Breeding for low nitrogen (N) input conditions has been shown to be more efficient under severe N stress than under high-input conditions for wheat (Brancourt-Hulmel et al., 2005), barley (Ceccarelli, 1996; Sinebo et al., 2002; Ryan, 2008), oat (Atlin and Frey, 1989), maize (Presterl et al., 2003), and rice (Mandal et al., 2010).

Crops bred for specific farming practices are key to sustainable management of resources and eco-functional intensification of organic farming (Schmid et al., 2009; Lammerts van Bueren et al., 2011; Lammerts van Bueren and Myers, 2012). However, few studies have examined differences between direct and indirect selection for organic farming. One study comparing wheat selection under organic and conventional conditions concluded that indirect selection of varieties bred under conventional conditions would not result in the best lines for organic farming, and that varieties selected under organic conditions would perform better on organic farms (Reid et al., 2009). Similar results were found with the direct and indirect selection of maize for organic farming (Messmer et al., 2009).

Another reason why international breeding research in developing countries has had limited impact is that culture, quality, and flavor are seldom considered by such programs. Breeding often follows a top-down approach (Figure 2.1) where breeders are disconnected from farmers (McGuire, 2008; Temudo, 2011). The seed replacement rate is especially low in unfavorable growing conditions (Aw-Hassan et al., 2008; McGuire, 2008). Smallholder farmers in stressful environments rely mainly on diverse cultivars to slow the spread of pests and diseases, and to adapt to unforeseen climatic changes (Jarvis et al., 2011; Mulumba et al., 2012). Yield stability is more important than yield potential to smallholders (Sthapit et al., 2008).

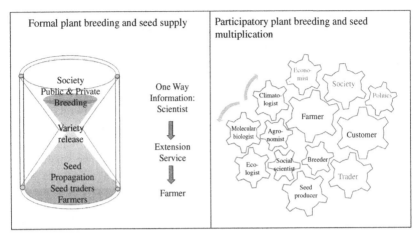

FIGURE 2.1   Formal and participatory plant breeding and seed supply.

Participatory cultivar selection (PCS) and participatory plant breeding (PPB) offer opportunities to involve farmers in cultivar selection and breeding (Ceccarelli et al., 2007; Dawson et al., 2008b; Desclaux et al., 2012). These approaches fully integrate the farmers' knowledge and skills, and resulting cultivars are more likely to be adopted in a short time (Vernooy, 2003). Large groups are able to deal with a wide range of different crops that need to be improved for different agroecological niches and cropping systems, resulting in greater biodiversity.

Genetic and agronomic improvement can be developed in parallel through PCS and PBB (Mayes et al., 2012). PPB is a transdisciplinary approach where different actors work together in a very dynamic and demand-driven process. This trans- and interdisciplinary approach requires new tools of group learning, innovation diffusion, and institutional changes (McGuire, 2008; Joshi et al., 2012).

Women involved in most field activities and cooking can contribute to breeding goals. Integrating women in the process will not only empower them in plant breeding and seed conservation but also improves their social status (Ceccarelli and Grando, 2007; Paris et al., 2008). In addition, they are very good multipliers not only of seed but also of information (McElhinny et al., 2007; Ashby 2009).

Murphy et al. (2005) combined PPB with evolutionary breeding based on natural selection for local adaptation, first proposed by Suneson (1956). Such integrated breeding approaches are known as evolutionary participatory breeding (EPB) and use the skills and knowledge of both breeders and farmers to develop heterogeneous landrace populations. EPB was shown to be an effective breeding method for both modern and traditional farmers (Murphy et al., 2005, Ceccarelli et al., 2010).

Farmers satisfied with new cultivars will multiply and distribute seeds among their social networks (Vernooy, 2003). Capacity building and training in seed multiplication, seed processing, seed testing and storage, as well as in the management of the startup of seed enterprises, will enable farmers to develop a local seed supply chain and become independent from global seed companies and volatile prices of hybrid seed (Aw-Hassan et al., 2008). Pautasso et al. (2012) highlighted the role of seed exchange networks in conservation of agrobiodiversity and their importance for local food security. Farm-saved seeds and local seed markets and seed fairs can all maintain and strengthen the formal and informal seed chain and thus improve smallholder farmers' access to seed (Sperling et al., 2008; Sthapit et al., 2008; da Silva Dias, 2010; Joshi et al., 2012).

Political dialog and public intervention in agricultural research are needed to enforce such farmer-driven seed initiatives (Vanloqueren and Baret, 2008, 2009). Collaborative participatory plant breeding and seed multiplication will improve biodiversity, resilience against environmental, economic, and food crises in the long term, and contribute to food security and sovereignty (Pretty, 2008; Mayes et al., 2012).

4. FUNCTIONAL BIODIVERSITY AND PEST MANAGEMENT IN ORGANIC FARMING

Traditional agricultural landscapes offered diverse habitats and succession stages; centuries of land-use have enabled a large number of species to survive in the agricultural landscape and to sustain pest control function (Winqvist et al., 2012). Modern agriculture is characterized by simplified landscapes, dominated by large-scale monocultures highly dependent on external inputs (Altieri, 1999). Agricultural intensification has resulted in less biodiversity (Postma-Blaauw et al., 2012). Intensification entails increased application of herbicides, insecticides, fungicides, and chemical fertilizer on local fields, to loss of natural and semi-natural habitats, and decreased habitat heterogeneity at the farm and landscape levels. The density and uniformity of crop cultivars offer locally concentrated food supplies for many pests and diseases to flourish, while depressing natural enemies' populations due to lack of food resources or shelter. Beneficial plants and insects might also be unintentionally harmed by the application of broad-spectrum pesticides. Diminished natural-enemy populations often result in pest outbreaks, which adversely affect crop yields (Geiger et al., 2010; Tscharntke et al., 2012). A key strategy in sustainable agriculture is to restore functional biodiversity of agroecosystems on crop, farm and landscape level and link agroecological intensification with biodiversity conservation (Brussaard et al., 2010; Tscharntke et al., 2012).

Functional biodiversity is defined by Moonen and Bàrberi (2008) as "that part of global diversity composed of clusters of elements (at the gene, species or habitat level) providing the same (agro)ecosystem service, that is driven by within-cluster diversity". The agroecosystem comprises managed fields, surrounding semi-natural or natural habitats, and human settlement and infrastructure. Production of food and other agricultural products is only one ecosystem service. Other ecosystem services include soil-related processes, soil food webs, and gene flow (Moonen and Bàrberi, 2008). Maximizing biodiversity *per se* does not necessarily increase the ecosystem functions, as often a few dominant species perform specific functions, while the remaining species are redundant for these functions (Moonen and Bàrberi, 2008; Postma-Blaauw et al., 2012). However, in high-input agricultural systems with low biodiversity on all levels, an increase in biodiversity is most likely adding complementary ecosystem services. More diverse communities have been shown to have higher and more stable ecosystem functioning, suggesting they should also have a consistently higher level of functioning over time. Diverse communities could maintain consistently high function because the species are driving function change over time (functional turnover) or because they are more likely to contain key species with temporally stable functioning. Agroecosystems should be redesigned in such a way that the system can develop mechanisms to recover disturbances autonomously in tune with the locally available biodiversity and with the existing environmental and socio-economic conditions (Altieri, 1999; Pfiffner and Wyss, 2004; Moonen and Bàrberi, 2008, Perfecto and Vandermeer, 2010).

Organic agriculture developed several strategies to improve crop production and other ecosystem services. Crop protection in organic agriculture is based on a multi-level approach (Figure 2.2) with a focus on indirect preventive plant protection measures (Step 1 to 3) followed by more-direct and curative measures (Step 4 and 5) only when needed at later stages (Wyss et al., 2005; Zehnder et al., 2007).

This multilevel strategy combines various tactics in farm management and cropping design in a holistic approach to limit pest populations below

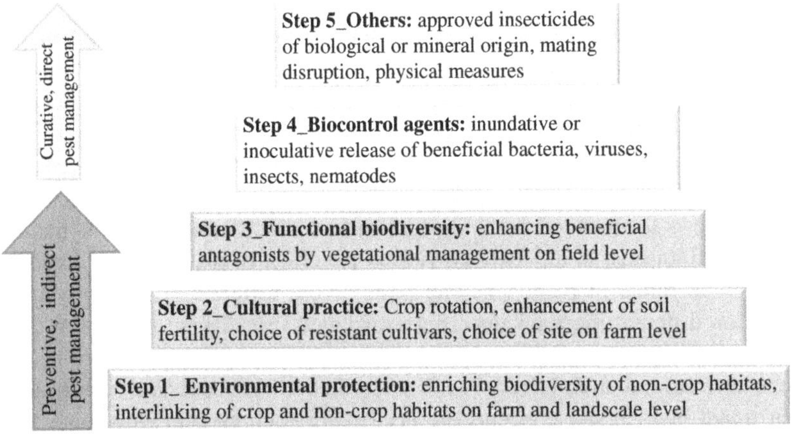

Step 5_Others: approved insecticides of biological or mineral origin, mating disruption, physical measures

Step 4_Biocontrol agents: inundative or inoculative release of beneficial bacteria, viruses, insects, nematodes

Step 3_Functional biodiversity: enhancing beneficial antagonists by vegetational management on field level

Step 2_Cultural practice: Crop rotation, enhancement of soil fertility, choice of resistant cultivars, choice of site on farm level

Step 1_ Environmental protection: enriching biodiversity of non-crop habitats, interlinking of crop and non-crop habitats on farm and landscale level

Curative, direct pest management

Preventive, indirect pest management

FIGURE 2.2   A five-step approach of arthropod pest management in organic agriculture based on the concept of Wyss et al. (2005) and Zehnder et al. (2007) modified by Hernyk Luka, FiBL 2012

damaging levels, thus minimizing the need for direct intervention (Pfiffner et al., 2005; Zehnder et al., 2007). Habitat management through farm site selection, crop isolation, and interlinking of natural and semi-natural ecosystems offers the first step. Non-crop habitats are known to play an essential role for reproduction and survival of natural enemies, offering food resources, overwintering sites, and refuges (Pfiffner and Luka, 2000). The second step involves adaptation of optimal cultural practices such as diverse crop rotation, selection of resistant or tolerant cultivars, and improved soil fertility through cover crops or soil amendments. On the individual field level, vegetational management in the form of hedgerows and flower strips enhances beneficial organism populations (step 3). Repellent or trap plants can be planted to divert the pest from the crop. Successful examples of preventive pest control include bio-fumigation effects of *Brassicaceae* crops on soil-borne pests and diseases, reduced preference of Colorado potato beetle for potato plants in manure-amended soils meeting optimal nutrients requirements, or reduced stem borer infestation through combined push-pull trap cropping in maize (for review see Zehnder et al., 2007, Cook et al., 2007).

Organic farming generally improved species richness and abundance compared with intensive conventional farming with higher inputs of energy, fertilizer, and pesticides; however, the surrounding landscape can either enhance or reduce its positive effects on functional biodiversity (Winqvist et al., 2012). Enriched non-production areas with diverse natural or semi-natural habitats are crucial for functioning pollination service and pest control, but these effects are less pronounced for soil biota (Smukler et al., 2010). Similar results were found by Batáry et al. (2012) for the response of plants, insects, and spiders to management intensity in cereals and grassland. Organic management promoted species richness of non-carnivore carabids and hunting spiders, while grasshoppers benefited from reduced management intensity. In general, reduction in field size by enlarging the edge area enhances functional biodiversity under organic and conventional farming. Functional groups that are not yet enhanced by organic farming need further improved management practices and strategies (Batáry et al., 2012).

Functional biodiversity specifically augments beneficial species relative to pest species by adding resources that selectively enhance only beneficials. The beauty of the functional biodiversity approach is that it starts a feedback loop: increased beneficial populations lead to decreased pest populations, which allows for reduced pesticide application, which in turn benefits natural enemies. Ratnadass et al. (2012) reviewed the effect of plant species diversity on a broad range of pathogens and pests across different cropping systems based on different mechanisms. They concluded that diverse vegetation does not necessarily reduce incidence of pests and diseases. Most pests and pathogens have several alternative hosts or reservoirs. To be successful, tailor-made functional biodiversity concepts are needed for each crop in each landscape (Pfiffner et al., 2005).

The Research Institute of Organic Agriculture (FiBL) in Frick, Switzerland, is currently investigating the efficiency of a functional biodiversity approach to control lepidopteran pests in cabbage. White cabbage (*Brassica oleracea* (L.), var. *capitata*) in central Europe is mainly attacked by the cabbage moth (*Mamestra brassicae*), the diamond-back moth (*Plutella xylostella*), and the small white cabbage moth (*Pieris rapae*), which all have specialized larval parasitoids. Because the endoparasitic wasp *Microplitis mediator* has different food requirements than its hosts, some plants benefit only the parasitoid and not any of the pests. Géneau et al. (2012) identified that cornflower (*Centaurea cyanus* L.), buckwheat (*Fagopyrum esculentum* Moench), and common vetch (*Vicia sativum* L.) improved the endoparasitic wasp's longevity and rate of cabbage moths parasitized, but had no effect on the longevity and fecundity of the cabbage moth. Behavioral assays showed that all plant species tested were attractive to *M. mediator* but cornflower was significantly more attractive than buckwheat (Belz et al., 2012).

Recent scientific studies on the complex interactions of crop, pest, and environment open new doors for preventive pest control by habitat management and functional biodiversity (Bàrberi et al., 2010; Petit et al., 2011; Farwig and Berens, 2012; Parolin et al., 2012). Many secondary metabolites and proteins were identified to be involved in plant defense as they are toxic or deterrent to herbivores or to other plants through allelopathy, whereas plant volatiles are released by the plant after attack, to attract natural enemies (Khan et al., 2010).

As the numerous combinations of species, environments, and practices are beyond the traditional factorial experimental approaches, a systems approach and dynamic modeling tools are required to test and verify the potential and limits of the management of functional biodiversity on plot as well as on landscape level (Ould-Sidi and Lescourret, 2011; Ratnadass et al., 2012). More complex cropping systems like permaculture or agroforestry need to be considered for more diversified agroecosystems (Malézieux et al., 2009; Tscharntke et al., 2011). Organic agriculture offers ideal conditions to integrate such a whole-system approach of functional biodiversity into innovative agroecosystems, that best benefits but also fosters biodiversity, thus minimizing the trade-off between production aims and biodiversity conservation (Simon, 2010).

5. AGRICULTURAL INNOVATION—THE NEED FOR TRANSDISCIPLINARY RESEARCH AND DEVELOPMENT

During the 20th century, the relation between science and society was based on an implicit agreement which stated that science was responsible for making discoveries and subsequently making them available to society (Gibbons, 1999). Disciplinary research in agriculture successfully developed and produced a multitude of innovations that increased productivity and contributed to food security. However, these knowledge-based solutions were usually communicated in a top-down approach. According to Pretty (1995), a one-way transfer of technology approach was used during the so-called production stage (1950–1975). Crop and animal breeding and genetics were predominant disciplines, while the farmers simply received the technology. In the following years (1975–1985) economists and agronomists gained importance and practiced a two-way communication approach. Farmers were included as informants and started to contribute to the design of new technologies. During the ecological stage (1985–1995) agroecologists, geographers, and anthropologists influenced agricultural research and development. The farmer was seen as cause and victim for unsustainable development. From 1995 onwards, a new understanding for institutional development emerged where psychologists, sociologists, political scientists, training specialists, and educators were involved. Society was calling for a revolution in science to enable itself to tackle the problems caused by the global industrial system of which science itself forms the basis.

A more systemic and holistic approach should replace the reductionist, disciplinary worldview (Funtowicz and Ravetz, 1993). Transdisciplinary research gained attention as it was seen as a conceptual approach oriented to solve problems of the life-world and comprising phases of problem identification, structuring, investigation, and bringing results to fruition (Hirsch Hadorn et al., 2008). Boundaries between scientific disciplines—and between science and society—are crossed in a recursive process, in which problems are not pre-determined, but jointly identified (Wiesmann et al., 2008). Disciplinary approaches are not excluded, but rather needed as the basic level of understanding. The two conceptual approaches are essentially complementary (Max-Neef, 2005). Additional knowledge and mutual understanding of each other's problems significantly improve the quality, acceptance, and sustainability of solutions developed for a particular problem. Thus, participation and integration are key elements of transdisciplinary research (Elzinga, 2008).

Despite the continuous development of new concepts in agricultural research and technology transfer, a broad movement into these new approaches is not visible. Participation in research and extension is still predominantly consultative, and farmers have little influence on the decision-making. Many research projects in food and agriculture are now positioned in distinct disciplines, and further narrowed down to particular segments of these (Francis et al., 2008). Specialization within disciplines even seems to pay off better than inter- and transdisciplinary research. Vanloqueren (Vanloqueren and Baret, 2009) compared two technological paradigms: genetic engineering—representative of specialization within disciplines—and agroecology—representative of inter- and transdisciplinary research. They identified the following determinants, which in combination lead to imbalances between both paradigms.

- *Complexity and framing of agricultural research.* While complexity in molecular biology and genetic engineering is applied to cell or gene level, complexity in agroecological engineering is referred to the ecosystems level.

The recent technology advances have helped to deal with the complexity at cell or gene level and led to the development of highly specialist research. In contrast, the agroecological system cannot be effectively addressed by a reductionist approach and requires a holistic view.

- *Performance of agricultural innovations.* Direct, local, and short-term benefits are favored through reductionist approaches leading in current sciences. Conversely, indirect, long-term or holistic benefits for agriculture are underestimated. Research focuses on simple quantifiable variables such as gross yield, while complex ones such as sustainability or externalities are left out.
- *Publication pressure.* Publications share scientific knowledge, and acceptance for publication should not be guided by market incentives. Nonetheless, large differences between the two paradigms with respect to appearance in high-ranking journals can be observed. Also the impact factor of journals representative of genetic engineering is far higher than that of journals for agroecological engineering.
- *Specialized research.* Molecular biology and genetic engineering profit from the growing specialization of science. Interdisciplinary research between medicine, animal, and plant sciences is gaining importance in biotechnology. However, in agroecological engineering, an interdisciplinary approach includes sciences such as agronomy, ecology, sociology, and economics. Interdisciplinarity in agroecology thus requires the crossing of boundaries between very distinctive disciplines.
- *Technology transfer.* Public sector research establishments aim to transfer scientific knowledge to the private sector. Thereby, spin-off companies with patented research results play a crucial role. As private sector investments into public universities increase, the number of patents and spin-off companies will likely grow as well. Extension and technology transfer to farmers may emphasize those technologies that farmers can use directly. Besides the private sector, genetically modified crops are more favorable for public sector research establishments than are agroecological innovations that require fundamental changes in cropping systems.

Under the current scientific paradigm, disciplinary, specialist approaches seem to have a clear advantage over inter- and transdisciplinary ones. Changing this trajectory leads to high transaction costs (Geels, 2004). Scientists are educated to develop specific competencies and to specialize. Crossing boundaries between scientific disciplines and between science and society implies, however, compromise on acquired specialization and the research environment and breaking out to new fields of research with considerable uncertainty (Vanloqueren and Baret, 2009). Thus, the question is how to move from the present unilateral reductionist system to a more holistic one that favors new inter- and transdisciplinary approaches?

Building up inter- and transdisciplinary agricultural innovation systems requires complementary institutions developed in parallel to the existing ones in basic and applied research. These institutions should be designed to undertake research and education in a problem-oriented way, where discipli-nary, inter- and transdisciplinary forms of education are equally supported (Max-Neef, 2005; Hadorn et al., 2006; Francis et al., 2008). The mutual recognition of different disciplines and the willingness to integrate disciplinary and interdisciplinary knowledge in a transdisciplinary framework is an impor-tant first step (Giri, 2002). Creation of a mutual learning environment is key to integration of transdisciplinary processes (Wiesmann et al., 2008).

Agricultural sustainability is probably best achieved through agroecosys-tem research and development. While Gliessman (1998) defines agroecology as the application of ecological concepts and principles to the design and man-agement of sustainable agroecosystems, Francis et al. (2003) goes a step fur-ther and sees agroecology as the integrative study of the ecology of the entire food system, including ecological, economic, and social dimensions. Whether agroecology is restricted to the ecology of the food system only, shall not be debated at this point. However, agroecology encompasses with certainty envi-ronmental (ecological), economic and social aspects—the three dimensions of sustainability. Agroecology is an inherently transdisciplinary process, as it links structure and functioning of agroecosystems and closes the gaps between different disciplines as well as between theory and practice (Caporali, 2011). Agroecosystems research and development create the mutual learning envi-ronment to develop appropriate solutions for future challenges associated with deterioration of natural resources and climate change.

Many farming systems claim to be sustainable. If agricultural sustainability can be ensured by the production system that makes best use of environmental goods and services while reducing and avoiding adverse system effects with regard to natural, social, and human capital, then organic agriculture can be considered to be truly sustainable (UNEP-UNCTAD, 2008). Organic agri-culture is the most important example of policy-regulated agriculture which explicitly integrates biophysical and socio-economic aspects of sustainability (Caporali, 2011). Driven by the aim to contribute to more holistic agricul-ture, the Research Institute of Organic Agriculture (FiBL) was an early adop-ter of participatory research processes. Agricultural research and extension became united under the single roof of FiBL, and henceforth allowed fruit-ful exchange between farmers and scientists. Research results were quickly transferred to practice and contributed significantly to the shaping of organic agriculture in Switzerland (Niggli, 2007). Meanwhile, FiBL is building on its experience in Europe and partners in multinational research and development projects in Asia, Africa, and Latin America. All of these projects have a com-mon feature—they aim to make substantial contributions to agroecology by inter- and transdisciplinary research and development.

In conclusion, if agricultural research is to provide sustainable solutions, inter- and transdisciplinary research is indispensable. Reductionist methodolo-gies need to be complemented by more holistic approaches. New institutions and education programs need to be introduced to foster a continuous dialogue leading to a change in perception of values in society. Students, scientists, and advisors need to be trained in a systems approach to address multi-dimensional problems of agroecosystems. Organic agriculture can be the point of departure to develop and advance inter- and transdisciplinary research and a driving force for a renewal of agriculture in the future (Kotschi, 2011).

6. OUTLOOK

Organic agriculture not only conserves resources and nature; it is also conservative as a farming system. Rooted in the traditional knowledge of farmers, driven by consumer expectations like "nature knows best", and guided by a precautionary approach, organic farming is skeptical of novel technologies.

The dos and don'ts of organic farming give direction for scientific innovation to benefit both the environment and society. However, organic farmers are confronted with real production problems that are already solved in conventional farming systems. The challenges can be found everywhere: in the insufficient control of weeds, plant diseases, and pests; in the management of soil and plant nutrients up to the optimal development of crops; in the availability of germplasm for low-fertilizer and low-pesticide input conditions, to name a few.

Consumers are often seen as the limiting factor in the transition to organic. Studies show that consumers are well informed about organic agriculture in most countries and a majority of them appreciate it as being the best option for agriculture and foods (Aertsens et al., 2009). In contrast with this positive marketing profile, organic food consumption is below 10% even in countries with developed organic markets.

Therefore, the bottleneck to bring organic agriculture into the mainstream is not consumption but agricultural production. Technical problems lead to high costs, to insufficient market supply, and to reluctance by farmers to accept high risks and greater labor requirements.

Organic food and farming systems might be very attractive fields for scientists to search for system-oriented solutions. On the other hand, systems research is not always rewarding in the short term, as results take longer to produce, and inter- and transdisciplinary research tends to overextend scientists. Investment in organic farming research promises an efficient return in contrast to conventional farming, as the marginal utility of organic research is still high. Research funding in organic agriculture is insufficient in most regions of the world, and a critical mass of research teams is needed to address the problems faced by organic farmers.

How do we tackle these constraints and opportunities? First of all, organic food and farming research should become viewed as excellent, effective, and targeted to the big challenges of the forthcoming decades. The development of organic agriculture needs to be driven more by science than by tradition. Scientists able to handle complexity, societal responsibility, and farm community knowledge need to drive the research agenda. Finally, organic agriculture should become more open towards innovation. The concept of eco-functional intensification (Niggli et al., 2008) includes the smart use of novel technologies. Well-selected elements of precision farming, ICT, nanotechnology, and molecular-based breeding might one day offer tools consistent with the principles of organic agriculture. These principles promote greater diversity on landscape, farm, field, and germplasm level. They care about ecosystems and their functioning, and they produce food and fiber with minimum waste, energy, and environmental impact. And most importantly, the principles of organic agriculture respect human and animal well-being.

ACKNOWLEDGEMENTS

We thank Christian Andres, Henryk Luka, and Oliver Balmer from FiBL for their valuable contributions and fruitful discussions.

REFERENCES

Abdullah, A., Lombin, L.G., 1978. Comparative Effectiveness of Separate and Combined Applications of Mineral Fertilizers and Fym in Maintaining Soil Productivity Under Continuous Cultivation in the Savannah (Long-Term Fertility Studies at Samuru, Nigeria). Institute of Agricultural Research, Ahmadu Bello University, Samaru, Zaira, Nigeria, (Misc. paper 75, 12).

Aertsens, J., Verbeke, W., Mondelaers, K., Van Huylenbroeck, G., 2009. Personal determinants of organic food consumption: a review. Br. Food J. 111, 1140–1167.

AFS, 2006. Achieving an African Green Revolution: A vision of sustainable agricultural growth in Africa. African Fertiliser Summit (AFS), Abuja, Nigeria, (Background papers 9–13 June).

Altieri, M.A., 1999. The ecological role of biodiversity in agroecosystems. Agric. Ecosyst. Environ. 74, 19–31.

Ashby, J.A., 2009. The impact of participatory plant breeding. In: Ceccarelli, S., Guimarães, E.P., Weltzien, E. (Eds.), Plant Breeding and Farmer Participation FAO, Rome, pp. 649–671.

Atlin, G.N., Frey, K.J., 1989. Predicting the relative effectiveness of direct versus indirect selection for oat yield in 3 types of stress environments. Euphytica 44, 137–142.

Aw-Hassan, A., Mazid, A., Salahieh, H., 2008. The role of informal farmer-to-farmer distribution in diffusion of new barley varieties in Syria. Exp. Agric. 44, 413–431.

Azontonde, H.A., 1993. Degradation et restauration des terres de barre au Sau-Benin Presentation to 10th Erosion Workshop. ORSTOM, Montpellier, (September, 1993, 15–18).

Bàrberi, P., Burgio, G., Dinelli, G., Moonen, A.C., Otto, S., Vazzana, C., et al., 2010. Functional biodiversity in the agricultural landscape: relationships between weeds and arthropod fauna. Weed Res. 50, 388–401.

Batáry, P., Holzschuh, A., Orci, K.M., Samu, F., Tscharntke, T., 2012. Responses of plant, insect and spider biodiversity to local and landscape scale management intensity in cereal crops and grasslands. Agric. Ecosyst. Environ. 146, 130–136.

Bationo, A., Kimetu, J., Vanlauwe, B., Bagayoko, M., Koala, S., Mokwunye, A.U., 2011. Comparative analysis of the current and potential role of legumes in integrated soil fertility management in West Africa and Central Africa. In: Bationo, A.E.A. (Ed.), Fighting Poverty in Sub-Saharan Africa: The Multiple Roles of Legumes in Integrated Soil Fertility Management. Springer, Science + Business Media B.V, pp. 117–150.

Bationo, A., Waswa, B., Abdou, A., Bado, B.V., Bonzi, M., Iwuafor, F., et al., 2012. Overview of Long term experiments in Africa. In: Bationo, Lessons Learned from Long Term Soil Fertility Management Experiments in Africa. Springer, Dordrecht, pp. 1–26.

Belz, E., Kölliker, M., Balmer, O., 2012. Olfactory attractiveness of flowering plants to the parasitoid *Microplitis mediator*: potential implications for biological control. BioControl. 13, 85–93.

Singh, B., Rengel, Z., 2007. The role of crop residues in improving soil fertility. In: Marschner, P., Rengel, Z. (Eds.), In Nutrient Cycling in Terrestrial Ecosystems. Springer, Berlin/Heidelberg, Germany, pp. 183–214.

Birkhofer, K., Bezemer, T.M., Bloem, J., Bonkowski, M., Christensen, S., Dubois, D., et al., 2008. Long-term organic farming fosters below and aboveground biota: implications for soil quality, biological control and productivity. Soil. Biol. Biochem. 40, 2297–2308.

Brancourt-Hulmel, M., Heumez, E., Pluchard, P., Beghin, D., Depatureaux, C., Giraud, A., et al., 2005. Indirect versus direct selection of winter wheat for low-input or high-input levels. Crop Sci. 45, 1427–1431.

Brussaard, L., Caron, P., Campbell, B., Lipper, L., Mainka, S., Rabbinge, R., et al., 2010. Reconciling biodiversity conservation and food security: scientific challenges for a new agriculture. Curr. Opin. Environ. Sustain. 2, 34–42.

Bumb, B., Baanante, C., 1996. The Role of Fertilizer in Sustaining Food Security and Protecting the Environment to 2020. IFPRI, Washington DC, 2020 Vision Discussion Paper 17.

Caporali, F., 2011. Agroecology as a transdisciplinary science for a sustainable agriculture biodiversity In: Lichtfouse, E. (Ed.), Biofuels, Agroforestry and Conservation Agriculture, vol. 5. Springer, Dordrecht, pp. 1–71.

Ceccarelli, S., 1996. Adaptation to low/high input cultivation. Euphytica 92, 203–214.

Ceccarelli, S., Grando, S., 2007. Decentralized-participatory plant breeding: an example of demand driven research. Euphytica 155, 349–360.

Ceccarelli, S., Grando, S., Baum, M., 2007. Participatory plant breeding in water-limited environments. Exp. Agri. 43, 411–435.

Ceccarelli, S., Grando, S., Maatougui, M., Michael, M., Slash, M., Haghparast, R., et al., 2010. Plant breeding and climate changes. J. Agric. Sci. 148, 627–637.

Chable, V., Louwaars, N., Hubbard, K., Baker, B., Bocci, R., 2012. Plant breeding, variety release, and seed commercialization: laws and policies applied to the organic sector. In: Lammerts van Bueren, E.T., Myers, J.R. (Eds.), Organic Crop Breeding Wiley-Blackwell Press, New York, pp. 139–159.

Chianu, J.N., Tsujii, H., 2005. Determinants of farmers' decision to adopt or not adopt inorganic fertilizer in the savannas of northern Nigeria. Nutr. Cycl. Agroecosys. 70, 293–301.

Cook, S.M., Khan, Z.R., Pickett, J.A., 2007. The use of push-pull strategies in integrated pest management. Annu. Rev. Entomol. 52, 375–400.

da Silva Dias, J.C.D., 2010. Impact of improved vegetable cultivars in overcoming food insecurity. Euphytica. 176, 125–136.

Danso, G., Drechsel, P., Fialor, S., Giordano, M., 2006. Estimating the demand for municipal waste compost via farmers' willingness-to-pay in Ghana. Waste Manage. 26, 1400–1409.

Dawson, J.C., Huggins, D.R., Jones, S.S., 2008a. Characterizing nitrogen use efficiency in natural and agricultural ecosystems to improve the performance of cereal crops in low-input and organic agricultural systems. Field Crop Res. 107, 89–101.

Dawson, J.C., Murphy, K.M., Jones, S.S., 2008b. Decentralized selection and participatory approaches in plant breeding for low-input systems. Euphytica 160, 143–154.

de Ponti, T., Rijk, B., van Ittersum, M.K., 2012. The crop yield gap between organic and conventional agriculture. Agric. Syst. 108, 1–9.

Deckers, J., 1993. Soil fertility and environmental problems in different ecological zones of the developing countries in sub-Saharan Africa. In: Van Reuler, H., Prins, W.H. (Eds.), The Role of Plant Nutrients and Sustainable Food Production in Sub-Saharan Africa. Vereniging van Kunstmest Producenten, Laidschendam, Netherlands.

Desclaux, D., Ceccarelli, S., Navazio, J., Coley, M., Trouche, G., Aguirre, S., et al., 2012. Centralized or decentralized breeding: the potentials of participatory approaches for low-input and organic agriculture. In: Lammerts van Bueren, E.T., James, R.M. (Eds.), Organic Crop Breeding Wiley-Blackwell Press, New York, pp. 99–123.

Diacono, M., Montemurro, F., 2010. Long-term effects of organic amendments on soil fertility. A review. Agron. Sustain. Dev. 30, 401–422.

Elzinga, A., 2008. Participation. In: Hadorn, H., Hoffmann-Riem, H., Biber-Klemm, S., Grossenbacher-Mansuy, W., Joye, D., Pohl, C., Wiesmann, U., Zemp, E. (Eds.), Handbook of Transdisciplinary Research. Springer, Dordrecht, pp. 433–441.

Engels, J.M.M., Dempewolf, H., Henson-Apollonio, V., 2011. Ethical considerations in agro-biodiversity research, collecting, and use. J. Agric. Environ. Ethics 24, 107–126.

ETC Group. (2011). Who Will Control the Green Economy? Retrieved July 27, 2012 from <www.etcgroup.org>.

Fageria, N.K., Baligar, V.C., Li, Y.C., 2008. The role of nutrient efficient plants in improving crop yields in the twenty-first century. J. Plant. Nutr. 31, 1121–1157.

FAO, 1996. Report on the State of the World's Plant Genetic Resources for Food and Agriculture. Food and Agriculture Organization, Rome, Italy.

FAO, 2010. The Second Report on the State of the World's Plant Genetic Resources for Food and Agriculture. FAO, Rome, Italy, Retrieved July 20, 2012 from <http://www.fao.org/docrep/013/i1500e/i1500e.pdf>.

FAO-RAF, 2000. Integrated soil management for sustainable agriculture and food security Case Studies from Four Countries in West Africa. FAO Regional Office, Accra-Ghana, pp. 146.

FAO/WHO (1999). Guidelines for the production, processing, labelling and marketing of organically produced foods. Joint FAO/WHO food standards programme codex alimentaries commission. Rome, CAC/GL 32. P. 49.

Farwig, N., Berens, D.G., 2012. Imagine a world without seed dispersers: a review of threats, consequences and future directions. Basic. Appl. Ecol. 13, 109–115.

Fliessbach, A., Oberholzer, H.-R., Gunst, L., Mäder, P., 2007. Soil organic matter and biological soil quality indicators after 21 years of organic and conventional farming. Agri. Ecosys. Environ. 118, 273–284.

Francis, C., Lieblein, G., Gliessman, S., Breland, T.A., Creamer, N., Harwood, R., et al., 2003. Agroecology: the ecology of food systems. J. Sustain. Agr. 22, 99–118.

Francis, C.A., Lieblein, G., Breland, T.A., Salomonsson, L., Geber, U., Sriskandarajah, N., et al., 2008. Transdisciplinary research for a sustainable agriculture and food sector. Agron. J. 100, 771–776.

Funtowicz, S.O., Ravetz, J.R., 1993. Science for the post-normal age. Futures 25, 739–755.

Garnett, T., Godfray, J.C.H., 2012. Sustainable intensification in agriculture. Navigating a Course Through Competing Food System Priorities. Food Climate Research Network – Oxford Martin Programme on the Future of Food, Oxford, UK.

Gattinger, A., Müller, A., Häni, M., Oehen, B., Stolze, M., Niggli, U., 2011. Soil Carbon Sequestration of Organic Crop and Livestock Systems and Potential for Accreditation by Carbon Markets "Organic Agriculture and Climate Change Mitigation. A Report of the Round Table on Organic Agriculture and Climate Change". FAO, Rome, Italy, (10–32).

Geels, F.W., 2004. From sectoral systems of innovation to socio-technical systems: insights about dynamics and change from sociology and institutional theory. Res. Policy 33, 897–920.

Geiger, F., Bengtsson, J., Berendse, F., Weisser, W., Emmerson, M., Morales, M., et al., 2010. Persistent negative effects of pesticides on biodiversity and biological control potential on European farmland. Basic and Applied Ecology 11, 97–105.

Géneau, C.E., Wäckers, F.L., Luka, H., Daniel, C., Balmer, O., 2012. Selective flowers to enhance biological control of cabbage pests by parasitoids. Basic and Applied Ecology. 13, 85–93.

Gibbons, M., 1999. Science's new social contract with society. Nature 402, C81–C84.

Gill, J.G. (1995). Major natural resource management concerns in South Asia. Food, Agriculture, and the environment discussion paper 8. Vision 2020. International Food Policy Research Institute, 1200 Seventeenth street, NW, Washington, DC, USA. 1–31.

Giri, A.K., 2002. The calling for creative transdisciplinarity. Futures 34, 737–755.

Gliessman, S.R., 1998. Agroecology: Ecological Processes In Sustainable Agriculture. Ann Arbor Press, Chelsea, Michigan.

Gotor, E., Irungu, C., 2010. The impact of bioversity international's African leafy vegetables programme in Kenya. Impact Assess. Proj. Appraisal 28, 41–55.

Hadorn, G.H., Bradley, D., Pohl, C., Rist, S., Wiesmann, U., 2006. Implications of transdisciplinarity for sustainability research. Ecol. Econ. 60, 119–128.

Hartmann, A., Schmid, M., Tuinen, D., Berg, G., 2009. Plant-driven selection of microbes. Plant Soil 321, 235–257.

Haussmann, B.I.G., Parzies, H., 2009. Methodologies for generating variability. Part 1: use of genetic resources in plant breeding. In: Ceccarelli, S., Guimarães, E.P., Weltzien, E. (Eds.), Plant Breeding And Farmer Participation FAO, Rome, pp. 107–128.

Henao, J., Baanante, C., 2006. Agricultural production and soil nutrient mining in Africa: implication for resource conservation and policy development International Fertilizer Development Centre. IFDC Tech. Bull, Muscle Shoals, AL, USA.

Hepperly, P., Douds, J.D., Seidel, R., 2006. The Rodale farming systems trial 1981 to 2005: long-term analysis of organic and conventional maize and soybean cropping systems. In: Raupp, J., Pekrun, D., Oltmanns, M., Köpke, U. (Eds.), Long-Term Field Experiments in Organic Farming International Society of Organic Agriculture Research (ISOFAR), Bonn, pp. 15–32.

Hirsch Hadorn, G., Hoffmann-Riem, H., Biber-Klemm, S., Grossenbacher-Mansuy, W., Joye, D., Pohl, C., et al., 2008. Handbook of Transdisciplinary Research. Springer, Dordrecht.

Hoeschle-Zeledon, I., and Jaenicke, H. (2007). A strategic framework for global research and development of underutilized plant species: a contribution to the enhancement of indigenous vegetables and legumes. In: Chadha, M.L. et al. (Eds.), "1st International conference on indigenous vegetables and legumes". 103–110.

Howard, P.H., 2009. Visualizing consolidation in the global seed industry: 1996–2008. Sustainability 1, 1266–1287.

IAASTD (2008). Reports from the International Assessment of Agricultural Knowledge, Science and Technology for Development. Available at: <http://www.agassessment.org/>.

IFOAM (2007). The IFOAM basic standards for organic production and processing version 2005. Retrieved July 27, 2012 from <http://www.ifoam.org/about_ifoam/standards/norms/norm_documents_library/IBS_V3_20070817.pdf>.

Jaggard, K.W., Qi, A., Ober, E.S., 2010. Possible changes to arable crop yields by 2050. Philos. Trans. R. Soc. Lond. B Biol. Sci. 365, 2835–2851.

Jarvis, D.I., Hodgkin, T., Sthapit, B.R., Fadda, C., Lopez-Noriega, I., 2011. An heuristic framework for identifying multiple ways of supporting the conservation and use of traditional crop varieties within the agricultural production system. Crit. Rev. Plant. Sci. 30, 125–176.

Joshi, K.D., Devkota, K.P., Harris, D., Khanal, N.P., Paudyal, B., Sapkota, A., et al., 2012. Participatory research approaches rapidly improve household food security in Nepal and identify policy changes required for institutionalisation. Field Crop Res. 131, 40–48.

Katyal, J.C., and Reddy, M.N. (2012). Fertilizer use in South Asia. Agricultural Science Vol. 2: Encyclopedia of Life Support Systems (EOLSS). Retrieved July 27, 2012 from <http://www.eolss.net/Eolss-sampleAllChapter.aspx.aspx>.

Keatinge, J.D.H., Easdown, W.J., Yang, R.Y., Chadha, M.L., Shanmugasundaram, S., 2011. Overcoming chronic malnutrition in a future warming world: the key importance of mungbean and vegetable soybean. Euphytica 180, 129–141.

Keneni, G., Imtiaz, M., 2010. Demand-driven breeding of food legumes for plant-nutrient relations in the tropics and the sub-tropics: serving the farmers; not the crops!. Euphytica 175, 267–282.

Khan, Z.R., Midega, C.A.O., Bruce, T.J.A., Hooper, A.M., Pickett, J.A., 2010. Exploiting phytochemicals for developing a 'push-pull' crop protection strategy for cereal farmers in Africa. J. Exp. Bot. 61, 4185–4196.

Kherallah, M., Delagade, C., Gabre-Madhim, E., Minot, N., Johson, M., 2002. Reforming Agricultural Markets in Africa. IFPRI and Johns Hopkins University Press, Baltimore, pp 201.

Kotschi, J. (2011). Less hunger through more ecology – what can organic farming research contribute? Retrieved July 27, 2012 from <http://www.ke.boell.org/downloads/Less_Hunger_through_mroe_ecology.pdf>.

Lal, R., 2009. Soils and world food security. Soils Tillage Res. 102, 1–4.

Lammerts van Bueren, E.T., Jones, S.S., Tamm, L., Murphy, K.M., Myers, J.R., Leifert, C., et al., 2011. The need to breed crop varieties suitable for organic farming, using wheat, tomato and broccoli as examples: a review. NJAS Wageningen J. Life Sci. 58, 193–205.

Lammerts van Bueren, E.T., Myers, J.R., 2012. Organic crop breeding: integrating organic agricultural approaches and traditional and modern breeding methods. In: Lammerts van Bueren, E.T., Myers, J.R. (Eds.), Organic Crop Breeding. Wiley-Blackwell Press, New York, pp. 3–13.

Lele, U. (1994). Structural adjustment and agriculture: a comparative perspective on Africa, Asia and Latin America. In: Heidhues, F., Knerr, B., Main, F., Lang, P. (Eds.), Food and Agricultural policies under structural adjustment, Seminar of the European Association of Agricultural Economists; 29 (Hohenheim): 1992.09.21–25, Lang, Peter, GmbH, Internationaler Verlag Der Wissenschaften (Juni 1994) 57–115.

Lynch, J.P., 2007. Roots of the second green revolution. Aust. J. Bot. 55, 493–512.

Mäder, P., Fliessbach, A., Dubois, D., Gunst, L., Fried, P., Niggli, U., 2002. Soil Fertility and Biodiversity in Organic Farming. Science 296, 1694–1697.

Mäder, P., Fliessbach, A., Dubois, D., Gunst, L., Jossi, W., Widmer, F., et al., 2006. The DOK experiment (Switzerland). In: Raupp, J., Pekrun, C., Oltmanns, M., Köpke, U. (Eds.), Long-Term Field Experiments in Organic Farming. Koester, Bonn, pp. 41–58.

Malézieux, E., Crozat, Y., Dupraz, C., Laurans, M., Makowski, D., Ozier-Lafontaine, H., et al., 2009. Mixing plant species in cropping systems: concepts, tools and models. A review. Agron. Sustain. Dev. 29, 43–62.

Mandal, N.P., Sinha, P.K., Variar, M., Shukla, V.D., Perraju, P., Mehta, A., et al., 2010. Implications of genotype × input interactions in breeding superior genotypes for favorable and unfavorable rainfed upland environments. Field Crop Res. 118, 135–144.

Max-Neef, M.A., 2005. Foundations of transdisciplinarity. Ecol. Econ. 53, 5–16.

Mayes, S., Massawe, F.J., Alderson, P.G., Roberts, J.A., Azam-Ali, S.N., Hermann, M., 2012. The potential for underutilized crops to improve security of food production. J. Exp. Bot. 63, 1075–1079.

McElhinny, E., Peralta, E., Mazon, N., Danial, D.L., Thiele, G., Lindhout, P., 2007. Aspects of participatory plant breeding for quinoa in marginal areas of Ecuador. Euphytica 153, 373–384.

McGuire, S.J., 2008. Path-dependency in plant breeding: challenges facing participatory reforms in the Ethiopian sorghum improvement program. Agric. Syst. 96, 139–149.

Meertens, H.C.C., 2003. The prospects for integrated nutrient management for sustainable rainfed lowland rice production in Sukumaland, Tanzania. Nutr. Cycl. Agroecosys. 65, 163–171.

Messmer, M.M., Burger, H., Schmidt, W., Geiger, H.H. (2009). Importance of appropriate selection environments for breeding maize adapted to organic farming systems. In: P. Ruckenbauer, A. Brandstetter, M. Geppner, H. Grausgruber, K. Buchgraber, (Ed.), 60. Tagung Der Vereinigung der Pflanzenzüchter und Saatgutkaufleute Österreichs 2009", vol. 60, Gumpenstein. 49–52.

Messmer, M., Hildermann, I., Thorup-Kristensen, K., Rengel, Z., 2012. Nutrient management in organic farming and consequences for direct and indirect selection strategies. In: Lammerts van Bueren, E.T., Myer, J.R. (Eds.), Organic Crop Breeding Wiley-Blackwell Press, New York, pp. 15–38.

Millennium Ecosystem Assessment. (2005). Natural assets and human well-being. Statement from the board. Retrieved July 27, 2012 from <http://www.maweb.org/en/index.aspx>.

Moonen, A.C., Bàrberi, P., 2008. Functional biodiversity: an agroecosystem approach. Agr. Ecosyst. Environ. 127, 7–21.

Mowo, J.G., Janssen, B.H., Oenema, O., German, L.A., Mrema, J.P., Shemdoe, R.S., 2006. Soil fertility evaluation and management by smallholder farmer communities in northern Tanzania. Agr. Ecosyst. Environ. 116, 47–59.

Mulumba, J.W., Nankya, R., Adokorach, J., Kiwuka, C., Fadda, C., De Santis, P., et al., 2012. A risk-minimizing argument for traditional crop varietal diversity use to reduce pest and disease damage in agricultural ecosystems of Uganda. Agri. Ecosys. Environ. (in press).

Mulvaney, R.L., Khan, S.A., Ellsworth, T.R., 2009. Synthetic nitrogen fertilizers deplete soil nitrogen: a global dilemma for sustainable cereal production. J. Environ. Qual. 38, 2295–2314.

Murphy, K., Lammer, D., Lyon, S., Carter, B., Jones, S.S., 2005. Breeding for organic and low-input farming systems: an evolutionary-participatory breeding method for inbred cereal grains. Renew. Agri. Food Syst. 20, 48–55.

Mwangi, S., and Kimathi, M. (2006). African leafy vegetables evolves from underutilized species to commercial cash crops., " Research Workshop on Collective Action and Market Access for Smallholders. 2–5th Oct. 2006 ", Cali, Colombia.

Niggli, U., 2007. FiBL and organic research in Switzerland. In: Lockeretz, W. (Ed.), Organic Farming: An International History CAB International, Wallingford.

Niggli, U. (2010). Organic Agriculture: A Productive Means of Low-carbon and High Biodiversity Food Production, Trade and Environment Review ed. UNCTAD (United Nations) Geneva. 112–118.

Niggli, U., Slabe, A., Schmid, O., Halberg, N., and Schlüter, M. (2008). Vision for an organic food and farming research agenda 2025. Organic Knowledge for the Future. Retrieved July 27, 2012 from <http://orgprints.org/13439/1/niggli-etal-2008-technology-platform-organics.pdf/>.

Niggli, U., Fließbach, A., Hepperly, P., and Scialabba, N. (2009). Low greenhouse gas agriculture: mitigation and adaptation potential of sustainable farming systems. Retrieved July 20, 2012 from <http://orgprints.org/15690/1/niggli-etal-2009-lowgreenhouse.pdf>.

Ogoke, I.J., Ibeawuchi, I.I., Ngwuta, A.A., Tom, C.T., Onweremadu, E.U., 2009. Legumes in the cropping systems of Southeastern Nigeria. J. Sustain. Agr. 33, 823–834.

Oldeman, L.R., Hakkeling, R.T.A., Sombroek, W.G., 1991. World Map of the Status of Human Induced Soil Degradation. An Explanatory Note. ISRIC, Wegeningen, The Netherlands, (Second revised version. Global Assessment of soil degradation).

Omotayo, O.E., Chukwuka, K.S., 2009. Soil fertility restoration techniques in sub-saharan Africa using organic resources. Afr. J. Agri. Resour. 4, 144–150.

Ould-Sidi, M.M., Lescourret, F., 2011. Model-based design of integrated production systems: a review. Agron. Sustain. Dev. 31, 571–588.

Paarlberg, R.L., 1993. Managing pesticides use in developing countries. In: Haas, P.M., Keohane, R.O., Levy, M.A. (Eds.), Institutions for the Earth MIT Press, Cambridge MA, pp. 309–350.

Paris, T.R., Singh, A., Cueno, A.D., Singh, V.N., 2008. Assessing the impact of participatory research in rice breeding on women farmers: a case study in eastern Uttar Pradesh, India. Exp. Agr. 44, 97–112.

Parolin, P., Bresch, C., Desneux, N., Brun, R., Bout, A., Boll, R., et al., 2012. Secondary plants used in biological control: a review. Int. J. Pest Manage. 58, 91–100.

Pautasso, M., Aistara, G., Barnaud, A., Caillon, S., Clouvel, P., Coomes, O., et al., 2012. Seed exchange networks for agrobiodiversity conservation. A review. Agron. Sustain. Dev. 1–25.

Perfecto, I., Vandermeer, J., 2010. The agroecological matrix as alternative to the land-sparing/ agriculture intensification model. Proc. Natl. Acad. Sci. U.S.A. 107, 5786–5791.

Petit, S., Boursault, A., Le Guilloux, M., Munier-Jolain, N., Reboud, X., 2011. Weeds in agricultural landscapes. A review. Agron. Sustain. Dev. 31, 309–317.

Pfiffner, L., Luka, H., 2000. Overwintering of arthropods in soils of arable fields and adjacent semi-natural habitats. Agric. Ecosyst. Environ. 78, 215–222.

Pfiffner, L., Wyss, E., 2004. Use of sown wildflower strips to enhance natural enemies of agricultural pests. In: Gurr, G.M., Wratten, S.D., Altieri, M.A. (Eds.), Ecological Engineering for Pest management. Advances in Habitat Manipulation for Arthropods CSIRO Publishing, Collingwood, VIC, Australia, pp. 165–186.

Pfiffner, L., Luka, H., Schlatter, C., 2005. Funktionelle Biodiversität: Schädlingsregulation gezielt verbessern. Oekologie und Landbau 134, 51–53.

Pimentel, D., Harvey, C., Resosudarmo, P., Sinclair, K., Kurz, D., McNair, M., et al., 1995. Environmental and Economic Costs of Soil Erosion and Conservation Benefits. Science 267, 1117–1123.

Pinstrup-Andersen, P., Pandya-Lorch, R., Rosengrant, M.W., 1999. World Food Prospect: Critical Issues for the Early 21st Century and Food Security: Problems, Prospects and Policies. International Food Policy Research Institute, Washington, DC.

Postma-Blaauw, M.B., de Goede, R.G.M., Bloem, J., Faber, J.H., Brussaard, L., 2012. Agricultural intensification and de-intensification differentially affect taxonomic diversity of predatory mites, earthworms, enchytraeids, nematodes and bacteria. Appl. Soil Ecol. 57, 39–49.

Presterl, T., Seitz, G., Landbeck, M., Thiemt, E.M., Schmidt, W., Geiger, H.H., 2003. Improving nitrogen-use efficiency in European maize: estimation of quantitative genetic parameters. Crop Sci. 43, 1259–1265.

Pretty, J.N., 1995. Regenerating Agriculture: Policies and Practice for Sustainability and Self-Reliance. Josef Henry Press, Earthscan, London.

Pretty, J.N., 2008. Agricultural sustainability: concepts, principles and evidence. Philos. Trans. Biolo. Sci. 363, 447–465.

Ramesh, P., Panwar, N.R., Singh, A.B., Ramana, S., Yadav, S.K., Shrivastava, R., et al., 2010. Status of organic farming in India. Curr. Sci. 98, 1190–1194.

Ratnadass, A., Fernandes, P., Avelino, J., Habib, R., 2012. Plant species diversity for sustainable management of crop pests and diseases in agroecosystems: a review. Agrono. Sustain. Dev. 32, 273–303.

Reid, T., Yang, R.-C., Salmon, D., Spaner, D., 2009. Should spring wheat breeding for organically managed systems be conducted on organically managed land? Euphytica 169, 239–252.

Ryan, J., 2008. Crop Nutrients for Sustainable Agricultural Production in the Drought-Stressed Mediterranean Region. J. Agri. Sci. Technol. 10, 295–306.

Sanchez, P.A., Swaminathan, M.S., 2005. Hunger in Africa: the link between unhealthy people and unhealthy soils. Lancet 365, 442–444.

Schader, C., Stolze, M., Gattinger, A., 2012. Environmental performance of organic agriculture. In: Boye, J., Arcand, Y. (Eds.), Green Technologies in Food Production and Processing. Springer, New York, pp. 183–210.

Scherr, S.J., 1999. Past and present effects of soil degradation. In: Scherr, S.J. (Ed.), Soil degradation – A Threat to Developing Country Food Security by 2020 International Food Policy Research Institute, Washington DC, pp. 13–30. (2020 Discussion Paper 27).

Schlecht, E., Buekert, A., Tielkers, E., Bationo, A., 2006. A critical analysis of challenges and opportunities for soil fertility restoration in Sudano-Sahelian West Africa. Nutr. Cycl. Agroecosys. 76, 109–136.

Schmid, O., Padel, S., Halberg, N., Huber, M., Darnhofer, I., Micheloni, C., et al., (2009). Strategic Research Agenda for organic food and farming. Retrieved July 27, 2012 from <http://orgprints.org/16694/1/tporganics_strategicresearchagenda.pdf>.

Scialabba El-Hage, N., and Hattam, C. (2002). Organic Agriculture, Environment and Food Security. Retrieved July 20. 2012 from <http://www.fao.org/docrep/005/y4137e/y4137e00.htm>.

Seufert, V., Ramankutty, N., Foley, J.A., 2012. Comparing the yields of organic and conventional agriculture. Nature 485, 229–232.

Simon, S., 2010. Biodiversity and organic farming – strengthening the interactions between agriculture and ecosystems. In: Kölling, A., Fertl, T., Schlüter, M. (Eds.), Organic Food and Farming Imprimerie Eschedé-Van Muysewinkel, Belgium, pp. 11–13. (Retrieved July 27, 2012 from) <http://orgprints.org/18489/1/18489.pdf>.

Sinebo, W., Gretzmacher, R., Edelbauer, A., 2002. Environment of selection for grain yield in low fertilizer input barley. Field Crop Res. 74, 151–162.

Smukler, S.M., Sanchez-Moreno, S., Fonte, S.J., Ferris, H., Klonsky, K., O'Geen, A.T., et al., 2010. Biodiversity and multiple ecosystem functions in an organic farmscape. Agric. Ecosyst. Environ. 139, 80–97.

Sperling, L., Cooper, H.D., Remington, T., 2008. Moving towards more effective seed aid. J. Dev. Stud. 44, 586–612.

Sthapit, B., Rana, R., Eyzaguirre, P., Jarvis, D., 2008. The value of plant genetic diversity to resource-poor farmers in Nepal and Vietnam. Int. J. Agri. Sustain. 6, 148–166.

Suneson, C.A., 1956. An evolutionary plant breeding method. Agrono. J. 48, 188–191.

Teasdale, J.R., Coffmann, C.B., Magnum, R.W., 2007. Potential long-term benefits of no-tillage and organic cropping systems for grain production and soil improvement. Agrono. J. 99, 1297–1305.

Temudo, M.P., 2011. Planting knowledge, harvesting agro-biodiversity: a case study of southern guinea-bissau rice farming. Hum. Ecol. 39, 309–321.

Thierfelder, C., Wall, P.C., 2011. Reducing the risks of crops failure for smallholder farmers in Africa through the adoption of conservation agriculture. In: Bationo, Innovations as Key to the Green Revolution in Africa Springer Science Business Media BV (doi:10.1007/978-90481-2543-2_129).

Tian, G., Hauser, S., Koutika, L.S., Ishida, F., Chianu, J.N., 2001. Pueraria cover crop fallow systems: benefits and applicability. In: Tian, Sustaining Soil Fertility in West Africa SSSA Special Publication, Madison, pp. 137–156. (Number 58).

Tscharntke, T., Clough, Y., Bhagwat, S.A., Buchori, D., Faust, H., Hertel, D., et al., 2011. Multifunctional shade-tree management in tropical agroforestry landscapes – a review. J. Appl. Ecol. 48, 619–629.

Tscharntke, T., Clough, Y., Wanger, T.C., Jackson, L., Motzke, I., Perfecto, I., et al., 2012. Global food security, biodiversity conservation and the future of agricultural intensification. Biol. Conserv. 151, 53–59.

United Nations Conference on Trade And Development (UNCTAD), 2006. Trade and Environment Review 2006. United Nations, Geneva, New York, Retrieved July 27, 2012 from <www.unctad.org/trade_env>.

United Nations Environmental Program (UNEP)-UNCTAD (2008). Organic Agriculture and Food Security in Africa. United Nations, New York and Geneva.

Vanloqueren, G., Baret, P.V., 2008. Why are ecological, low-input, multi-resistant wheat cultivars slow to develop commercially? A Belgian agricultural 'lock-in' case study. Ecol. Econ. 66, 436–446.

Vanloqueren, G., Baret, P.V., 2009. How agricultural research systems shape a technological regime that develops genetic engineering but locks out agroecological innovations. Res. Policy 38, 971–983.

Vernooy, R. (Ed.), 2003. Seeds that give – participatory plant breeding. International Development Research Centre. Retrieved July 27, 2012 from <http://www.scribd.com/doc/30166294/Seeds-That-Give-Participatory-Plant-Breeding-In-Focus-Collection>.

Wiafe, E.K. (2010). Evaluation of some leguminous cover crops for restoration of degraded lands. A Case Study in the Kwaebibirem District in the Eastern Region of Ghana., University of Ghana, Legon, Phd thesis.

Wiesmann, U., Biber-Klemm, S., Grossenbacher, W., Hirsch Hadorn, G., Hoffmann-Riem, H., Joye, D., et al., 2008. Enhancing transdisciplinary research: a synthesis in fifteen propositions. In: Hirsch Hadorn, G., Hoffmann-Riem, H., Biber-Klemm, S., Grossenbacher-Mansuy, W., Joye, D., Pohl, C., Wiesmann, U., Zemp, E. (Eds.), Handbook of Transdisciplinary Research Springer, Dordrecht, pp. 433–441.

Willer, H., Kilcher, L., 2012. The World of Organic Agriculture: Statistics and Emerging Trends. FiBL and IFOAM, Frick and Bonn, 331.

Winqvist, C., Ahnström, J., Bengtsson, J., 2012. Effects of organic farming on biodiversity and ecosystem services: taking landscape complexity into account. Ann. N. Y. Acad. Sci. 1249, 191–203.

Wissuwa, M., Mazzola, M., Picard, C., 2009. Novel approaches in plant breeding for rhizosphere-related traits. Plant Soil 321, 409–430.

Wyss, E., Luka, H., Pfiffner, L., Schlatter, C., Uehlinger, G., and Daniel, C. (2005). Approaches to pest management in organic agriculture: a case study in European apple orchards In: CAB International: Organic-Research.com May 2005, 33N–36N.

Zehnder, G., Gurr, G.M., Kühne, S., Wade, M.R., Wratten, S.D., Wyss, E., 2007. Arthropod pest management in organic crops. Annu. Rev. Entomol. 52, 57–80.

Chapter 3

Guar: An Industrial Crop from Marginal Farms

N.A. Kuravadi*, S. Verma*, S. Pareek*, P. Gahlot*, S. Kumari*,
U.K. Tanwar*, P. Bhatele*, M. Choudhary*, K.S. Gill†, V. Pruthi*,
S.K. Tripathi**, K.S. Dhugga‡, and G.S. Randhawa*

*Department of Biotechnology, Indian Institute of Technology Roorkee, India, †Department of
Crop & Soil Sciences, Washington State University, Pullman, WA, USA, **Department of Water
Resource Development and Management, Indian Institute of Technology Roorkee, India,
‡Trait Discovery & Technology, DuPont Pioneer, IA, USA

1. INTRODUCTION

Guar (*Cyamopsis tetragonoloba* [L.] Taub.), also known as cluster bean, has traditionally been used as a fodder and vegetable crop. Because of its ability to fix nitrogen and because it has a smaller seed than other legumes, it is also used as green manure. Galactomannan gum, a polysaccharide derived from guar seeds that results in highly viscous aqueous solutions at low concentrations, has long been used in many industrial applications (Dhugga et al., 2004).

Guar is a drought-hardy, deep-rooted, annual legume grown in the summer season. The name guar is believed to have arisen from the Sanskrit words *gau* and *ahaar*, which mean cow and fodder respectively. In the earlier literature, *C. tetragonoloba* was known as *Dolichos fabaeformis* or *C. psoralioides* (Stephens, 1998). It belongs to the tribe Galegae of the family fabaceae; which has three species, of which *C. tetragonaloba* is the only economically important one. The haploid chromosome number of guar is 7 (Patil, 2004). In India, guar is mainly grown in the dry habitats of Rajasthan, Haryana, Gujarat, and Punjab and to a limited extent in Uttar Pradesh and Madhya Pradesh. Guar is also grown in Pakistan as a cash crop, and to a limited extent in other parts of the world such as Australia, Bangladesh, Brazil, Myanmar, South Africa, southwestern USA, and Sri Lanka.

Two guar varieties, one dwarf and the other tall, are commonly grown in India. A smooth vegetable and a hairy fodder type are both dwarf. Guar is generally 50–100 cm tall and bears 4–10 branches (branched types); however, unbranched types are also available (Kumar and Singh, 2002). The flowers are borne as axillary racemes on long pedicels. The papilionaceous flowers are

Agricultural Sustainability. DOI: http://dx.doi.org/10.1016/B978-0-12-404560-6.00003-4

2–5 mm in size. The pods, which are oblong and 5–12 cm long, normally contain 5–12 oval or cube-shaped seeds (Poats, 1961).

Guar can tolerate high temperature and drought. It grows well in sandy loam soil with pH in the range 7.5–8.0 (Douglas and Routley, 2004). Its immature pods are a source of green vegetable (Kumar and Singh, 2002). In the early varieties the first pod picking starts in 45–50 days and in late varieties in 70–90 days after planting. The crop duration for the early types is 80–90 days and in the late types 135–145 days. The average seed yield is about 5–6 tons per hectare (Sharma et al., 2011). Guar seed consists of seed coat (14–17%), endosperm (35–42%) and germ (43–47%) (Sharma et al., 2007).

2. ORIGIN, GENETICS, AND BREEDING

Guar is considered to have been originated by the domestication of the African wild species, *C. senegalensis*. The domestication process most likely took place in the dry areas of the northwestern region of the Indo-Pakistan subcontinent (Hymowitz, 1972).

In 1903, guar was introduced into the USA for experimentation in the southwestern region, where the hot climate and long growing seasons were suitable for its adaptation (Hymowitz and Matlock, 1963). The objective was to use guar as a soil-improving legume and forage for cattle. Before World War II the carob seed (*Ceratonia siliqua*) from the Mediterranean was used to extract gum for extensive use in the paper industry. During World War II, when the supply of imported carob seed was cut off, a search for a domestic source of galactomannan gum was initiated in the USA by the Institute of Paper Chemistry, Appleton, Wisconsin. Guar filled the role of an alternative crop for galactomannan (Anderson, 1949).

Crossing leads to the hybridization of DNA from two plants belonging to the same genus with a different genetic makeup, but it becomes difficult with small flowers and results in less-consistent pollen production. In guar the flower is only 8 mm long and requires a magnifying lens for emasculation. Once the anthesis begins, ten anthers can be seen encircling the stigma. The pollen is viable from 2 hours before anthesis up to 11 hours after anthesis (Stafford, 1982). The flower morphology leads to self-pollination although an outcrossing rate of 9% has been observed (Gill, 2009). Chaudhary et al. (1974) reported that the optimal time for pollinating the flowers is between 8 a.m. and 9 a.m. and obtained a success rate of 7% for pollination on emasculated flowers.

Stafford (1989) studied the inheritance of a partial male sterility system in guar. Recently, other methods, for example caging, have also been applied in guar to speed up the process of hybrid production (Gill, 2009). However, making crosses in guar still remains an inefficient process.

Mutation breeding is considered to be a useful tool to enrich the variation in crops where useful genetic variability is lacking (Arora and Pahuja, 2008). Vig (1965, 1969) reported reciprocal translocations in a 20 kR gamma-irradiated population of guar with a low fertility rate. Singh et al. (1981) developed early flowering and determinate mutants by gamma- irradiation. Many mutants with useful traits like early flowering, increased yields and gum content have been produced by the mutation breeding approach (Singh, 1986). Rao and Rao (1982) isolated a stabilized early-flowering mutant with an increased number of pods by irradiating the seeds with a 10 kR dose of Co^{60} gamma rays.

Inheritance of branching behavior, clustering pattern, growth habit, and leaf size and hairiness in cluster bean has been studied (Chaudhary and Lodhi, 1981). Each of these traits is controlled by a single gene, except branching behavior which exhibited a two-gene inheritance. The study also showed that the alleles governing branching, discontinuous clustering, indeterminate growth habit, small leaf size and hairiness were dominant over the alleles controlling alternative traits.

Genetic diversity for fodder nutritive value has been assessed by Pathak et al. (2009). They found that genotypes with common geographic origin belonged to different clusters. The extent of genetic variation in guar has also been studied using nuclear rDNA and RAPD markers (Pathak et al., 2011). Intra-specific polymorphism approaching 90% was detected by multilocus genotyping using nine RAPD markers. A single band of uniform size was obtained in all samples after PCR-based amplification of ITS regions encompassing the 5.8 S rDNA. This study suggested wide genetic distribution of guar genetic diversity across agro-climatic zones.

3. WATER AND SALT STRESS

The sustainability of guar cultivation is affected by water and salt stress. Garg et al. (1986) reported adverse effects at pre-flowering stage when water-stressed soil was supplied with different concentrations of saline water as compared with well-watered plants fed with the same concentrations of saline water. The varieties RGC 1076 and GUAG 004 showed stable seed yield as compared with other assessed varieties under rain-fed conditions (Pathak et al., 2010). Studies by Vinisky and Ray (1985) have shown that increase in salt concentration decreases the germination rate, but no effect of increase in temperature was observed on germination in guar. Garg et al. (1997) reported that supplementing guar with calcium significantly countered the adverse effect of NaCl on guar by increasing potassium uptake and reducing sodium uptake. Francois et al. (1990) determined the effect of salinity on vegetative growth and seed yield of two guar cultivars: *Kinman* and *Esser*. Lahiri et al. (1996) reported effects of soil salinity on dry weight, seed yield, leaf area, mineral composition, concentration of leaf metabolites, and enzyme activities in guar.

Salt stress affects mostly the plant height, seed yield, plant growth, and dry and fresh weight of root and shoot in guar (Ashraf et al., 2002). Ashraf et al. (2005) concluded that plants with better rooting systems had a higher seed yield and were more salt tolerant. Igino et al. (2009) reported that the germination range of 42 guar accessions varied from 7.7% to 90.3% at a stress of 200 mM NaCl, but in a greenhouse experiment 100 mM salt was as effective in reducing germination as 200 mM in the field.

4. SEED COMPOSITION

Greater than 90% (w/w) of the endosperm of guar seeds consists of Golgi-synthesized galactomannan (Dhugga et al., 2004). Guar seed is a potential source of additional phytochemicals (Wang and Morris, 2007). Kays et al. (2006) found that the protein, fat, carbohydrate, and ash content of *C. tetragonoloba* L. seeds were in the range 22.9–30.6%, 2.9–3.4%, 50.2–59.9% and 3.0–3.5%, respectively.

The phenolics vary in the range 0.74–1.24% for total phenols, 0.74–1.24% for flavonols, and 0.05–0.24% for hydroxycinnamic acid in the dry matter of guar leaves (Kaushal and Bhatia 1982). As reported by Singh and Misra (1981), the total lipid content of guar meal was found to be 7% by weight of guar seed meal. Guar seed sterols include campesterol, avenasterol, stigmasterol, sitosterol, and traces of delta-7-avenasterol, stigmast-7-enol, brassicasterol, and cholesterol (Ali et al., 1977).

In the powdered endosperm of 11 commercial varieties of *C. tetragonoloba*, the most abundant amino acids were glycine, glutamate, aspartate, serine, and alanine; however, their relative proportions varied considerably. Moreover the proportions of other amino acids, such as histidine, isoleucine, phenylalanine, threonine, tyrosine, and valine, were found to be remarkably constant (Anderson et al., 1985). Kobeasy et al. (2011) carried out biochemical analysis of general chemical components of *Plantago major* L. and *C. tetragonoloba* L. They found that guar seeds contained high contents of proteins, fats, and total hydrolysable carbohydrates as compared with *Plantago* seeds and leaves.

5. GALACTOMANNAN PROPERTIES, BIOSYNTHESIS, AND DEGRADATION

The galactomannan gum from guar is known to have a high molecular weight as compared with other water-soluble polysaccharides. It is a hydrocolloid and shows very high viscosity in polar solvents: ranging from 7000 to 8000 cP (mPa s) in 1% solution at 25°C (Gupta et al., 2009). A viscosity increase of 2000, 3000, 3300, and 4000 cP was observed, respectively, after ½, 1, 2, and 24 hours in a 1% solution of guar gum. A high viscosity value at low concentrations makes guar gum an excellent thickener in the food industry, in preparations like soups, desserts, and pie fillings (Venugopal and Abhilash, 2010).

Galactomannans are deposited as massive thickenings of the endosperm cell walls during seed development (Reid and Meier, 1973). Galactomannan is a product of combined action of two enzymes: mannan synthase (ManS), which makes β-1, 4-linked mannan backbone, and α-galactosyltransferase, which adds galactosyl residues to the mannan backbone by an α-1,6 linkage (Edwards et al., 1992; Reid et al., 1995).

The ratio of mannose to galactose (M:G) in galactomannans is genetically controlled (Dey and Dixon, 1985; Sandhu et al., 2009). Galactosyltransferase specificity regulates the statistical distribution of galactosyl substituents along the mannan backbone as well as the degree of galactose substitution in the primary product formed during galactomannan biosynthesis (Reid et al., 1995). Edwards et al. (1999), following detergent solubilization of fenugreek galactosyltransferase activity and native gel electrophoresis, identified a 51 kDa protein associated with the activity. The corresponding cDNA upon expression in yeast produced a protein that carried the galactosyltransferase activity.

Edwards et al. (1989) proposed a model whereby the transfer of D-galactosyl residues from UDP-galactose to galactomannan was absolutely dependent upon the simultaneous transfer of D-mannosyl residues from GDP-mannose. D-mannan sequences pre-formed *in situ* using the mannosyltransferase in the absence of UDP-galactose could not become galactose-substituted in a subsequent incubation either with UDP-galactose alone or with UDP-galactose plus GDP-mannose. The degree of galactosylation is also believed to be determined by the action of an α-galactosidase in natural galactomannans during later stages of seed development (Joersbo et al., 2001). The degree of galactose-substitution in galactomannans could be regulated either by the selective removal of galactosyl residues after deposition or at the level of the biosynthetic process itself (Reid et al., 1987).

The precursors for galactomannan biosynthesis, GDP-D-mannose and UDP-D-galactose, are formed by the actions of GDP-mannose pyrophosphorylase and UDP-glucose 4-epimerase, respectively. The relative concentrations of these precursors have been found to affect the M:G ratio of the galactomannan polymer (Edwards et al., 1992).

Enzymes involved in the synthesis of polymer backbones of plant cell wall polysaccharides have shown recalcitrance towards biochemical purification (Dhugga, 2012). Dhugga et al. (2004) isolated a cDNA clone of the gene mannan synthase (ManS) from a cDNA library constructed from developing guar endosperm. They found that the transformed soybean somatic embryos expressing ManS cDNA contained high levels of mannan synthase activity that was localized to the Golgi compartment. Naoumkina et al. (2007) determined that cDNA libraries from developing seeds of guar possessed widely differing sets of genes that were activated at the early and late developmental stages. Recently, Naoumkina and Dixon (2011) characterized a ~1.6 kb guar ManS promoter by fusing the promoter sequence with the *GUS* reporter gene and expressing it in alfalfa (*Medicago sativa*).

Enzymes involved in the hydrolysis of galactomannans during seed germination are: β-mannosidase, responsible for hydrolysing the oligomannans released by prior endo β-mannanase activity; β-mannanase, which cleaves the mannan backbone; and α-galactosidase, which removes the galactose side-chain units (Naoumkina et al., 2007).

Three isomorphs of α-galactosidase have been known: α-galactosidase A, α-galactosidase C1, and α-galactosidase C2 (Shivanna and Ramakrishna, 1985). Hughes et al. (1988) reported that synthesis of an α-galactosidase is directed by messenger RNA isolated from aleurone cells during germination of guar seed. Subsequent study by Overbeeke et al. (1989) gave the complete nucleotide sequence of the guar α-galactosidase cDNA.

6. PREPARATION AND APPLICATIONS OF GUAR GUM

The production of guar is dominated by India, as it contributes 80% of the world's total production of guar gum. The country exported a total of 240,000 tons of guar gum worth Rupees (INR) 11.3 billion (over 209 million USD) and 445,000 tons worth INR 28.1 billion (over 520 million USD) in years 2010 and 2011, respectively. Major guar gum importing countries are USA, China, Germany, Russia, and Italy. Also, in the rest of the world, the trend of consumption has increased with time, leading to the introduction of this crop in many countries (www.apeda.gov.in).

Guar gum is obtained by grinding the seed endosperm. After harvesting, the pods are dried in the sun and then threshed mechanically. The endosperm contains most of the pure galactomannan, and thus needs to be separated from the seed coat (Vandamme et al., 2002). On a commercial scale, seeds are roasted in a furnace to loosen the seed coat. After heat treatment, it becomes easy to separate the hull by attrition milling (Kawamura, 2008). The endosperm is powdered with pulverizers. The pulverization step is carried out in humid conditions and special care is taken to avoid degradation of high molecular weight polymers. The gum so obtained is called crude gum or commercial grade gum (Vandamme et al., 2002).

Crude gum obtained by the milling process contains proteins and fibers as impurities. To determine physico-chemical properties, the gum is dissolved in water and partially precipitated with polar solvents. Gum can also be purified by complexation with Cu^{2+} and Ba^{2+} salts or by dialysis for advanced studies (Srivastava and Kapoor, 2005).

6.1 Medicinal Uses

Guar is useful as a cooling agent, digestive aid, appetizer, and laxative (Mukhtar et al., 2006). Guar gum holds potential for improving glycemic control and decreasing serum LDL-cholesterol concentrations in mildly hypercholesterolemic insulin-dependent diabetic patients (Markkola et al., 1992). It has also been found to be effective in the treatment of irritable bowel syndrome (Parisi et al., 2002; Giannini et al., 2006). Guar is also considered helpful in weight loss (Butt et al., 2007). Dietary intake of high viscosity guar gum decreases protein efficacy (Poksay and Schneeman, 1983). Guar gum also decreases lipid utilization by interfering with digestion and absorption of nutri-ents when dissolved in water. This results in slower gastric emptying (Yoon et al., 2008). The gum has hypolipidemic effects (Sharma et al., 2011) and is also used as a synthetic cervical mucus (Burruano et al., 2002) and viscosup-plementation agent in osteoarthritis treatment (Cunha et al., 2005). In cancer chemopreventive activity studies by Eldeen et al. (2006), a *C*-glycosylated derivative of guar gum was found to inhibit the carcinogen activator enzyme cytochrome P450 1A (CYP1A) as well as to induce the carcinogen detoxifi-cation enzymes glutathione-*S*-transferases (GSTs), while its sulfated derivative inhibited both CYP1A and GSTs. Partially hydrolyzed guar gum (PHGG) is produced by controlled partial enzymatic hydrolysis of gum and has a smaller molecular weight and less viscosity than native guar gum. PHGG has been found to be effective in the treatment of diarrhea, constipation, and irritable bowel syndrome (Slavin and Greenberg, 2003). Guar gum and its derivatives in various forms such as coatings, matrix tablets, hydrogels, and nano- or micro-particles have been exploited as potential carriers for colon-specific, antihy-pertensive, protein and transdermal targeted drug delivery (Prabaharan, 2011). Guar gum solubilizes well in aqueous solutions, but issues such as cloudiness, alcohol insolubility, and thermal instability have led to the development of a number of chemically modified guar gums. Also, the substitution of hydroxyl groups with ethers such as hydroxylpropyl allows side groups extension, which may change the solubility and other characteristics of the guar gum. The decrease in viscosity of these derivatized products, which could bestow an advantage in processing pumping and mixing, is credited to a polysurfactant effect imparted by the polyoxyalkylene grafts (Bahamdan and Daly, 2007).

According to Moorhouse et al. (1998), the most widely known derivatives of guar gum are: carboxymethyl guar (CMG), hydroxypropyl guar (HPG), car-boxymethyl hydroxypropyl guar (CMHPG). The remaining part of the seed (except endosperm), which is a by-product of the guar gum industry, is known as guar meal and is a rich source of proteins, essential amino acids (other than methionine, threonine, and tryptophan), and 5–6% fatty acids. Guar meal can be used as a feed supplement for ruminants (Kaur et al., 1981).

Presence of toxic constituents is common in leguminous plants. Bakshi et al. (1964) reported that residual gum accounts for about 18% toxicity of guar meal and is not digested and absorbed from the gastrointestinal tract. They found that when chicks were given 1.8–2.7% guar gum, their growth was depressed. This most likely results from an increase in the viscosity of the chyme, slowing down digestion (Dhugga, 2007). Presence of other toxic anti-growth factors, such as trypsin inhibitor, has also been reported in guar meal (Vidyasagar et al., 1978).

Guar gum is used as a binder, disintegrant, suspending agent, thickening agent and stabilizing agent. The applications of guar gum are summarized in Table 3.1.

TABLE 3.1 Applications of Guar Gum and its Modified Forms in Various Industries

Form of gum	Functions
Explosives Industry	
Crosslinked guar gum	Water remover
Guar gum	Plasticity improver
Hydratable guar gum and self-complexing guar gum	Gelling agent
Nitrate ester of guar gum	Thickening agent
Petroleum Industry	
Sulfonated guar gum and $Me_3\ N^+$ guar gum	Thickener
Alkali refined guar gum crosslinked with borax	Gelling agent
Textiles Industry	
Guar gum and xanthomonas gum	Printing paste thickener
Guar gum derivatives (polyacrylic acid, polyacrylamide, carboxy methyl guar gum, and guar gum mixture grafted with acrylic acid or acrylonitrile or acrylamide)	Thickener
Paper Industry	
Guar gum	Increase fines retention
Quaternary ammonium guar gum	Imparts dry strength
Cationic guar gum	Retention and drainage-providing agent
Guar gum formate	Flocculent and sizing agent
Sodium salt carboxy-methylated guar gum	Dry strength
Ore-Refining/Metal Industry	
Amino ethyl gum	In settling fine colloidal particles
Guar gum	Flocculent
Guar gum with polyacrylamide	Reduces mechanical work
Guar gum and its derivatives	Binder
Coal-Mining Industry	
Esterified guar gum	Stabilizer
Guar gum	Dispersant
Guar gum with boric acid/borax	For shock impregnation of coal seams
Food Industry	
Guar gum	Thickener, binder
Guar gum	Stabilizer and gelling agent
Guar gum with carrageenan and O-carboxy methyl cellulose	Stable thixotropic stabilizer, emulsifier system
Pharmaceutical and Cosmetic Industry	
Guar gum	Stabilizer, suspending agent, binder/disintegrant
Sodium carboxymethyl guar	Binds medicine

TABLE 3.1 (Continued)

Form of gum	Functions
Guar gum	Drug targeting to colon
Modified guar gum	Sustained release of diltiazem-hydrochloride
Guar gum/urea and sulfite	Thickener
Partially hydrolyzed guar gum	Blood glucose and serum cholesterol-lowering agent and decreases transit time of colon
Agriculture	
Guar gum or guar gum + polyvinyl alcohol + borax	Increased water retention capacity of soil
Guar gum	Anticrusting agent and adhesive of azotobacter
Guar gum polyacrylamide	Water-retaining agent
Guar gum with fatty acid derivatives and kaolin	Prevention of granules
Other Industries	
Guar gum derivatives (glycinhydroxamate in guaran, acetic acid hydroxamate in guaran)	Separation of metal ions
Guar gum	Purification of lectins
Guar gum with isobutylene and carboxymethyl cellulose	Lubricant for installation of electric and telephone cable and also as electrical insulator
Guar gum	Adhesive in tobacco products
Guar gum	Water proofing in building/construction material

Modified from Kumar and Singh (2002).

7. BIOTECHNOLOGY FOR GUAR GUM MODIFICATION

Transformation and regeneration of guar has been challenging. Single-cell culture has emerged as an important application of plant tissue culture in sexually incompatible crosses. Saxena et al. (1986) isolated protoplasts of *C. tetragonoloba* from leaf tissue of *in vitro* grown plants and reported their sustained division in modified Kao and Michayluk (1975) nutrient medium. In a separate study Saxena and Gill (1986) also studied the role of polyvinylpolypyrrolidone in enhancement of growth of protoplast cultures of guar.

The recalcitrant nature of leguminous crops has always been a problem for developing an efficient regeneration system for various *in vitro* manipulations. Ramulu and Rao (1991) reported formation of embryoids from callus in guar using naphthaleneacetic acid (NAA) and benzyladenine (BA); however, the development of these embryoids into plantlets and subsequent regeneration has not been elucidated. In another study, Ramulu and Rao (1993) found kinetin to be essential for callus induction and maintenance. Attempts have been made by Prem et al. (2003) to use meristematic tissue as a source of totipotent cells. They induced direct shoot morphogenesis using cotyledonary node explants of guar. Prem et al. (2005) have also studied the effect of different concentrations of various growth regulators on *de novo* regeneration in guar. They achieved shoot regeneration in medium containing NAA (13.0 μM) and BA (5.0 μM).

Optimization of the transformation parameters in guar using a marker (*npt II*) and a reporter gene (*uidA*) was carried out by Joersbo et al. (1999). Analysis of the transformants showed stable expression and inheritance of the genes. The alpha-galactosidase gene of the senna species was transformed into guar by Joersbo et al. (2001). Endosperm with reduced galactose content was produced by 30% of the transformants.

That the expression of guar ManS in soybean led to a significant increase in the mannose content of the seeds opened the door for the production of guar gum in species other than the galactomannan gum-producing native species (Dhugga et al., 2004). An alternative is to modify the galactomannan composition in guar itself.

The quality of a galactomannan gum is determined by the mannose/galactose (M:G) ratio, which is approximately 2 in guar and close to 4 in locust bean. Comparatively longer stretches of unsubstituted mannan backbone in the locust bean gum allow for the formation of micelles because of interchain hydrogen bonding. The result is a more viscous solution than with the guar gum at lower concentrations. Locust bean gum thus commands a much higher market price than guar gum. Down-regulation of galactosyltransferase or an overexpression of a galactosidase in guar endosperm could lead to an increased M:G ratio, which will increase the value of guar.

8.   SUMMARY AND FUTURE PROSPECTS

Guar offers an environmentally friendly alternative to undesirable chemicals which can leak into underground aquifers, contaminating the water. With increasing global demand for guar gum, mainly because of fracking, which involves deep drilling for natural gas, prices have increased by 50% in the past two years (Dezember, 2011). Despite its economic importance, very few genomic resources exist for guar. Development of genetic markers and tissue transformation capability will help improve guar yield in general and gum quality in particular, further improving the lives of marginal farmers in developing countries, besides contributing towards environmental health.

REFERENCES

Ali, A.M., Hussain, N., Haq, S.A., 1977. Sterol composition of guar seed oil. Pak. J. Sci. Ind. Res. 20, 279–281.

Anderson, E., 1949. Endosperm mucilages of legumes. Ind. Eng. Chem. Res. 41, 2887–2890.

Anderson, D.M.W., Howlett, J.F., McNab, C.G.A., 1985. The amino acid composition of the protein-aceous component of guar gum (*Cyamopsis tetragonolobus*). Food Addit. Contam. 2, 225–230.

Arora, R.N., Pahuja, S.K., 2008. Mutagenesis in Guar [*Cyamopsis tetragonoloba* (L.) Taub.]. Plant Mutation Reports 2, 7–9.

Ashraf, M.Y., Akhtar, K., Sarwar, G., Ashraf, M., 2002. Evaluation of arid and semi-arid eco-types of guar (*Cyamopsis tetragonoloba* L.) for salinity (NaCl) tolerance. J. Arid Environ. 52, 473–482.

Ashraf, M.Y., Akhtar, K., Sarwar, G., Ashraf, M., 2005. Role of the rooting system in salt toler-ance potential of different guar accessions. Agron. Sustain. Dev. 25, 243–249.

Bahamdan, A., Daly, W.H., 2007. Hydrophobic guar gum derivatives prepared by controlled graft-ing processes. Polymer. Adv. Tech. 18, 652–659.

Bakshi, Y.K., Greger, C.R., Couch, J.R., 1964. Studies on guar meal. Poult. Sci. 43, 1302.

Burruano, B.T., Schnaare, R.L., Malamud., D., 2002. Synthetic cervical mucus formulation. Contraception 66, 137–140.

Butt, M.S., Shahzadi, N., Sharif, M.K., Nasir, M., 2007. Guar gum: a miracle therapy for hyper-cholesterolemia, hyperglycemia and obesity. Cr. Rev. Food Sci. 47, 389–396.

Chaudhary, B.S., Lodhi, G.P., 1981. Studies on the inheritance of five qualitative characteristics in clusterbean (*Cyamopsis tetragonoloba* (L.) Taub). Euphytica 30, 161–165.

Chaudhary, B.S., Paroda, R.S., Solanki, K.R., 1974. A new crossing technique in cluster bean (*Cyamopsis tetragonoloba* [L.] Taub.). Curr. Sci., 43.

Cunha, P.L.R., Castro, R.R., Rocha, F.A.C., de Paula, R.C.M., Feitosa, J.P.A., 2005. Low vis-cosity hydrogel of guar gum: Preparation and physicochemical characterization. Int. J. Biol. Macromol. 37, 99–104.

Dey, P.M., Dixon, R.A., 1985. Biochemistry of storage carbohydrates in green plants. Academic Press.

Dezember, R., 2011. Farmer says: hitch your wagons to some guar. Wall St. J. November (25).

Dhugga, K.S., 2007. Maize biomass yield and composition for biofuels. Crop Sci. 47, 2211–2227.

Dhugga, K.S., 2012. Biosynthesis of non-cellulosic polysaccharides of plant cell walls. Phytochemistry 74, 8–19.

Dhugga, K.S., Barreiro, R., Whitten, B., Stecca, K., Hazebroek, J., Randhawa, G.S., et al., 2004. Guar seed beta-mannan synthase is a member of the cellulose synthase super gene family. Science 303, 363–366.

Douglas, C.A., Routely, R., 2004. Guar in the new crop industries - Handbook. Salvin, S., Bourke, M., Byrne, T. (Eds.), Rural Industries Research and Development Corporation RIRDC. Publication No. 04/125.

Edwards, M., Bulpin, P.V., Dea, I.C.M., Reid, J.S.G., 1989. Biosynthesis of legume-seed galactoman-nans in vitro. Cooperative interactions of a guanosine 5'-diphosphate-mannose-linked (1~4)-13-D-mannosyltransferase and a uridine 5'-diphosphategalactose-linked Ct-D-galactosyltransferase in particulate enzyme preparations from developing endosperms of fenugreek (*Trigonella foenum graecum* L.) and guar (*Cyamopsis tetragonoloba* [L.] Taub.). Planta 178, 41–51.

Edwards, M.E., Dickson, C.A., Chengappa, S., Sidebottom, C., Gidley, M.J., Reid, J.S.G., 1999. Molecular characterisation of a membrane-bound galactosyltransferase of plant cell wall matrix polysaccharide biosynthesis. Plant J. 19, 691–697.

Edwards, M., Scott, C., Gidley, M.J., Reid, J.S.G., 1992. Control of mannose/galactose ratio dur-ing galactomannan formation in developing legume seeds. Planta 187, 67–74.

Eldeen, A.M.G., Amer, H., Helmy, W.A., 2006. Cancer chemopreventive and anti-inflammatory activities of chemically modified guar gum. Chem. Biol. Interact. 161, 229–240.

Francois, L.E., Donovan, T.J., Maas, E.V., 1990. Salinity effects on emergence, vegetative growth, and seed yield of guar. Agron. J. 82, 587–592.

Garg, B.K., Kathju, S., Vyas, S.P, Lahiri, A.N., 1986. Effect of saline waters on drought affected cluster bean. Proc. Indian Acad. Sci. Anim. Sci. 96, 531–538.

Garg, B.K., Kathju, S., Vyas, S.P., Lahiri, A.N., 1997. Alleviation of sodium chloride induced inhibition of growth and nitrogen metabolism of clusterbean by calcium. Biol. Plant. 39, 395–401.

Giannini, E.G., Mansi, C., Dulbecco, P., Savarino, V., 2006. Role of partially hydrolyzed guar gum in the treatment of irritable bowel syndrome. Nutrition 22, 334–342.

Gill, S.L., 2009. Evaluation of reciprocal hybrid crosses in guar. Texas Tech University, Texas, USA.

Gupta, S.C., Hooda, K.S., Mathur, N.K., Gupta, S., 2009. Tailoring of guar gum for desert and stabilization. Indian J. Chem. Technol. 16, 507–512.

Hughes, S.G., Overbeeke, N., Robinson, S., Pollock, K., Smeets, F.L.M., 1988. Messenger RNA from isolated aleurone cells directs the synthesis of an alpha-galactosidase found in the endosperm during germination of guar (Cyamopsis tetragonaloba) seed. Plant Mol. Biol. 11, 783–789.

Hymowitz, T., 1972. The trans-domestication concept as applied to guar. Econ. Bot. 26, 49–60.

Hymowitz, T., Matlock, R.S., 1963. Guar in the United States. Okla. Aes. Tech. Bull. 611, 1–34.

Igino, T., Weixin, L., Ellen, B.P., 2009. Salinity effects on seed germination and plant growth of guar. Crop Sci. 49, 637–642.

Joersbo, M., Brunstedt, J., Marcussen, J., Okkels, F.T., 1999. Transformation of the endospermous legume guar (Cyamopsis tetragonoloba) and analysis of transgene transmission. Mol. Breed. 5, 521–529.

Joersbo, M., Pedersen, S.G., Nielsen, J.E., Marcussen, J., Brunstedt, J., 1999. Isolation and expression of two cDNA clones encoding UDP-galactose epimerase expressed in developing seeds of the endospermous legume guar. Plant Sci. 149, 147–154.

Joersbo, M., Marcussen, J., Brunstedt, J., 2001. In vivo modification of the cell wall polysaccharide galactomannan of guar transformed with α-galactosidase gene cloned from senna. Mol. Breed 7, 211–219.

Kao, K.N., Michayluk, M.R., 1975. Nutrient requirements for growth of Vicia hajastana cells and protoplasts at a very low population density in liquid media. Planta 126, 105–110.

Kaur, A., Arora, S.K., Joshi, U.N., 1981. Nutritive value of guar meal, proteins isolate and concentrate. Guar News Lett. 2, 41–43.

Kaushal, G.P., Bhatia, I.S., 1982. A study of polyphenols in the seeds and leaves of Guar (Cyamopsis tetragonoloba L. Taub). J. Sci. Food Agric. 33, 461–470.

Kawamura, Y., 2008. Guar gum chemical and technical assessment [Online]. Available at: <www.fao.org/ag/agn/agns/jecfa/cta/69/Guar_gum_CTA_69.pdf>.

Kays, S.E., Morris, J.B., Kim, Y., 2006. Total and soluble dietary fiber variation in cyamopsis tetragonoloba (L.) Taub.(guar) genotypes. J. Food Qual. 29, 383–391.

Kobeasy, M.I., Fatah, O.M.A., Salam, S.M.A., Mohamed, Z.E.M., 2011. Biochemical studies on Plantago major L. and Cyamopsis tetragonoloba L. Int. J. Biodivers. Conservation. 3, 83–91.

Kumar, D., Singh, N.B., 2002. Guar in India Jodhpur. Scientific Publishers Jodhpur.

Lahiri, A.N., Garg, B.K., Vyas, S.P, Kathju, S, Mali, P.C., 1996. Genotypic differences to soil salinity in clusterbean. Arid Soil Res. Rehab. 10, 333–345.

Markkola, H.V., Sinisalo, M., Koivisto, V.A., 1992. Guar gum in insulin-dependent diabetes: effects on glycemic control and serum lipoproteins. Am. J. Clin. Nutr. 56, 1056–1060.

Moorhouse, R., Harry, D.N., Merchant, U.D., 1998. Society of Petroleum Engineering 39531, 253–269.

Mukhtar, H.M., Ansari, S.H., Bhat, Z.A., 2006. Hypoglycemic activity of psidium guajava Linn. Bark. Tradit. Syst. Med., 288.

Naoumkina, M., Dixon, R.A., 2011. Characterization of the mannan synthase promoter from guar (Cyamopsis tetragonoloba). Plant Cell Rep. 30, 997–1006.

Naoumkina, M., Torres-Jerez, I., Allen, S., He, J., Zhao, P.X., Dixon, R.A., et al., 2007. Analysis of cDNA libraries from developing seeds of guar (*Cyamopsis tetragonoloba* (L.) Taub. BMC Plant Biol. 7, 62.

Overbeeke, N., Fellinger, A.J., Toonen, M.Y., Wassenaar, D., Verrips, C.T., 1989. Cloning and nucleotide sequence of the α-galactosidase cDNA from *Cyamopsis tetragonoloba* (guar). Plant Mol. Biol. 13, 541–550.

Parisi, G.C., Zilli, M., Miani, M.P., Carrara, M., Bottona, E., Verdianelli, G., et al., 2002. High-fiber diet supplementation in patients with irritable bowel syndrome (IBS): a multicenter, randomized, open trial comparison between wheat bran diet and partially hydrolyzed guar gum (PHGG). Dig. Dis. Sci. 47, 1697–1704.

Pathak, R., Singh, M., Henry, A., 2009. Genetic diversity for fodder nutritive aspects among guar [*Cyamopsis tetragonoloba* (L.) Taub.] genotypes. Emerg. Trends Forage Res. Livestock Prod., 160–163.

Pathak, R., Singh, S.K., Singh, M., Henry, A., 2010. Performance and stability of *Cyamopsis tetragonoloba* (L.) Taub. genotype under rainfed conditions. Indian J. Dryland Agric.Res. Dev. 25, 82–90.

Pathak, R., Singh, S.K., Singh, M., 2011. Assessment of genetic diversity in clusterbean using nuclear rDNA and RAPD markers. J. Food Legumes 24, 180–183.

Patil, C.G., 2004. Nuclear DNA amount variation in *Cyamopsis* DC (Fabaceae). Cytologia 69, 59–62.

Poats, J.J., 1961. Guar a summer row crop for southwest. Econ. Bot. 14, 241.

Poksay, K.S., Schneeman, B.O., 1983. Pancreatic and intestinal response to dietary guar gum in rats. J. Nutr. 113, 1544.

Prabaharan, M., 2011. Prospective of guar gum and its derivatives as controlled drug delivery systems. A Review. Int. J. Biol. Macromolec 49, 117–124.

Prem, D., Singh, S., Gupta, P.P., Singh, J., Yadav, G., 2003. High frequency multiple shoot regeneration from cotyledonary nodes of guar (*Cyamopsis tetragonoloba* L. Taub.). *In Vitro* Cell Dev. Biol. 39, 384–387.

Prem, D., Singh, S., Gupta, P.P., Singh, J., Pal, S., Kadayan, S., 2005. Callus induction and *de novo* regeneration from callus in guar (*Cyamopsis tetragonoloba*). Plant Cell Tiss. Org. 80, 209–214.

Ramulu, C.A., Rao, D., 1991. Tissue culture studies of differentiation in a grain legume *Cyamopsis tetragonoloba* (L.) Taub. J. Physiol. Res. 4 (2), 183–185.

Ramulu, C.A., Rao, D., 1993. *In vitro* effect of phytohormones on tissue cultures of cluster bean (*Cyamopsis tetragonoloba* L. Taub.). J. Physiol. Res. 20, 7–9.

Rao, S., Rao, D., 1982. Studies on the effect of X-irradiation on *Cyamopsis tetragonoloba* (L.) Taub. Proc. Indian Nat. Sci. Acad. (Biol. Sci.) 48, 410–415.

Reid, J.S.G., Meier, H., 1973. Formation of the endosperm galactomannan in leguminous seeds: preliminary communications. Caryologia Suppl. 25, 219–222.

Reid, J.S., Edwards, M., Dea, I., 1987. Biosynthesis of galactomannan in the endosperms of developing fenugreek (*Trigonella foenum-graecum* L.) and guar (*Cyamopsis tetragonoloba* [L.] Taub.) seeds. Food Hydrocolloids 1, 381–385.

Reid, J.S.G., Edwards, M., Gidley, M.J., Clark, A.H., 1995. Enzyme specificity in galactomannan biosynthesis. Planta 195, 489–495.

Sandhu, A.P.S., Randhawa, G.S., Dhugga, K.S., 2009. Plant cell wall matrix polysaccharide biosynthesis. Mol. Plant 2, 840–850.

Saxena, P.K., Gill, R., 1986. Removal of browning and growth enhancement by polyvinylpolypyrrolidone in protoplast cultures of Cyamopsis tetragonoloba (L.). Biol. Plant 28, 313–315.

Saxena, P.K., Gill, R., Rashid, A., 1986. Isolation and culture of protoplasts from mesophyll tissue of legume Cyamopsis tetragonoloba (L.). Plant Cell Tiss. Org. 6, 173–176.

Sharma, B.R., Chechani, V., Dhuldhoya, N.C., Merchant, U.C., 2007. Guar gum Science Tech Entrepreneur. Lucid Colloids Limited, Jodhpur, Rajasthan, India.

Sharma, P., Dubey, G., Kaushik, S., 2011. Chemical and medico-biological profile of Cyamopsis tetragonoloba (L.) Taub: an overview. J. Appl. Pharmaceut. Sci. 01, 32–37.

Shivanna, B.D., Ramakrishna, M., 1985. α-Galactosidase from germinating guar (Cyamopsis tetragonolobus) seeds. J. Biosci. 9, 109–116.

Singh, S.P., Misra, B.K., 1981. Lipids of guar seed meal (Cyamopsis tetragonoloba L. Taub). J. Agricult. Food Chem. 29, 907–909.

Singh, V.P., 1986. Induced high yielding mutants in clusterbean. Indian J. Agr. Sci. 56, 695–700.

Singh, V.P., Yadav, R.K., Chowdhury, R.K., 1981. Note on a determinate mutant of clusterbean. Indian J. Agri. Sci. 51, 682–683.

Slavin, J.L., Greenberg, N.A., 2003. Partially hydrolyzed guar gum: clinical nutrition uses. Nutrition 19, 549–552.

Srivastava, M., Kapoor, V.P., 2005. Seed galactomannans: an overview. Chem Biodivers. 2, 295–317.

Stafford, R.E., 1982. Yield stability of guar breeding lines and cultivars. Crop Sci. 2, 1009–1011.

Stafford, R.E., 1989. Inheritance of partial male-sterility in guar. Plant Breed. 103, 43–46.

Stephens, J.M., 1998. Guar – Cyamopsis tetragonoloba (L.) Taub. HS608. Horticultural Sciences Department, Florida Cooperative Extension Service, Institute of Food and Agricultural Sciences, University of Florida.

Vandamme, E.J., De Baets, S., Steinbuchel, A., 2002. Polysaccharides II, polysaccharides from eukaryotes. Wiley-VCH Weinheim, Germany. Biopolymers 6.

Venugopal, K.N., Abhilash, M., 2010. Study of hydration kinetics and rheological behaviour of guar gum. Int. J. Pharma Sci. Res. 1, 28–39.

Vidyasagar, P.D., Thakur, R.S., Pradhan, K., 1978. Nutritional evaluation of processed guar (Cyamopsis tetragonoloba) meal for broilers. Indian J. Poult. Sci. 13, 155–160.

Vig, B.K., 1965. Effect of reciprocal translocation on cytomorphology of guar. Sci. and Cult. 31, 532–533.

Vig, B.K., 1969. Studies with Co[60] radiated guar [Cyamopsis tetragonoloba (L.) Taub.]. Ohio J. Sci. 69, 18.

Vinisky, I., Ray, D.T., 1985. Effects of various salts and temperatures on germination of guar. Forage and Grain Reports 64, 102–103.

Wang, M.L., Morris, J.B., 2007. Flavonoid content in seeds of guar germplasm using HPLC. Plant Genet. Resour. 5, 96–99.

Yoon, S.J., Chu, D.C., Juneja, L.R., 2008. Chemical and physical properties, safety and application of partially hydrolyzed guar gum as dietary fiber. J. Clin. Biochem. Nutr. 42, 1–7.

Effective Management of Resources (Nutrients and Water) and Crop Modelling

Nitrogen Use as a Component of Sustainable Crop Systems

Amritbir Riar and David Coventry

School of Agriculture, Food and Wine, The University of Adelaide, Glen Osmond, Australia

1. INTRODUCTION

In early 2012 the population of the world reached 7 billion, and given the present growth rate, where about 1 billion people will be added every 12 years, food production must be an increasing concern (FAO, 2012). As the impacts of present agricultural practice on land, water, and other resources are already under question environmentally, the issue of food production increasingly becomes more complex: i.e., can we feed nine billion people (the projected population in 2050) sustainably? Although food production has consistently increased since the 1950s, the definitions of limiting and non-limiting resources have also changed as increasing production has occurred, with recognition of the costs of soil degradation, groundwater pollution, eutrophication, emission of greenhouse gases, and ammonia and nitrate losses.

Resources are finite to meet these challenges, and land and water suitable for agriculture are ultimately limited on a global scale. Countries like China and India are already on the threshold for the use of arable land, with little scope to expand. There are some projections in other developing countries for expansion of arable land but by only up to 8%, and mostly at the cost of tropical forests (Eickhout et al., 2006). So globally there is not much scope to increase food production by introducing new arable land. Thus the focus for efforts to expand food production in both developing and developed countries must be on raising crop yields on existing arable lands and improving production efficiencies, outcomes that can only be achieved by using improved cultivars together with improved agronomic practices. A key part of this process of agronomic management is the improvement of resource use efficiencies, especially of the key nutrient nitrogen (Cassman et al., 2003).

Nitrogen is a unique element as we have it in abundance in our environment but its availability in forms suitable for plant use has made it an expensive input in crop production systems. Historically, agriculture has been sustained

Agricultural Sustainability. DOI: http://dx.doi.org/10.1016/B978-0-12-404560-6.00004-6

without externally added N, using N from various sources like organic material inputs, dry depositions from the atmosphere, and biological N_2-fixation of soil (Fischer, 2000). A dramatic escalation has occurred in the last five decades in consumption of synthetic N at the global level, as it has been increased from 11.6 million to 104 million tonnes in the period 1961-2006 (Mulvaney et al., 2009; Hoang and Alauddin 2010). Nitrogen input and its efficient use by plants is integral both to increasing crop production as well as addressing issues of sustainability. In this chapter, we will review the role of nitrogen in relation to sustainable crop production and discuss possible strategies for improving the efficiency of N use.

2. PRINCIPLES OF CROP SUSTAINABILITY IN RAINFED FARMING

A general understanding of sustainable practices is that they meet the needs of the present generation (e.g., yield, resource, and environment integrity) without compromising resources for future needs. However it is recognized that sustainability may mean different things for developed and developing countries: for example, in developing countries the priority may be to establish the capacity to produce enough food first, with considerations of sustainability coming when this is achieved. However, the development of higher input and higher output cropping systems must be achieved with the dual goals of higher yield and sustainable practice. The past concerns of researchers remained bilateral: on one side, seeking development of alternative methods for conventional agricultural practices often with high external inputs and low resource use efficiencies (Matson et al., 1997); and, in contrast, developing systems that target low external inputs (Altieri, 1995). Today it is increasingly recognized that both approaches can be applied selectively and in an integrated manner. Such integrated approaches must necessarily consider the individual components of the production system. For example, within cropping systems, which are the focus of this chapter, tillage systems, fertilization and other agronomic inputs can be improved in tandem. But consideration also has to be given to understanding the long-term impacts of these system components and their interactions with the other components, and also the complexity involved with and within these components. Here, we will briefly discuss principles of crop sustainability as they are expressed within the key components of the cropping system. In particular we will highlight both the direct benefits that occur from a given practice or input, and also the associated or additional benefits that occur as interactions arising from the combinations of practices or inputs.

2.1 Nitrogen Use Efficiency (NUE)

NUE is a universal parameter that is used to describe how efficiently the crop plant uses available nitrogen. Crop yields have increased in the last 40 years by 2.4 fold as compared with a 7.4 fold increase in synthetic N use (Tilman et al., 2002). This reflects a sharp decrease in the NUE by crop production. Several parameters are used to describe and identify the N pathways and how they can be used efficiently. Agronomic efficiency and the apparent recovery are the two parameters that are considered most important from the sustainability point of view. Agronomic efficiency is an indicator of the ability of a crop to increase the grain yield in response to applied N, whereas the apparent recovery reflects the ability of the crop to recover N from soil. Fageria and Baligar (2005) describe NUE as the maximum economic yield per unit of applied N. But maximum economic yield, which is the outcome pursued by farmers, cannot be achieved with low efficiency levels. So, in order to achieve maximum economic yield without compromising sustainability, the only option is to improve resource use efficiency.

Current values of NUE show that recovery of N for most crops is less than 50% (Fageria and Baligar, 2005). For cereals like rice, wheat, sorghum, millet, barley, corn, oat, and rye the recovery of N is reported to be about 33% (Raun and Johnson, 1999). The Australian trends for dryland wheat are also in line with world statistics, and are in the range 40–50% (McDonald, 1989; Fillery and McInnes, 1992; Van Herwaarden et al., 1998). The authors cited above all indicate that there is a considerable potential and need to improve the NUE, especially when recognizing sustainability outcomes in crop production systems. The low values of NUE may also be due to physiological limits of crop plants, but the N dynamics within the farming system can be much improved and are as yet not completely understood. On the basis of present values of Harvest Index, Anderson (1985) argues that the agronomic efficiency of N can be doubled (i.e., from 20–30kg/kg to 50kg/kg) in Australian wheat by breeding genotypes more efficient in N uptake and utilization. Conversely, studies from North America, UK, and elsewhere in Australia indicate that much higher values for agronomic efficiency and apparent recovery for N can be achieved with existing genotypes (Craswell and Godwin, 1984; Angus and Fischer, 1991; Sylvester-Bradley et al., 2001). These studies make it clear that, apart from appropriate genotypes, there are agronomic and environmental factors that are limiting the NUE of existing farming systems.

2.2   Water–N Relationship

The low recovery values of N indicate the loss of N or inability of the plant to take up the N. Crop-available water plays a very important role in N recovery by plants. Crop water when present in abundance leads to various losses of N, and its absence inhibits the uptake of N. So the first goal of farming systems is to achieve maximum return from the available water; i.e., to maximize water use efficiency. Water use efficiency (WUE) is generally defined as the yield per unit of crop water use. Total water use may not be the main limitation on crop growth in the Australian rainfed cropping system (Cornish and Murray, 1989; Norton and Wachsmann 2006; Passioura, 2006), but plant-available water and timing of rainfall in pre-anthesis or post-anthesis growth period is considered to be critical (Angus and Van Herwaarden, 2001). Thus the aim of rainfed cropping systems from an economic, environmental (and sustainability) point of view is to optimize the balance between water and N-use efficiencies. In a rainfed farming system it can be challenging to make a balance between N-use and water use as crop N demand remains at low levels in the early period of growth when usually there is adequate availability of stored water, whereas after post-anthesis, when the crop requires more N, water often remains at the limiting end (Angus 2001; López-Bellido et al., 2005). N-use and water use efficiencies can be explained by the co-limitation theory (Sadras, 2004, 2005). Theoretically, the maximum biological yield can be obtained only when all the factors are equally limiting (Bloom et al., 1985). And in accordance with this concept, maximum yield can only be obtained when water and N co-limit equally, as water is the primary limiting factor at the dry end and N is limiting at the wet end of the growth cycle (Sadras, 2004). In rainfed cropping systems it is difficult to control the water supply at critical crop development stages, so the only management option to be considered is to manage N effectively. In contrast, water availability is not a limitation in areas with irrigation facilities, and these areas constitute 17% of the total cultivated land and contribute 40% of world food (Rhoades and Loveday, 1990). Low NUE values in these areas are mainly due to mismanagement of N causing losses of N to the environment. These losses are in the form of denitrification, immobilization, leaching, and volatilization individually or combined (Craswell and Godwin, 1984; Sowers et al., 1994; Raun and Johnson, 1999). To overcome these losses and to improve the NUE, there is a greater need to fine-tune N supply and irrigation according to the crop demand.

2.3   Crop Rotations

In the past three to four decades the use of synthetic fertilizers (especially N) and pesticides have become the basis for increases in the yield of crops in developed and developing countries. The use of N fertilizers changed the natural cycling of C and N in the field, which was previously managed by the legume rotations, fallow periods, and applications of manure. Crop rotations reduce the risk of disease, insect attack, weeds, and crop failure risk associated with monoculture (Helmers et al., 1986; Karlen et al., 1994), with resulting improvements in crop yield (Pierce and Rice, 1988). The mechanism and factors that result in yield increases attributed to this rotation effect are mostly understood, and where legumes are involved the benefits can be associated with increased N availability (Bullock, 1992). Many studies have shown that legumes contribute N to the rotation, but often not enough to maintain high yields (Pierce and Rice, 1988). Whilst the inclusion of such legumes in the crop rotation may reduce the dependence on externally applied N, there still may be a requirement for externally applied N to optimize yield. The success of legumes in any crop rotation depends upon the amount of N_2 fixed by it for the subsequent crop, which will be affected by environmental and edaphic factors. Where the legume crop is to be used for human consumption there is need to consider that this crop will be removing significant amounts of N from the field.

2.4   Fertilization to Optimize Yields

Over-application of fertilizers (inorganic and organic) has created problems with water and soil health in some situations in both developed and developing countries. Conversely, in other situations there is over-extraction of nutrients, hence poor soil fertility leading to lower yields and lower resource use efficiency. This raises the concept of balanced fertilization, particularly with respect to the use of N, P, and K. The cited ratios of N, P and K use are 1.0:0.56:0.50 in developed countries and 1.0:0.39:0.19 in developing countries (IFA (2012)). The disparity between these ratios (i.e., away from a balance of 1.0:1.0:1.0) show that N is by far the largest contributing nutrient used to drive yield outcomes. The low recovery of N might be related to the high application of N, which is generally applied at the early growth stage of the crop when the capacity of the crop plant to take up N is small (Zhu and Chen, 2002). The elemental imbalance in application of fertilizer also suggests there is less understanding about the interacting effects of these nutrients at the farm level. Interactions of N with the different nutrients may be different and could be affected by amounts and availability of other nutrients. N has synergistic interactions with P, K, and S, and with many other micronutrients, and is known to be often related to yield increases with additional application of N in the presence of other nutrients (Fageria and Baligar, 2005). For example, Jackson (2000) reported a linear relationship in the yield of canola with addition of 22 kg/ha S, and by increasing N from 200 to 250 kg/ha. In another study showing a typical example of the need to balance fertilizers, rice did not respond to the application of N, NP, and NK individually, but does to the combined application of NPK (Dobermann et al., 1998). The imbalance of nutrient use can cause antagonistic interactions that may lead to the loss of any synergistic advantage and hence result in lower NUE (Roy et al., 2006). The studies cited above make it clear that all nutrients are equally important for the better use of the others, so there is great need to take into account the concept of nutrient balance when considering nutrient efficiencies and sustainability of crop production systems. As Cooke (1982) reports, increases in yield potential will result from such interaction effects and will contribute towards sustainable agriculture.

2.5   No-Tillage Systems

The shift from a system that uses cultivations as the basis of cropping practices to a system using no-tillage technology is seen as an integral part of sustainable agriculture (Llewellyn et al., 2012). The adoption of no-tillage is slightly different in different parts of the world, with modification and fine tunings done in accordance with local socio-economic and cultural environments. Consequently, here we define no-tillage as all tillage practices that aim to minimize soil disturbance. The direct benefits of no-tillage are lower cost of cultivation—resulting from saving in time, fuel, labour, and machinery cost—and, in some situations, opportunities to intensify cropping through improved trafficability (Doran and Linn, 1994; Llewellyn et al., 2012). Yield improvements can also be achieved with no-tillage systems associated with more water availability and hence water/precipitation use efficiency (Zentner et al., 2002), and possibly other indirect benefits associated with higher soil organic matter availability. Whilst the success of no-tillage relies on both direct and indirect benefits, in some situations these benefits may not be apparent. There are reports of benefits arising from better N-use efficiencies associated with no-tillage practices, but in some cases higher inputs of N may be required. Wood and Edwards (1992) suggested increasing the N application rate in no-tillage systems to overcome problems of short-term N immobilization, whereas Torbert et al. (2001) reported no N limitation in no-tillage compared with other tillage practices. In general, no-tillage has greater benefits for N on low organic level and in well drained soils (Cameron and Haynes, 1986), but given the increasing need for N-fertilizer inputs to manage yield improvements, more understanding of the interactions with N availability and any additional benefits associated with no-tillage is required.

3.   IMPROVING NUE AND ON-SITE N MANAGEMENT

Notwithstanding the overall position that NUE is low, it is recognized that NUE must be improved significantly. Initially it was thought that improvement would most likely be achieved through physiological efficiency, by reducing biomass relative to that of straw (Fischer and Wall, 1976; Craswell and Godwin, 1984; Oritz-Monasterio, 2002). N utilization and physiological efficiency are largely related to the genetic character of the crop plant, and this improvement of NUE is from better utilization of N rather than greater N uptake. It seems that in the present circumstances NUE is restricted by limits associated with uptake rather than utilization. It is possible to improve N uptake by agronomic management, and in this section we will discuss some strategies to improve NUE by agronomic management.

3.1   Pre- and Post-Anthesis Water Use and N Uptake

N uptake by roots and translocation of N to and from leaves is a complex and integral determinant of crop yield (Imsande and Touraine, 1994). N uptake will be related to limited N absorption by roots, which is affected by both external and internal factors (physiological factors). Plant available water (PAW) acts as the driver for such factors in the field: for example, PAW enhances N uptake from deeper soil layers by increasing the absorption and translocation of N in the plant (McDonald, 1989). Plant-available water and timing of rainfall in the pre-anthesis or post-anthesis growth period for dryland and irrigated crops is critical (Angus and van Herwaarden, 2001). Early N application leads to high pre-anthesis water use that sometimes limits the N response and causes reduction in crop yield (Fischer and Kohn, 1966; Fischer, 1979; McDonald, 1989; Asseng et al., 2001). Adequate early N and water reduce the evaporation and increase the transpiration by enhancing the Leaf Area Index (Van Harwaarden et al., 1998). Adequate post-anthesis water use helps to provide the N required at the grain-filling stage and reduces the limit imposed by water soluble carbon re-translocation, the greatest single contributor to yield reduction (Van Harwaarden et al., 1998). Pre-anthesis water use helps in good sink establishment, whereas sufficient post-anthesis water use provides more time for the translocation of N from leaves to grains. Balance is therefore required in pre- and post-anthesis water use for a crop to optimize yield, WUE, and NUE (Passioura, 1976; Fischer, 1979). Achieving the balance between pre- and post-anthesis water use is very challenging in both rainfed and irrigated systems. Various agronomic strategies can be used to achieve this, such as no-tillage and stubble retention to improve soil water relations, adjusting the time of sowing, and managing N supply according to water availability. Moreover, in irrigation systems, N supply can be associated with irrigation schedule, and the number of pre- and post-anthesis irrigations can be adjusted to use water and N more efficiently.

3.2 Agronomic Strategies to Improve NUE

To optimize economic yield from crop plants, the source of N fertilizer input, method of application, rate and timing of fertilizer all are critical issues to consider. In this section we will discuss strategies relating to each of these and related management options.

3.2.1 Source and Method of N Application

The source of applied N and method of application is more important for N than for any other nutrient, because of variable mobility of N in the soil–plant system. Ammonium sulfate and urea are two major and widely used sources of N amongst other sources available in the market (Fageria and Baligar, 2005). In the USA, anhydrous ammonia (NH_3) is also widely used and is a cheap source of N. Anhydrous ammonia helps to avoid losses through volatilization as it is supplied in liquid form under pressure and injected into the soil. But this fertilizer is not used in developing countries, because of lack of equipment and infrastructure for the process. In developing countries, urea is commonly used because of availability and flexibility of methods of application (often it is applied by hand broadcast). Urea is susceptible to ammonium loss by volatilization, and ideally urea should be beneath the soil surface soon after its application. Urea ammonium nitrate (UAN) is another popular source of N fertilizer because of easy handling and wide compatibility with other chemicals. The source of fertilizer, time of application, and cost are important driving factors in choosing the method of application. Several studies have reported the benefits of applying N into soil or subsurface before sowing, particularly in no-tillage systems (Whitaker et al., 1978; Beyrouty et al., 1986). Pre-sowing application of N can increase the availability at early growth stages and hence improve uptake and NUE (Rao and Dao, 1996). The effectiveness of pre-sowing application over topdressing and increases of 20–40% in recovery are also associated with subsurface N application (Whitaker et al., 1978; Beyrouty et al., 1986). Yield increases in corn crop of 5–40% are also reported by Stecker et al. (1993) for knife-injected UAN compared with broadcasting. Nitrogen losses associated with broadcasting can be heavy and are reported to be 9% to 30% by UAN and urea, respectively, under no-till conditions (Keller and Mengel, 1986). Losses through volatilization are not the only concern, as immobilization of N can also occur in no-tilled soil surface, resulting in yield reduction and poor NUE (Kitur et al., 1984). In many situations, side dressing of N can be helpful to maximize efficiency (Piekkielek and Fox, 1992). Clearly consideration must be given to the most appropriate method for N application, combined with the source, the amount and the time of crop growth, and also soil fertility (available soil N) and crop type, so that the crop can easily access and use N when it is required.

3.2.2 Amount of N

Over-fertilization is another important issue related to economics, losses of N and NUE, and hence sustainability. Problems of over-fertilization are common in developing countries, where the farmer relies upon N input to obtain maximum crop yield. Excessive application of N increases the residual NO_3-N in soil (Raun and Johnson, 1995; Porter et al., 1996) and produces favorable conditions for N leaching below the root zone. Dinnes et al. (2002) write on the importance of rate and time of N application to avoid these losses. Setting target yields and calculation of these yield goals can help farmers to limit uncertainty about using the proper rate of N for economic (and sustainable) yield. Yield goals must be calculated by taking account of experience over several years, as well as economic and seasonal circumstances. Mullen et al. (2003) suggest calculating the N rate by considering the data of N uptake of the crop from previous years. For example in Oklahoma a simple equation is recommended to calculate N rate for wheat crop:

$$N \text{ rate} = \text{yield goal (kg/ha)} \times 0.33$$

where the recommendations are 33 kg-N/ha for 1000 kg grains. A general rule of thumb for the N rate is that high-yielding crops require high amounts of N. But the responsiveness of the crop to N varies with location and from year to year, as driven by the soil and environmental factors. On the other hand, most N rate recommendations are made prior to crop sowing and sometimes need to be changed according to seasonal variability; in these situations, tools like leaf colour charts, chlorophyll meters, and optical sensors like GreenSeeker (Normalized Difference Vegetation Index) can be very useful for matching crop demands (Coventry et al., 2011b).

3.2.3 In-Season Application of N

High application rates of N as a basal dose can be toxic to the seed and have greater potential to expose the emerging crop to losses. Splitting or delaying the N application to meet the demand of the crop and/or to respond to water availability can thus be used as a tool to manage economic and environmental yield. Recommendations are therefore made for such split applications, with usually 30–50% application at sowing and sometimes as low as 10% in dry conditions, and the rest can be applied according to favorable events (or otherwise) in the growing season (Fageria and Baligar, 2005; Potter, 2009). Many studies have shown the benefits of splitting N application in both irrigated and rainfed farming systems (Wuest and Cassman, 1992; Coventry et al., 2011a,b). In non-water-limited environments, growers use this strategy to control the early vigour of the crop, and to overcome the significant problem of lodging as well as to limit losses through leaching. It is well understood that splitting the N application for a wheat crop increases yield and grain protein, and in achieving this desired outcome the recovery of applied N is much enhanced (Ellen and Spietz, 1980; Mercedes et al., 1993; López-Billido et al., 2007; Coventry et al., 2011b). For example, Wuest and Cassman (1992) have shown recoveries of 55–80% N when applied at anthesis, as compared with 30–55% at planting. Further, Coventry and colleagues (2011b) showed that a three-way split of N is more effective in terms of yield and wheat grain quality, and N recovery, than a two-way split N application. This latter study shows a classic example of water-nitrogen co-limitation in an irrigated crop system. Split application is known to be beneficial in densely tillering wheat crops (Baethgen and Alley, 1989) and also with sparsely tillering wheat crop (Weisz et al., 2001). Thus, regarding the management of N, there are many studies that have shown that split application is probably the most appropriate way to increase NUE. But also obvious from these studies is the conclusion that it can be risky to be too prescriptive or rigid in strategy setting, as yield potential and other factors are variable and site specific. When targeting optimum N use in both irrigated and rainfed conditions, flexibility and caution are required when developing site-specific management approaches.

3.2.4   N Management in No–Tillage Systems

Fertilizer practices and rates are roughly the same between no-till and conventional till systems, although some special care may be required for N input in no-tillage systems. Decomposition of stubble and residues can immobilize soil N in no-tillage systems, and application of more N may be required in the first few years of no-till practice. To minimize immobilization and volatilization, N can be applied with banding below the soil surface at the time of sowing, with a higher rate of N as an option when adequate soil moisture is present. In addition, strategies regularly used are the injection of liquid solutions (such as UAN), incorporation of granular fertilizer with irrigation or rain when possible, and application of urea prior to seeding for partial incorporation with the seeding tool. Where surface broadcasting is chosen, a source of N that becomes quickly available, such as ammonium nitrate or urea, and surface application during cool periods is recommended (Jones et al., 2007). Although the timing of N application does not change with the tillage system used, extra benefits, such as improvement in grain quality, have been obtained under no-till conditions (Coventry et al., 2011a).

4.   TARGETING CROP SUSTAINABILITY

In this chapter we highlight the importance of N in relation to production sustainability. Sustainability as recognized here is a wide concept consisting of various components that work together as a system, so there is a need to understand and maximize the operation and efficiency of the individual components of the system.

Whilst strict rotations are mostly followed on farms in developing countries, farmers should recognize the need for flexibility in practices within the crop (and pasture) sequences. These practices will be based on rules that account for short-term goals and long-term environmental considerations. The rules will consider product prices, cost of production and diversification of income, breaks in disease and pest cycles, soil water, mineral nutrients, and tillage and weed factors. When dealing with flexibility, both knowledge and understanding of the production system are required in making decisions that target optimum economic yield. In this chapter we have argued that, when targeting economic yield, it is likely that one will also be targeting sustainable yield, particularly with respect to N management. Thus yield and sustainability can be increased in parallel by adhering to underlying agronomic principles. These principles relate to careful choice of varieties to match the environment and product requirement, crop rotations, early sowing where appropriate, stubble management, zero tillage, and fertilizer balance. An example of system development where consideration is given to completely rethinking and restructuring a cropping system is given in Piggin et al. (2011).

Almost by definition the suite of management practices used will be site specific, taking into account soil conditions (reflecting the history of past management) and available infrastructure. Thus the "package of practices" will vary from situation to situation and cannot be prescriptive, but must only be seen as a guide to best practice opportunities. Conservation agriculture is the foundation of best agronomic practices, and no-tillage, crop rotations, and management of N fertility by relating to crop demand will remain the basis of a sustainable system. This emphasis on conservation agriculture will underpin food production to meet the needs of future generations.

ACKNOWLEDGEMENT

The principal author acknowledges the John Allwright fellowship award from ACIAR that is enabling him to undertake PhD degree study at The University of Adelaide.

REFERENCES

Altieri, M., 1995. Agroecology: The Science of Sustainable Agriculture, second ed. Westview Press, Boulder, CO.

Anderson, W., 1985. Differences in response of winter cereal varieties to applied nitrogen in the field II. Some factors associated with differences in response. Field Crops Res. 11, 369–385.

Angus, J., 2001. Nitrogen supply and demand in Australian agriculture. Aust. J. Exp. Agri. 41, 277–288.

Angus, J., Fischer, R., 1991. Grain and protein responses to nitrogen applied to wheat growing on a red earth. Aust. J. Agric. Res. 42, 735–746.

Angus, J., Van Herwaarden, A., 2001. Increasing water use and water use efficiency in dryland wheat. Agron. J. 93 (2), 290–298.

Asseng, S., Turner, N., Keating, B., 2001. Analysis of water-and nitrogen-use efficiency of wheat in a Mediterranean climate. Plant Soil 233 (1), 127–143.

Baethgen, W., Alley, M., 1989. Optimizing soil and fertilizer nitrogen use by intensively managed winter wheat. II. Critical levels and optimum rates of nitrogen fertilizer. Agron. J. 81 (1), 120–125.

Beyrouty, C., Nelson, D.W., Sommers, L.E., 1986. Transformations and losses of fertilizer nitrogen on no-till and conventional till soils. Nutr. Cycl. Agroecosys. 10 (2), 135–146.

Bloom, A., Chapin, F.S., Mooney, H.A., 1985. Resource limitation in plants—an economic analogy. Annu. Rev. Ecol. Syst. 16, 363–392.

Bullock, D., 1992. Crop rotation. Crit. Rev. Plant Sci. 11 (4), 309–326.

Cameron, K., Haynes, R., 1986. Retention and movement of nitrogen in soils. Mineral Nitrogen in the Plant-Soil System. Academic Press, Inc. (166–241).

Cassman, K., Dobermann, A., Walters, D.T., Yang, H., 2003. Meeting cereal demand while protecting natural resources and improving environmental quality. Annu. Rev. Environ. Resour. 28 (1), 315–358.

Cooke, G., 1982. Fertilizing for Maximum Yield, third ed. Macmillan, New York.

Cornish, P., Murray, G.M., 1989. Low rainfall rarely limits wheat yields in southern New South Wales. Aust. J. Exp. Agri. 29, 77–83.

Coventry, D., Gupta, R., Yadav, A., Poswal, R., Chhokar, R., Sharma, R., et al., 2011a. Wheat quality and productivity as affected by varieties and sowing time in Haryana, India. Field Crops Res. 123, 214–225.

Coventry, D., Yadav, A., Poswal, R., Sharma, R., Gupta, R., Chhokar, R., et al., 2011b. Irrigation and nitrogen scheduling as a requirement for optimising wheat yield and quality in Haryana, India. Field Crops Res. 123, 80–88.

Craswell, E., Godwin, D.C., 1984. The efficiency of nitrogen fertilizers applied to cereals in different climates. Adv. Plant Nutr. 1, 1–55.

Dinnes, D., Karlen, D.L., Jaynes, D.B., Kaspar, T.C., Hatfield, J.L., Colvin, T.S., et al., 2002. Nitrogen management strategies to reduce nitrate leaching in tile-drained midwestern soils. Agron. J. 94 (1), 153–171.

Dobermann, A., Cassman, K.G., Mamaril, C.P., Sheehy, J.E., 1998. Management of phosphorus, potassium, and sulfur in intensive, irrigated lowland rice. Field Crops Res. 56 (1-2), 113–138.

Doran, J., Linn, D.M., 1994. Microbial ecology of conservation management systems. In: Hatfield, J, Stewart, B. (Eds.), Soil Biology: Effects on Soil Quality Lewis Publications, Boca Raton, pp. 1–27.

Eickhout, B., Bouwman, A.F., Van Zeijts, H., 2006. The role of nitrogen in world food production and environmental sustainability. Agric. Ecosyst. Environ. 116 (1), 4–14.

Ellen, J., Spiertz, J.H.J., 1980. Effects of rate and timing of nitrogen dressings on grain yield formation of winter wheat (T. aestivum L.). Nutr. Cycl. Agroecosys. 1 (3), 177–190.

Fageria, N., Baligar, V.C., 2005. Enhancing nitrogen use efficiency in crop plants. Adv. Agron. 88, 97–185.

FAO (2012) Committee on world food security Rome, Italy. Retrieved April 26, 2012 from <http://www.fao.org/cfs/en/>.

Fillery, I., McInnes, K.J., 1992. Components of the fertiliser nitrogen balance for wheat production on duplex soils. Aust. J. Exp. Agri. 32, 887–899.

Fischer, K., 2000. Frontier project on nitrogen fixation in rice: looking ahead. In: Ladha, J.K., Reddy, P.M. (Eds.), The Quest for Nitrogen Fixation in Rice International Rice Research Institute, Los Banos, Philippines, pp. 25–31.

Fischer, R., 1979. Growth and water limitation to dryland wheat yield in Australia: a physiological framework. J. Aust. Inst. Agric. Sci. 45 (2), 83–94.

Fischer, R., Kohn, G.D., 1966. The relationship between evapotranspiration and growth in the wheat crop. Aust. J. Agric. Res. 17, 255–267.

Fischer, R., Wall, P.C., 1976. Wheat breeding in Mexico and yield increases [spring wheat]. J. Aust. Inst. Agric. Sci. 42, 139–148.

Helmers, G., Langemeier, M.R., Atwood, J., 1986. An economic analysis of alternative cropping systems for east-central Nebraska. Am. J. Altern. Agric. 1 (4), 153–158.

Hoang, V., Alauddin, M., 2010. Assessing the eco-environmental performance of agricultural production in OECD countries: the use of nitrogen flows and balance. Nutr. Cycl. Agroecosys. 87 (3), 353–368.

IFA (2012) Statistics International Fertilizer Industry Association, Paris, France. Retrieved May 30, 2012 from <http://www.fertilizer.org/ifa/ifadata/search30/5/2012>.

Imsande, J., Touraine, B., 1994. N demand and the regulation of nitrate uptake. Plant. Physiol. 105 (1), 3–7.

Jackson, G., 2000. Effects of nitrogen and sulfur on canola yield and nutrient uptake. Agron. J 92 (4), 644–649.

Jones, C., Koenig, R.T., Ellsworth, J.W., Brown, B.D., Jackson, G.D., 2007. Management of Urea Fertilizer to Minimize Volatilization. Montana State University Extension Service, Washington State University Extension Service, Bozeman, Montana.

Karlen, D., Varvel, G.E., Bullock, D.G., Cruse, R.M., 1994. Crop rotations for the 21st century. Adv. Agron. 53, 1–45.

Keller, G., Mengel, D.B., 1986. Ammonia volatilization from nitrogen fertilizers surface applied to no-till corn. Soil Sci. Soc. Am. J 50 (4), 1060–1063.

Kitur, B., Smith, M.S., Frye, W.W., Blevins, R.L., 1984. Fate of ^{15}N-depleted ammonium nitrate applied to no-tillage and conventional tillage corn. Agron. J. 76 (2), 240–242.

Llewellyn, R., D'Emden, F.H., Kuehne, G., 2012. Extensive use of no-tillage in grain growing regions of Australia. Field Crops Res. 132, 204–212.

López-Bellido, L., López-Bellido, R.J., Redondo, R., 2005. Nitrogen efficiency in wheat under rainfed Mediterranean conditions as affected by split nitrogen application. Field Crops Res. 94 (1), 86–97.

López-Bellido, R.J., López-Bellido, L., Benitez-Vega, J., López-Bellido, F.J., 2007. Tillage system, preceding crop, and nitrogen fertilizer in wheat crop. Agron. J. 99 (1), 66–72.

Matson, P., Parton, W.J., Power, A.G., Swift, M.J., 1997. Agricultural intensification and ecosystem properties. Science 277, 504–509.

McDonald, G., 1989. The contribution of nitrogen fertiliser to the nitrogen nutrition of rainfed wheat crops in Australia: a review. Aust. J. Exp. Agri. 29, 455–481.

Mercedes, M., Alcoz, F.M., Haby, V.A., 1993. Nitrogen fertilization timing effect on wheat production, nitrogen uptake efficiency, and residual soil nitrogen. Agron. J. 85 (6), 1198–1203.

Mullen, R., Solie, J.B., Johnson, G.V., Stone, M.L., Raun, W.R., Freeman, K.W., 2003. Identifying an in-season response index and the potential to increase wheat yield with nitrogen. Agron. J. 95 (2), 347–351.

Mulvaney, R., Khan, S.A., Ellsworth, T.R., 2009. Synthetic nitrogen fertilizers deplete soil nitrogen: A global dilemma for sustainable cereal production. J. Environ. Qual. 38 (6), 2295–2314.

Norton, R., Wachsmann, N.G., 2006. Nitrogen use and crop type affect the water use of annual crops in south-eastern Australia. Aust. J. Agric. Res. 57 (3), 257–267.

Ortiz-Monasterio, I., 2002. Nitrogen management in irrigated spring wheat. FAO Plant. Prod. and Prot. Ser.

Passioura, J., 1976. Physiology of grain yield in wheat growing on stored water. Funct. Plant Biol. 3 (5), 559–565.

Passioura, J., 2006. Increasing crop productivity when water is scarce—from breeding to field management. Agric. Water. Manage. 80 (1), 176–196.

Piekkielek, W., Fox, R.H., 1992. Use of a chlorophyll meter to predict sidedress nitrogen. Agron. J. 84 (1), 59–65.

Pierce, F., Rice, C.W., 1988. Crop rotation and its impact on efficiency of water and nitrogen use In: Hargrove, W. (Ed.), Cropping Strategies for Efficient Use of Water and Nitrogen, vol. 51 ASA special publication, American Society of Agronomy, pp. 21–42.

Piggin, C., Haddad, A., Khalil, Y. (2011). Development and promotion of zero tillage in Iraq and Syria. 5th World Congress of Conservation Agriculture incorporating 3rd Farming system Design Conference, Brisbane, Australia, 304-305.

Porter, L, Follett, R.F., Halvorson, A.D., 1996. Fertilizer nitrogen recovery in a no-till wheat-sorghum-fallow-wheat sequence. Agron. J. 88 (5), 750–757.

Potter, T., 2009. The canola plant and how it grows. In: McCaffery, D., Potter, T., Marcroft, S., Pritchard, F. (Eds.), Canola Best Practice Management Guide for South-Eastern Australia Grains research and development corporation, Kingston, pp. 11–14.

Rao, S., Dao, T.H., 1996. Nitrogen placement and tillage effects on dry matter and nitrogen accumulation and redistribution in winter wheat. Agron. J. 88 (3), 365–371.

Raun, W., Johnson, G.V., 1995. Soil-plant buffering of inorganic nitrogen in continuous winter wheat. Agron. J. 87 (5), 827–834.

Raun, W., Johnson, G.V., 1999. Improving nitrogen use efficiency for cereal production. Agron. J. 91 (3), 357–363.

Rhoades, J., Loveday, J., 1990. Salinity in irrigated agriculture. Agron. J. 30, 1089–1142.

Roy, R., Finch, A., Blair, G.J., Tandon, H.L.S., 2006. Plant nutrition for food security: a guide to integreated nutrient management "Fertiliser and Plant Nutrition", Bull. FAO, Rome, (16).

Sadras, V., 2004. Yield and water-use efficiency of water-and nitrogen-stressed wheat crops increase with degree of co-limitation. Eur. J. Agron. 21 (4), 455–464.

Sadras, V., 2005. A quantitative top-down view of interactions between stresses: theory and analysis of nitrogen-water co-limitation in Mediterranean agro-ecosystem. Aust. J. Agric. Res. 56 (11), 1151–1157.

Sowers, K., Pan, W.L., Miller, B.C., Smith, J.L., 1994. Nitrogen use efficiency of split nitrogen applications in soft white winter wheat. Agron. J. 86 (6), 942–948.

Stecker, J., Buchholz, D.D., Hanson, R.G., Wollenhaupt, N.C., McVay, K.A., 1993. Application placement and timing of nitrogen solution for no-till corn. Agron. J. 85 (3), 645–650.

Sylvester-Bradley, R., Stokes, D.T., Scott, R.K., 2001. Dynamics of nitrogen capture without fertilizer: the baseline for fertilizing winter wheat in the UK. J. Agri. Sci. 136 (1), 15–33.

Tilman, D., Cassman, K.G., Matson, P.A., Naylor, R., Polasky, S., 2002. Agricultural sustainability and intensive production practices. Nature 418, 671–677.

Torbert, H., Potter, K.N., Morrison, J.E., 2001. Tillage system, fertilizer nitrogen rate, and timing effect on corn yields in the Texas Blackland Prairie. Agron. J. 93 (5), 1119–1124.

Van Herwaarden, A., Angus, J.F., Richards, R.A., Farquhar, G.D., 1998. Haying-off. the negative grain yield response of dryland wheat to nitrogen fertiliser II. Carbohydrate and protein dynamics. Aust. J. Agric. Res. 49, 1083–1094.

Weisz, R., Crozier, C.R., Heiniger, R.W., 2001. Optimizing nitrogen application timing in no-till soft red winter wheat. Agron. J. 93 (2), 435–442.

Whitaker, F., Heinemann, H.G., Burwell, R.E., 1978. Fertilizing corn adequately with less nitrogen. J. Soil. Water. Conserv. 33, 28–32.

Wood, C., Edwards, J.H., 1992. Agroecosystem management effects on soil carbon and nitrogen. Agric. Ecosyst. Environ. 39 (3), 123–138.

Wuest, S., Cassman, K.G., 1992. Fertilizer-nitrogen use efficiency of irrigated wheat: I. Uptake efficiency of preplant versus late-season application. Agron. J. 84 (4), 682–688.

Zentner, R., Lafond, G.P., Derksen, D.A., Campbell, C.A., 2002. Tillage method and crop diversification: effect on economic returns and riskiness of cropping systems in a thin black Chernozem of the Canadian Prairies. Soil Tillage Res. 67 (1), 9–21.

Zhu, Z., Chen, D.L., 2002. Nitrogen fertilizer use in China—Contributions to food production, impacts on the environment and best management strategies. Nutr. Cycl. Agroecosys. 63 (2), 117–127.

Chapter 5

Potential of Management Practices and Amendments for Preventing Nutrient Deficiencies in Field Crops under Organic Cropping Systems

Sukhdev S. Malhi*, Tarlok S. Sahota†, and Kabal S. Gill**

*Agriculture and Agri-Food Canada, Melfort, Saskatchewan, Canada, †Thunder Bay Agriculture Research Station, Thunder Bay, Ontario, Canada, **Smoky Applied Research and Demonstration Association, Falher, Alberta, Canada

1. INTRODUCTION

The interest and demand for organically grown food and fiber is increasing, because of (i) possible high economic returns due to price premiums (Zentner et al., 2011) and (ii) a belief that organically produced food is tastier and healthier than that from conventional farming. Optimum soil fertility is one of the top three (weeds, nutrients, crop species/rotations) challenges for sustainable organic farming systems. Adequate supply of nutrients from soil is essential for sustainable high yields, produce quality, water, nutrients, and energy use efficiency, and soil quality; and to prevent accumulation of nitrate-N in the soil profile (Table 5.1). Most soils under organic cropping systems can be deficient in available nitrogen (N) and phosphorus (P), and some soils may contain insufficient amounts of available sulfur (S) and potassium (K) for optimum crop growth and yield (Entz et al., 2001b; Watson et al., 2002). Returning crop residues alone cannot replenish all the nutrients exported, and in the long run nutrients in soils are depleted, resulting in poor productivity. Therefore, maintaining soil fertility is a key production challenge facing organic farmers.

Most organic producers usually focus on minimizing N deficiency or increasing N availability in soil–crop systems by including legume crops—for grain, forage, or green manure (GM)—in the rotations (Zentner et al., 2004). However, if soils are deficient in available P, K, S, or other essential nutrients,

Agricultural Sustainability. DOI: http://dx.doi.org/10.1016/B978-0-12-404560-6.00005-8
2013 Published by Elsevier Inc.

TABLE 5.1 Effects of Balanced Fertilization on Yield (Seed or Forage Dry Matter Yield), Partial Factor Productivity/N Fertilizer Use Efficiency (kg Seed or Forage Dry Matter kg^{-1} of Applied Nutrient/N), Seed Quality (Oil Content, %), Water Use Efficiency (kg Yield kg^{-1} mm^{-1} of Water), and Soil Organic C (Mg-C ha^{-1}) and nitrate-N (kg-N ha^{-1})

Parameter/Study	N Alone (Unbalanced)	Balanced Fertilization
NUE—seed yield/DMY		
Wheat Cu—seed yield (PFP/NFUE)	9.3 to 12.0	19.6 to 20.1 (N + Cu)
Canola S—seed yield (PFP/NFUE)	1.2	10.2 (N + S)
Grass S—DMY (PFP/NFUE)	10.0	42.3 (N + S)
		47.5 (N + S+K)
Alfalfa P—DMY (5-yr)	2164 (0 P)	6036 (Broadcast P)
		7192 (Disc-banded P)
Seed quality		
Canola seed oil content (%)	37.3	41.4 (N + S)
WUE—ACS Scott (1995–2006)		
First 6-yr cycle	5.5 (Organic—no fertilizer)	8.5 (Conventional—N + P)
Second 6-yr cycle	3.1 (Organic—no fertilizer)	4.9 (Conventional—N + P)
Soil quality		
Grass S—TOC (LFOC), 0–10 cm	44.1 (6.71)	51.0 (12.13) (N + S)
		55.7 (14.19) (N + S + K)
Canola S—TOC (LFOC), 0–15 cm	23.59 (2.758)	25.77 (3.776) (N + S)
Soil nitrate-N—ACS Scott (1995–2006)		
First 6-yr cycle (0–90 cm)	123 (Organic—no fertilizer)	42 (Conventional—N + P)
Second 6-yr cycle (0–90 cm)	94 (Organic—no fertilizer)	45 (Conventional—N + P)
Canola S (0–60 cm)	194	19

ACS, alternative cropping systems; DMY, dry matter yield; LFOC, light fraction organic carbon; PFP/NFUE, partial factor productivity / N fertilizer use efficiency; TOC, total organic carbon; WUE, water use efficiency.
Source: Malhi et al. (2010).

the only alternative is to use external sources. Manure can provide these nutrients to organic crops, but often there is not enough manure, especially in remote areas where the cost of transporting manure over long distances can be uneconomical (Freeze and Sommerfeldt, 1985). Also, most organic producers being grain farmers, they cannot rely on manure to replace nutrients exported. Crop yields on organic farms may also be increased by intercropping non-legume and legume crops (Getachew et al., 2006). Diversification/mix of shallow- and deep-rooted crops is another option to tap different soil layers for nutrients at the same or different times.

Thus, there is a great challenge to find suitable nutrient sources to replace nutrients harvested and exported from organic farms. The efficacy of permitted nutrient sources and management practices in increasing yield by preventing nutrient deficiencies in organic crops could vary with soil and climate (agro-ecological regions). This chapter summarizes results of assessment of management practices and amendments for their potential to prevent nutrient deficiencies in organic crops for sustainable yield of quality produce, while enhancing soil quality/health and minimizing negative environmental impacts: e.g., contamination of ground water with nitrate-N, surface water with labile P, and air with greenhouse gas (GHG) emissions.

2. MANAGEMENT PRACTICES

2.1 Crop Diversification/Rotation

Cropping systems are generally cereals and oilseeds based. Various management practices to diversify these are discussed below.

2.1.1 Legumes in Crop Rotation

Legumes—if properly inoculated with respective rhizobia—can fix sufficient amounts of atmospheric N_2 for their requirements and also provide N to subsequent or companion crops. Inclusion of annual and perennial legumes with different rooting patterns in crop rotations could play an important role in integrated nutrient management (Izaurralde et al., 1993; Entz et al., 1995). Annual grain legumes as cash crops are an option for reducing external N input. They can fix atmospheric N_2 for their own requirement, in soils that are supplied adequately with all other nutrients, and also provide N and other benefits to subsequent non-legume crops to improve their yields (Welty et al., 1988; Beckie and Brandt, 1997). Field pea has been found to provide the largest and most consistent N benefit to succeeding wheat, as compared with soybean, chickpea, and dry bean (Przednowek et al., 2004). Grain legumes in rotations also improve organic matter, mineral N content, and N mineralization potential of soil compared with non-legume crops. Therefore, crop rotations that include grain legumes are recommended for improving sustainability of soil productivity and fertility/quality/health. However, annual grain legumes cannot be relied on for substitution of entire N removed by other crops.

Perennial forage legumes do not need any external N input to produce high hay/pasture yields, conserve energy, increase economic returns and sustainability of crop production, improve soil organic C/microbial biomass N, and minimize soil erosion and damage to the environment (Juma et al., 1997). For example, alfalfa (*Medicago sativa* Leyss.) is grown for forage/hay as a livestock food/feed and the dehydrated alfalfa pellet market. Perennial legumes in rotations usually provide fairly large yield and N benefits to succeeding cereals (Wani et al., 1994b). Overall, diversified annual–perennial cropping systems that include alfalfa have the potential to (i) enhance yields of succeeding non-legume annual crops, (ii) improve soil quality/health, and (iii) save energy. In sub-humid regions, inclusion of perennial forage legumes in organic cropping systems can play an important role in integrated nutrient management; though in relatively dry regions, crop yields of the succeeding grain crops may be depressed. Despite N contribution from forage legumes to subsequent crops, they could increase risk of nitrate leaching in sub-humid environments, if N released in the soil is not synchronized with the crop N uptake.

2.1.2 Deep-rooted and Shallow-Rooted Crops in Rotation

Loss of nitrate-N by leaching, an economic loss to farmers, may cause environmental damage. Inclusion of deep-rooted crops in the rotations can tap nutrients from deeper soil layers and return those nutrients in the form of residues to the topsoil for subsequent crops, resulting in effective nutrient cycling. This can improve economic productivity when surface soil has poor fertility (especially available P). The recycling enhances retention of nutrients in the soil–plant system and minimizes environmental impact. Research has shown that excess nitrate-N in the soil profile can be decreased by including deep-rooted perennial forages in the cropping sequence (Entz et al., 2001a). Inclusion of both deep-rooted and shallow-rooted crops in the rotations or their mixed cultivation in hay/pastures could be conducive to sustainability of organic cropping systems, especially when surface soils are low in available P or other nutrients and sub-soils contain considerable amounts of these nutrients.

2.2 Crop Species/Cultivars

Crop species/cultivars vary in their demand for nutrients and/or ability to tolerate nutrient deficiencies. Nutrient management practices may therefore be tailored to suit different crop species/cultivars and optimize crop yields and quality. For example, S requirements of canola are much higher than those of cereals, and yield loss due to S deficiency may occur in canola grown on soils with marginal availability of S that may otherwise be sufficient for optimum wheat production. However, canola may deplete S and accelerate occurrence of S deficiency in subsequent cereal crops grown on such soils.

Research has revealed that wheat genotypes differ considerably in their P requirement, P use efficiency, and grain yield response to P application (Yaseen and Malhi, 2009). Some wheat genotypes were much better than others in exploring P, tolerating P deficiency stress, and producing higher grain yield per kg of P absorbed under P stress conditions. In some cases, these cultivars without applied P produced grain yields that were very close to the maximum grain yields from other cultivars with applied P. Such findings suggest a potential of screening/adapting some location-specific cultivars on low P soils, and also suggest the rotation of crop species/cultivars that have low requirements for S or P, especially on soils low in these nutrients, for sustainable production.

2.3 Crop Residue Return

Approximately 25% of N and P, 50% of S, and 75% of K taken up by the cereals are retained in crop residues. Contribution of crop residues to supply nutrients to subsequent crops is important in effective nutrient management, more so in relatively humid regions where mineralization of large amounts of crop residues could enhance supply of nutrients from the soils, and gradually improve soil productivity and quality (Malhi et al., 2011a,b), depending on crop type and C:N ratio of the residues. Legume crops residues benefit subsequent crops in rotation both by increasing N supply and by improving soil properties. Overall, crop residues, especially from legume crops, contribute appreciable amounts of plant nutrients and energy for the proliferation and activity of soil organisms to enhance soil quality/tilth/health (organic matter, nutrient supplying power, microbial and faunal activity) and productivity.

2.4 Intercropping Non-Legumes with Legumes

The use of legumes as intercrops with cereals may be a viable approach to sustainable cropping, because legumes contribute N through biological N fixation (BNF), and also reduce weed competition and increase input of root mass to soil. Crop yields on organic farms can be increased by intercropping non-legume and legume annual crops (Getachew et al., 2006; Malhi, 2012c; Table 5.2). Intercropping systems reduce the land requirements compared to mono cropping (Getachew et al., 2006; Malhi, 2012c). In addition to added crop yields, crop residues from barley–pea intercropping could return more N to soil than barley–barley residues (Izaurralde et al., 1990). The findings from the above-mentioned studies suggest the potential of intercropping annual non-legume and legume crops for improving yields and economic returns from the organic crops.

Intercropping is quite common in perennial forage production systems. Grasses grown in association with legumes use part of the N fixed by legumes, leading to higher forage dry matter and protein yield (Tomm et al., 1995). Recycling of N derived from BNF to grasses lends sustainability to forage production and can result in minimizing N requirements from external sources. Forage yield response and net returns for mixed stands from applied N are influenced by percentage of legumes in such stands. Malhi et al. (2002)

TABLE 5.2 Seed Yield, Net Returns (NR) and Land Equivalency Ratio (LER) for Barley or Canola and Pea Grown as Sole Crops and in Various Intercrop Combinations at Star City, Saskatchewan (Average of 2009 to 2011)

Treatment	Barley–Pea			Canola–Pea		
	Seed Yield (kg ha^{-1})	NR[a] ($ ha^{-1})	LER[b]	Seed Yield (kg ha^{-1})	NR ($ ha^{-1})	LER
Barley or canola, 0 kg-N ha^{-1}	2062	309		834	375	
Barley or canola, 40 kg-N ha^{-1}	3065	400		1167	465	
Barley or canola, 80 kg-N ha^{-1} Does not match N rates for inter crops	3975	476		1596	598	
Pea, 0 kg-N ha^{-1}	3097	929		2742	823	
(Barley or canola)–pea in alternate rows, 0 kg-N ha^{-1}	3717	826	1.50	2702	881	1.45
(Barley or canola)–pea in alternate rows, 20 kg-N ha^{-1} to barley	3815	752	1.24	3047	971	1.50
(Barley or canola)–pea in alternate rows, 40 kg-N ha^{-1} to barley	4067	756	1.15	2965	905	1.31
(Barley or canola)–pea in same row, 0 kg-N ha^{-1}	3764	824	1.54	2887	953	1.56
(Barley or canola)–pea in same row, 20 kg-N ha^{-1}	3741	736	1.22	3114	1006	1.51
(Barley or canola)–pea in same row, 40 kg-N ha^{-1}	3827	682	1.07	3023	962	1.40

[a] The cost of N fertilizer was $1500 Mg^{-1} of N. The price was $150 Mg^{-1} for barley, $450 Mg^{-1} for canola, and $300 ha^{-1} for pea.
[b] LER = (Intercrop1/Sole Crop1) + Intercrop2/Sole Crop2.
Source: Malhi (2012c).

found only a marginal increase in forage yield and net returns from N application to bromegrass–alfalfa mixtures (Table 5.3). The findings suggest that the use of alfalfa in mixed stands could increase profitability by reducing or eliminating N fertilizer requirement and sustaining high forage production, while fertilizer N application to pure grass stands could increase the risk of nitrate-N leaching.

TABLE 5.3 Dry Matter Yield (DMY) and Net Returns above N Fertilizer Costs (NR) of Hay from Various Bromegrass–Alfalfa Compositions, Treated Annually with Different Rates of N as Ammonium Nitrate at Lacombe and Eckville in Central Alberta, Canada (Average of 1993–1995 and Two Sites)

N Rates (kg-N ha^{-1})					
Composition	0	50	100	150	200
DMY (Mg ha^{-1})					
Pure bromegrass	4.95	7.61	10.41	12.22	14.63
Bromegrass: alfalfa (2:1)	10.98	12.76	13.57	14.24	15.09
Bromegrass: alfalfa (1:1)	11.30	12.84	13.38	14.54	14.81
Bromegrass: alfalfa (1:2)	11.30	13.03	14.15	14.49	15.15
Pure alfalfa	10.47	10.48	10.97	10.44	10.53
NR ($ ha^{-1}) at N rates (kg-N ha^{-1})[a]					
Pure bromegrass	192	261	388	496	615
Bromegrass: alfalfa (2:1)	649	673	672	686	715
Bromegrass: alfalfa (1:1)	714	716	716	741	731
Bromegrass: alfalfa (1:2)	732	788	802	778	798
Pure alfalfa	817	752	755	689	665

[a] The cost of N was $770 Mg^{-1}. The market price of bromegrass was $70 Mg^{-1} for first and $77 Mg^{-1} for second cut. The market price of alfalfa was $90 Mg^{-1} for first and $99 Mg^{-1} for second cut.
Source: Malhi et al. (2002).

2.5 Mixed Farming (Dairy, Beef Cattle or Swine, and Cropping) Systems

In mixed farming systems, cropping systems usually revolve around forage or feed requirements of the livestock, which contribute manure for crop production. Because of manure and also because a part of the crop acreage is under soil-protective perennial forage cropping, mixed farming is considered to be more sustainable than cropping without the livestock component. In most cropping systems, plant residues contribute most of the organic matter in soils, unless manure is bought from livestock producers. In mixed farming systems there can be a substantial contribution of animal manure, in addition to crop residues. The manure and perennial crops in mixed farming systems significantly improve soil biological, physical, and chemical properties and thereby increase crop yields as well as the quality of annual (Beauchamp and Voroney, 1994) and perennial forages (Warman, 1986).

2.6   Agroforestry (Integration of Trees with Field Crops or Animal Production Systems)

Agroforestry, which combines production of trees along with field crops or animal production systems, is practiced in many countries. Products of such systems include timber and livestock/crop produce as well as other valuable products such as mushrooms, florals, medicinals, herbs, and craft products. Tree species tap nutrients and water from deeper soil layers and add to organic matter at soil surface through litter fall. In these systems, shade tolerant plants are grown under a closed tree canopy, and sunshine loving plants are grown in open areas between trees with open canopies. Winter annual crops such as wheat are intercropped in winter deciduous trees such as poplar, especially in sub-temperate and sub-tropical parts of the world. The knowledge on the interactions of crops and trees for integrated nutrient management will be helpful to determine nutrient input rates and times to the field crops and trees for organic crop production. The information on this aspect is scarce.

2.7   Summer Fallow

The practice of summer fallow increases availability of nutrients (especially N from mineralization) and soil water storage and helps to control weeds. Therefore, it is often used to improve crop production on organic farms where manure is not available or applied (Zentner et al., 2004). Decomposition of soil organic matter due to tillage during summer fallow can increase the potential of GHG (CO_2, N_2O) emissions and contamination of ground water by nitrate leaching deep into the soil profile as a result of heavy rain or irrigation. The practice leaves soil surface exposed for erosion and can result in loss of productive surface soil, and deterioration of soil quality and productivity in the long run. However, in very dry areas it may not be economically feasible to eliminate summer fallowing from cropping systems.

3.   AGRICULTURAL ORGANIC AMENDMENTS

Most organic farms are either without livestock or do not have sufficient livestock to supply the entire farm with optimal nutrients from manure; thus they rely on live plant inputs such as green manure. Various amendments not only ameliorate soil alkalinity or acidity, but could also supply essential plant nutrients to build up soil fertility to sustain productivity in the long run. Research has shown potential beneficial effects on crop yields, produce quality, and nutrient uptake from organic amendments (Zentner et al., 2004; Malhi, 2012a,b), and soil activators/inoculants (Kucey, 1987). Nutrient management through organic sources can improve soil chemical, biological and physical properties and optimize crop production (Watson et al., 2002). Soil organic matter can be managed through direct input of organic matter (e.g., animal

TABLE 5.4 Yield of Canola, Barley and Crested Wheatgrass from Injected Liquid Swine Manure (LSM) and Urea in East-Central Saskatchewan, Canada

Treatment	Seed Yield (Mg ha^{-1})		Hay Yield (Mg ha^{-1})
	Canola	Barley	Crested Wheatgrass
Control (no fertilizer)	0.56	2.04	1.06
LSM @ 37,059 L ha^{-1} (84 kg-N ha^{-1})	1.29	4.03	2.72
LSM @ 74,118 L ha^{-1} (168 kg-N ha^{-1})	1.74	4.30	4.99
LSM @ 148,236 L ha^{-1} (336 kg-N ha^{-1})	1.62	3.98	4.89
Urea (112 kg-N ha^{-1})	1.46	4.09	

Source: Schoenau et al. (2000).

manures, humic acids) or through live plant materials (e.g., green manures, cover crops, intercrops, crop residues, etc.).

3.1   Compost/Manure

Composts/manures are slow-release organic fertilizers and excellent sources of most plant nutrients to increase crop yields (Schoenau et al., 2000; Table 5.4). Manures are also beneficial in improving productivity and/or quality of degraded soils (Izaurralde et al., 1998; Table 5.5). In field studies, composted manure and alfalfa pellets were found effective in increasing crop yield and nutrient uptake of cereals (Malhi, 2012a; Table 5.6; S. S. Malhi, unpublished results; Table 5.7).

Application of organic manures at optimum rates could improve soil fertility/quality/health and at the same time reduce nitrate-N accumulation in the soil profile as compared with conventional fertilizers (Yanan et al., 1997). In a long-term crop rotation study in Alberta, Canada, application of manure showed higher concentrations of total N, water soluble C, and microbial C than plots supplied with conventional N, P, K, and S fertilizers (Juma et al., 1997). Long-term application of cattle manure increased potentially mineralizable N (N_{min}) and P in soil (Whalen et al., 2001). In Germany, Heitkamp et al. (2011) observed a significant increase in labile N with four annual applications of farm yard manure (FYM). In Saskatchewan, Canada, applications of compost manure increased total and light fraction organic C, organic N, and N_{min} in soil (Malhi, 2012b).

The perception that organic farming practices can reduce accumulation and leaching of nitrate-N in the soil profile because of zero input of chemical N fertilizers does not always hold true. For example, Kirchmann and Bergstrom

TABLE 5.5 Effects of Manure and Fertilizer on Seed Yield of Wheat (Average of 1991 to 1994), and Total Organic C, Total Organic N, Light Fraction Organic C, and Light Fraction Organic N in Soil (0–20 cm) Under Different Levels of Simulated Erosion (Artificial Top Soil Removal) at Cooking Lake, Alberta, Canada

Treatments		Organic C and N in 0–20 cm Soil (kg-C or kg-N ha^{-1})				
Erosion (cm)	Amendments	Seed Yield (Mg ha^{-1})	TOC (Mg-C ha^{-1})	TON (Mg-N ha^{-1})	LFOC (kg-C ha^{-1})	LFON (kg-N ha^{-1})
0	Control	2.05	67.72	7.07	2940	150
	Fertilizer	3.32	75.83	7.67	3470	180
	Manure	3.07	76.34	7.97	4620	280
	M + F	3.75	82.92	8.43	5750	360
10	Control	0.81	39.77	4.83	910	40
	Fertilizer	2.66	47.42	5.30	1860	90
	Manure	2.27	55.47	6.12	3000	190
	M + F	3.20	59.11	6.32	4380	270

F, fertilizer; LFOC, light fraction organic carbon; LFON, light fraction organic nitrogen; M, manure; TOC, total organic carbon; TON, total organic nitrogen
Source: Izaurralde et al. (1998).

(2001) could not find any research evidence in favor of such a perception in their review on the effects of organic farming practices on nitrate leaching. Scheller and Vogtmann (1995) claimed that the continuous use of organic manures alone might result in asynchronicity between N supply and demand by crops, which may lead to a greater potential of nitrate leaching, especially under wet/humid conditions. Repeated excessive applications of animal manures were observed to adversely affect soil quality and crop yield (Chang et al., 1990), increase accumulation of P in soil (Zhang et al., 2004) and nitrate leaching (Yuan et al., 2000), and lead to high gaseous N losses and ammonia volatilization (Beauchamp et al., 1982).

A substantial increase in extractable P and total P in 0–15 cm soil with short-term annual applications of compost manure in Saskatchewan, Canada (Malhi, 2012b) and with long-term annual applications of FYM in China (Yang et al., 2007) is reported. Excessive build-up of P in surface soil with repeated/regular applications of manure can become a potential risk for contamination of surface waters with P from surface run-off of water, due either

TABLE 5.6 Seed Yield of Barley in 2010 After Three Annual Applications of Various Amendments (2008 to 2010) in Field Experiments on Certified Organic Farms at Spalding (Dark Brown Chernozem) and Star City (Gray Luvisol), Saskatchewan, Canada

Treatment	Seed Yield (kg ha^{-1})	
	Experiment 1	Experiment 2
Control (no amendment)	1253	2233
Amendments		
Compost @ 20 Mg ha^{-1}	2576	3570
Wood ash @ 2 Mg ha^{-1}	1779	2705
Alfalfa pellets @ 4 Mg ha^{-1}	2174	3859
Control + Inoculate seed with *Penicillium bilaiae*	1306	2128
Rock phosphate granular @ 20 kg P ha^{-1}	1526	2323
Rock phosphate granular @ 20 kg P ha^{-1} + Inoculate seed with *Penicillium bilaiae*	1534	2227
Rock phosphate finely ground @ 20 kg P ha^{-1}	1368	2170
Rock phosphate finely ground @ 20 kg P ha^{-1} + Inoculate seed with *Penicillium bilaiae*	1462	2133
MykePro	1412	2202

Source: Malhi (2012a).

to quick snow melt or heavy rain. Research in China has shown that substantial decline in exchangeable K in soil was ameliorated to some extent with continuous long-term annual applications of FYM (Yang et al., 2007). In a study in Saskatchewan, Canada, three annual applications of compost manure increased exchangeable K and sulphate-S in the soil profile (Malhi, 2012b).

3.2   Green Manure

On organic farms where compost manures are not available, green manuring with N_2-fixing legumes not only increases availability and recycling of N and other nutrients, but could also act as short-term storage for N that might otherwise be lost (Zentner et al., 2004). Green manuring increased availability of P, K, S, and some micronutrients through biocycling, increased chelation capacity, and improved soil properties (Rick et al., 2011). In an alternative cropping systems study in Alberta, Canada, barley grain yield was greater (by 0.6 Mg ha^{-1}) after faba bean green manure with no applied N than for

TABLE 5.7 Seed Yield of Canola in 2011 After Three Annual Applications of Various Amendments from 2009 to 2011 in a Field Experiment on a Gray Luvisol Soil at Star City, Saskatchewan, Canada

Treatment	Seed yield (kg ha^{-1})	
	2010	2011
Control (no amendment)	463	410
Amendments		
Compost @ 20 Mg ha^{-1}	534	651
Alfalfa pellets @ 2 Mg ha^{-1}	563	628
Distiller grain (wheat)—dry @ 1 Mg ha^{-1}	541	989
Thin stillage @ 20,000 L ha^{-1}	853	1088
Fish food additive @ 1 Mg ha^{-1}	541	832
N only—80 kg-N ha^{-1} (using 34-0-0)	230	247
N + P—80 kg-N ha^{-1} (using 34-0-0) + 20 kg-P ha^{-1} (using 0-45-0)	491	854
N + S—80 kg-N ha^{-1} (using 34-0-0) + 20 kg-S ha^{-1} (using 0-0-51-17)	680	1083
N + P + S—80 kg-N ha^{-1} + 20 kg-P ha^{-1} + 20 kg-S ha^{-1}	835	1262
Gypsum @ 20 kg-S ha^{-1} + 80 kg-N ha^{-1} + 20 kg-P ha^{-1}	714	1184
Rapid release elemental S @ 20 kg-S ha^{-1} + 80 kg-N ha^{-1} + 20 kg-P ha^{-1}	615	1187
Glycerol @ 1 Mg ha^{-1}	512	321
Glycerol @ 1 Mg ha^{-1} + 80 kg-N ha^{-1}	308	497
Wood ash @ 2 Mg ha^{-1}	453	493
Wood ash @ 2 Mg ha^{-1} + 80 kg-N ha^{-1}	834	1250
Penicillium bilaiae + 80 kg-N ha^{-1} + 20 kg-S ha^{-1}	616	986
Rock phosphate granular (BC Mines) @ 20 kg-P ha^{-1} + 80 kg-N ha^{-1} + 20 kg-S ha^{-1}	638	1010
Rock phosphate granular (BC Mines) @ 20 kg-P ha^{-1} + 80 kg-N ha^{-1} + 20 kg-S ha^{-1} + *Penicillium bilaiae*	568	974
Rock phosphate finely ground (BC Mines) @ 20 kg-P ha^{-1} + 80 kg-N ha^{-1} + 20 kg-S ha^{-1}	698	1247
Rock phosphate finely ground (BC Mines) @ 20 kg-P ha^{-1} + 80 kg-N ha^{-1} + 20 kg-S ha^{-1} + *Penicillium bilaiae*	702	1218
Rock phosphate [powder] (BC Mines) @ 20 kg-P ha^{-1} + 80 kg-N ha^{-1} + 20 kg-S ha^{-1}	689	1310

Source: S. S. Malhi, unpublished results.

continuous cereals supplied with $90\,kg\ N\ ha^{-1}$; potentially mineralizable and total N in soil, and plant N uptake were also greater with green manure in crop rotation than with continuous cereals (Wani et al., 1994a,b). It is possible that properly inoculated legume green manure in rotation could reduce reliance on non-renewable energy, enhance soil quality, sustain productivity, and improve grain quality. The amount of N fixed by green manure and the overall success of various crops depends on soil-climatic conditions. Amongst annual legumes evaluated, field pea (*Pisum sativum*), chickling vetch (*Lathyrus sativus*), and Indian Head lentil (*Lens culinaris* Medik.) were rated as superior green manure crops; lentil was considered most economical due to its low seed cost and small seed size. Chickling vetch could perform reasonably well in biomass production and N contribution, though due to its large seed size and prohibitive seed cost it might be the least economical. Pea accumulated the highest total N and could thus be the best choice on organic farms, also due to its greatest potential impact on improving soil fertility.

4. INDUSTRIAL ORGANIC PRODUCTS/BYPRODUCTS

4.1 Alfalfa Pellets

Alfalfa pellets used to be exported from Canada, as feed for horses, to Japan. After the collapse of the export market, the feasibility of alfalfa pellets as an alternate nutrient source to prevent nutrient deficiencies in organic crops was tested in Saskatchewan, Canada. Alfalfa pellets were generally found effective in increasing crop yield and nutrient uptake in cereals, by improving nutrient-supplying power of soil and its quality (Malhi, 2012a,b; Table 5.7). These findings indicated an improvement in soil quality, fertility, and productivity by the application of alfalfa pellets, and their potential for preventing nutrient deficiencies in organic crops.

4.2 Thin Stillage, Distiller Grain (Byproduct of Ethanol), Fish Food Additive, and Glycerol (Byproduct of Biodiesel)

Some industrial byproducts, containing nutrients in both organic and inorganic forms, could supply adequate amounts of nutrients for optimum crop growth in the first growing season. In a 4-year study in Saskatchewan, Canada, thin stillage (one of the byproducts after alcohol distillation from a fermented cereal grain mash) was found to produce the highest seed yield and nutrient uptake of canola in all the years, and it was similar to (or even slightly better in some years) the balanced fertilization treatment supplying optimum amounts of N, P, and S (S. S. Malhi – unpublished data; Table 5.7). Fish food additive and distiller grain produced only moderate increase in canola seed yield and nutrient uptake. On the contrary, application of glycerol resulted in a reduction in yield and nutrient uptake, most likely due to N immobilization.

TABLE 5.8 Seed Yield of Crops in Different Years and Soil Properties/ Nutrients After Harvest of 2008 Crop with Wood Ash, and Fertilizer Nitrogen and Phosphorus Applications in 2006 and 2007

Crop/Soil Property	Check	$N^a + P^a$	Ash[b]	Ash + N
Crop	Seed yield (kg ha^{-1})			
Pea 2006[c]	3977	4923	4870	5237
Barley 2006	3753	5849	4730	6447
Oat 2007	3461	4646	3842	5197
Wheat 2008	1624	1601	1856	1854
Wheat 2010	1796	1890	2107	2056
Property	Soil properties/nutrients after harvest of 2008 crop			
pH, water	6.20	5.65	6.85	6.70
Phosphorus (ppm)	23.5	29.5	35.0	39.0
Potassium (ppm)	96.5	93.5	161.5	175.5
Calcium (ppm)	1485	1345	1655	1715
Zinc (ppm)	3.55	3.50	8.60	9.55
Manganese (ppm)	8.0	9.0	10.5	11.0
Copper (ppm)	0.55	0.60	0.80	0.75
Sodium (ppm)	75	102	118	117

[a] Fertilizer rates were 90 kg-N + 34 kg-P_2O_5 ha^{-1} in 2006 and 27 kg-N + 39 kg-P_2O_5 ha^{-1} in 2007.
[b] Application of wood ash (3360 kg ha^{-1} in 2006 and 4368 kg ha^{-1} in 2007) supplied 39 kg-P_2O_5 ha^{-1} + other nutrients.
[c] There was no N fertilizer applied to pea, but it received granular Rhizobium inoculant at a proper rate.
Source: K. S. Gill, unpublished results.

4.3 Wood Ash (Byproduct of Forest Industry)

Wood ash (a waste product of forest industry) contains large amounts of Ca and Mg, about 0.44% P, 4.2% K, 1% S, and small amounts of other essential plant nutrients. Application of wood ash increased crop yield on an acidic soil in Alberta (K. S. Gill, unpublished data; Table 5.8), most likely due to the increased soil pH and supply of P and other essential plant nutrients. In another field study, wood ash increased barley yield on organic farms in Saskatchewan, Canada (Malhi, 2012a). After three annual applications, wood ash increased the pH, extractable P, exchangeable K, and sulphate-S in soil (Malhi, 2012b). A comparative long-term study on lime and wood ash at Thunder Bay, Canada, revealed that both lime and wood ash increased soil pH

TABLE 5.9 Effects of Wood Ash and Lime (Experiment 1) or Wood Ash and Manure (Experiment 2) on Total Dry Matter Yield of Alfalfa (2005–2007) and Grain Yield of Barley (2008–2010) at Thunder Bay, North-Western Ontario, Canada

Treatment	Alfalfa DMY (kg ha^{-1})			Barley GY (kg ha^{-1})			Treatment	Barley GY (kg ha^{-1})		
	2005	2006	2007	2008	2009	2010		2008	2009	2010
Control	6561	5177	3852	7991	6515	5369	Control	8270	8061	7446
Wood ash	6606	6550	4578	9447	7367	7178	Wood ash	10179	8680	8603
Lime	6334	6617	4249	7810	7166	6329	Manure	8725	10397	8839
WA + L	6365	4961	4285	8976	7474	6909	WA + M	9751	10516	9665

DMY, dry matter yield; GY, grain yield; L, lime; M, manure; WA, wood ash.
Source: T. S. Sahota, unpublished results.

from 5.9 to neutral (T. S. Sahota, unpublished results). Wood ash improved availability of P, K, Zn, Mn, and B more than lime, whereas the reverse was true for available Ca. Wood ash improved soil organic matter more than lime. Increase in total crop productivity (3 years alfalfa followed by barley for 3 years) was 0.52 Mg ha^{-1} yr^{-1} with lime and 1.08 Mg ha^{-1} yr^{-1} with wood ash (Table 5.9). Application of wood ash increased protein content in first cut alfalfa by ~2% compared with 1% for lime only. Wood ash can be an excellent substitute for lime in increasing soil pH and a better amendment for crop production on acidic soils, apparently because of the presence of other plant nutrients in wood ash compared with only Ca and Mg in lime.

5. MINERAL AMENDMENTS

5.1 Phosphate Amendments

Maintaining an adequate level of plant-available P in soil is the prime concern of organic producers, because most soils are inherently low in available P, and there is a lack of suitable amendments rich in available P. Long-term production of organic crops without adding adequate amounts of P can deplete available P in soils, resulting in poor crop yields and produce quality (Miller et al., 2008).

Taproot crops could absorb nutrients from deeper soil layers (Entz et al., 2001a). However, if the surface and sub-surface soil is low in available P, then it would be very difficult to increase P availability and sustain organic farming systems at a high level of productivity by using deep taproot crops in the rotation. Thus another alternative appears to be to add P from external mineral sources, such as rock phosphate, that are approved for improving P fertility in organic farming systems. Likewise, S deficiencies in organic crops could be corrected by the use of elemental S.

Growth responses to P application through rock phosphate are usually slight to non-existent due to inadequate dissolution of P from this amendment (Bolland and Gilkes, 1990). Effectiveness of rock phosphate to increase soil available P, and consequently improve crop yields, may vary with source and formulation of phosphate rock (Haque et al., 1999), severity of P deficiency in soil (Malhi, 2012a; S. S. Malhi, unpublished results), soil type/pH, and climatic conditions (Haque et al., 1999; Rick et al., 2011). Application of P-enriched compost, prepared by aerobic decomposition of crop/organic residues with low-grade rock phosphate and phosphate-solubilizing microorganisms increased available P in soil and sustained high crop yield (Manna et al., 2001). Fine grinding of rock phosphate should increase solubilization and availability of P, by increasing its surface area for solubilization by the soil microorganisms and chemical conditions. A field study is underway in Saskatchewan, Canada, to determine the feasibility of finely ground rock phosphate in increasing availability of P to crops (S. S. Malhi, unpublished results). Although rock phosphate is not useful as a quick/readily available P source in organic farming, it could play a role to gradually replace P removed from the system and build up soil P in the long run. This suggests the need for long-term experiments to determine the efficacy of rock phosphate in preventing deficiency of P in organic cropping systems for sustaining optimum crop yields.

Use of bone meal to prevent P deficiency and thus increase crop yields in P-deficient soils is also suggested (Jeng et al., 2009). However, like rock phosphate, bone meal too is not a source of readily available P.

5.2 Lime, Gypsum, and Elemental S

Crop growth is adversely affected by low pH on acidic soils. Liming of an acidic soil increased seed yield of pea, which could be attributed to improved pH, NO_3-N, available P, and bulk density of soil as well as reduced exchangeable Al level (Arshad and Gill, 1996).

If a soil on an organic farm is low in available S, and organic manures are not available, then a practical alternative is to add S from mineral sources that are allowed under organic farming. Earlier research by Malhi et al. (2000) had suggested that gypsum could be a suitable source of S to prevent S deficiency in crops (Tables 5.7 and 5.10). In many areas, gypsum may be unavailable/inaccessible or expensive because of the need to transport it over long distances. There is a wide variety of granular elemental S (ES) fertilizers commercially available, but granular ES fertilizers may not be effective in the year of application, especially when applied at seeding in cold spring. Increase in crop yields has been achieved by minimizing S deficiency through broadcast or spray application of fine-particle elemental S (as suspension or powder) on the surface of S-deficient soils (Malhi et al., 2005; Table 5.10). However,

TABLE 5.10 Effects of Elemental S, and Gypsum or other Sulfate-S Fertilizers on Yield of Canola Seed or Grass Forage Dry Matter Yield

Experiment/Treatment	Year	Grass DMY Increase (kg ha^{-1})		
		ES Granular	Gypsum	Potassium Sulfate
15 kg-S ha^{-1}	1	357	1696	2401
	2	195	1256	722
	3	1533	4646	4271

ES Formulation	Seed Yield Increase of Canola (kg ha^{-1})			
	Porcupine Plain, Saskatchewan		Canwood, Saskatchewan	Legal, Alberta
	2000	2001	2001	2000
ES-90 granular	0	127	1296	299
Biosul-ES90 granular	143	256	1518	75
Biosul-ES50 suspension	784	593	1710	637
Sulfate-S	861	581	1788	430

DMY, dry matter yield; ES, elemental sulfur.
Source: Malhi et al. (2000, 2005).

powder or suspension formulations of elemental S may not be convenient to use on a commercial scale. Thus, there is a need for future research to determine the feasibility of granular elemental S as an organic amendment in preventing S deficiency in organic crops for optimum crop growth and yield. Recent findings from an ongoing research study suggest the potential of rapid-release elemental S (Sulvaris Inc., Calgary, Alberta, Canada) in preventing S deficiency in organic crops, provided other nutrients are not limiting in the soil (S. S. Malhi, unpublished results; Table 5.7).

5.3   Biological Fertilizers/Biofertilizers (Microbial Products/Inoculants)

Biofertilizer technology has shown promise for integrated nutrient management through biological N fixation (BNF). Biofertilizers may also be used to improve P availability to crops. The efficacy of inoculants can vary with inoculant type, crop species, formulation, soil nutrient level, soil pH/type, existence of relevant microbes in the soil, and weather conditions.

5.3.1 Rhizobium, Azospirillum, Azotobacter, Azolla, and Blue Green Algae for Biological N Fixation

Rhizobium inoculation of legume crops is commonly practiced in most countries, and the importance of using rhizobial inoculants specific to the legume being planted is well established. *Rhizobium* inoculation has shown beneficial effects on nodulation, grain yield, and protein content of peas (McKenzie et al., 2001; Kutcher et al., 2002). In warm and humid regions, inoculation of *Azospirillum, Azotobacter*, blue green algae (BGA), and *Azolla* (a water fern) was found to increase N supply through BNF in rice-based cropping systems (Singh, 1981). *Azolla* is used as green manure in rice production systems in many Asian countries.

5.3.2 Penicillium Bilaiae

Phosphate-solubilizing microorganisms (PSM) increase availability of sparingly soluble (fixed) P in soil. *Penicillium bilaiae* inoculation could increase available P in soil, and improve dry matter yield, grain yield, and P uptake of wheat (Kucey, 1987). In a study on pea in Manitoba, Canada, *Penicillium bilaiae* inoculation increased crop growth and P uptake in soils with low available P (Vessey and Heisinger, 2001). Beckie et al. (1998) reported that inoculation of alfalfa with *Penicillium bilaiae* usually increased forage yield and P uptake by 3–18%, in the establishment year. However, some other studies did not find any beneficial effect of the fungus on crop yield (Goos et al., 1994; Malhi, 2012a). The efficacy of *Penicillium bilaiae* in increasing crop yield and P uptake depends probably on soil P status, soil type, and climatic conditions. Perhaps because of uncertainty of likely beneficial effects from inoculation with *Penicillium bilaiae*, it is not widely practiced by organic farmers. This emphasizes the need for further in-depth research on the use of *Penicillium bilaiae* for consistent improvement in the yield of organic field crops.

5.3.3 Arbuscular Mycorrhizal Fungi (AMF) Inoculants

Mycorrhizal association is a symbiotic relationship between mycorrhizal fungi and a plant. With the exception of *Brassica* species, AMF (also known as VAM, i.e., vesicular-arbuscular mychorrhiza) improves nutrients uptake, especially P. Research has shown that long-term fertilization with P can reduce AMF colonization (Clapperton et al., 1997; Joner, 2000). Root colonization by AMF in flax was found to be higher in organic treatments than in conventional treatments (Entz et al., 2004), most likely because of lower available P in the organic treatments. The use of VAM has been shown to prevent P deficiency in P-deficient soils by enhancing release of fixed soil P and increase in crop yields (Chalk et al., 2006). However, in field experiments conducted on organic and conventional farms in Saskatchewan, Canada, there was no consistent beneficial effect of AMF (marketed as MykePro) on seed yield and P uptake of barley, wheat, or pea (Malhi, 2012a). Similarly at Thunder Bay,

Canada, the mycorrhizal fungi (MykePro) seed treatment had no positive effect on grain yield, protein content, N removal, or straw yield of the organic spring wheat cultivars in any of the 2009–2011 years in a field not deficient in available P (T. S. Sahota, unpublished results). It is possible that the efficacy of AMF inoculants depends on the level of labile P in soil or soil type. Several AMF inoculants for broad crop production have been developed, but there is very little use of these products on a commercial scale. Therefore, further medium- to long-term research is required to determine the efficacy of all commercially available AMF/VAM inoculants in preventing P deficiency in organic crops grown on soils varying in texture/pH in different agroecological regions.

6. SUMMARY OF RESEARCH FINDINGS, GAPS, AND FUTURE NEEDS

Deficiencies of N and P are widespread on organic farms, and some soils may also contain insufficient amounts of K, S, and some micronutrients for optimum crop growth and sustainable yields. Crop rotations with legumes generally increased yield and protein content. Fertility/quality and productivity of soils could be improved through efficient nutrient cycling by growing deep-rooted legume crops in the rotation. Legumes for green manure could replace summer fallowing, to minimize nutrient inputs, enhance soil quality and increase sustainability of crop production. Because green manure legumes compete with cash crops for space, time, and water, it may not be easy to fit green manure legumes into cropping systems, especially in dry areas. In order to assess the future scope for green manures, there is a need to critically analyse factors that can limit the use of this technology on a commercial scale. Alternatively, intercropping non-legume and legume annual crops or perennial forage grass–legume mixtures may be adopted to improve and sustain crop yields and net returns on organic farms.

On mixed animal and crop farms, application of animal manure/compost and inclusion of perennial forage legumes in the rotation can be a good strategy to prevent multi-nutrient deficiencies in organic crops and improve physical, biological, and chemical properties of the soils. However, there is a need for long-term research studies in a wide range of agro-climatic conditions with a view to (i) minimize the impacts of these systems on soil quality, surface water contamination with labile P, groundwater contamination with nitrate-N, and greenhouse gas emissions including N losses, and (ii) sustain a high level of production from such systems.

Agroforestry could be another eco-friendly option on mixed organic farms. Both shade-tolerant plants grown under a closed tree canopy and sun-loving plants in open areas between trees can be utilized. But in-depth knowledge of the interactions between crops, trees, livestock, and other components appears to be lacking.

Amongst organic amendments/industrial organic products/byproducts, thin stillage was found very effective in improving crop yields. Alfalfa pellets, distiller grain and fish food additives resulted in a moderate increase in crop yields. Because of their residual effects on soil fertility and productivity, it may be important to have long-term information on the economic feasibility of such amendments.

If organic farm soils are deficient in P and organic manures are not available, wood ash, an industrial byproduct, is a good source of P, K, S, and micronutrients to prevent multiple nutrient deficiencies in organic crops. Wood ash, due to its basic nature, could substitute lime to ameliorate acidic soils. However, because of large volumes of application, it is not always convenient to apply, and therefore it needs to be granulated for easy use on a commercial scale.

Biofertilizers/microbial inoculants can play an important role in sustaining productivity, but their effectiveness varies with the source, nutrient type, and level as well as soil type and climatic conditions. *Rhizobium* inoculation is very effective in increasing BNF and crop yields in most legumes, and the practice is well adopted by the producers. It may be possible to enhance availability of native soil P to crops by the use of some microbial inoculants, such as *Penecillium bilaiae*, to increase availability of sparingly soluble soil P. However, the beneficial effects of these microbial inoculants on crop yields have been usually small and inconsistent. Long-term studies are required to assess the visible economic benefits from P-solubilizing microbial inoculants in soils varying in labile P level, pH, and texture, under different climatic conditions.

In warm/humid regions, inoculation of *Azospirillum*, *Azotobacter*, blue green algae (BGA), and *Azolla* has been found to increase N supply through BNF in rice-based cropping systems. However, high labor requirements, cold climate, short growing season, lack of ability to survive over winter, and the need of available P for optimum growth could limit its adoption on a commercial scale in colder countries such as Canada.

Commercially available mineral amendments can be used to prevent/correct deficiency of a single nutrient. Rock phosphate fertilizer is being promoted as a natural source of P to correct P deficiency. In most field studies, the amount of P that becomes available from rock phosphate in the first crop season is very limited, especially on high pH soils. In only a few cases, however, rock phosphate was found effective to increase herbage/pasture yield on P-deficient soils. In a few other cases, rock phosphate was more effective when it was used in combination with *Penicillium bilaiae*, but the yield increase was usually not economical. The results on the feasibility of rock phosphate in improving crop yields on P-deficient soils are inconsistent and not very conclusive, suggesting the need for long-term studies on the feasibility of these amendments on extremely P-deficient soils under variable agroclimatic conditions.

On organic farms with S-deficient soils, gypsum is an excellent mineral amendment to supply S in readily available form (sulfate). In addition, wood ash can be used to supply S to organic crops on S-deficient soils. A granular rapid release elemental S fertilizer has also shown good potential to prevent S deficiency in organic crops, though information on its use in crops at this stage is too limited to allow valid conclusions to be made.

Important strategies to maintain or enhance organic matter in soils include increasing biomass production and return of plant/root residues to soil per unit area by preventing nutrient deficiencies in organic crops and by decreasing loss of soil and organic matter by preventing soil erosion with crop residue cover. However, long-term studies are needed under organic farming systems to investigate the beneficial effects of integrated use of crop management practices, organic manures, and amendments on sustainable crop yields, produce quality, net returns, nutrient uptake and use efficiency, nutrient accumulation and distribution in the soil profile or nutrient balances, nitrate leaching and ground water contamination, soil physical, chemical, and biological properties, soil organic matter dynamics, and greenhouse gas emissions.

7. CONCLUSIONS

Animal and compost manures are an important source of plant nutrients. Often though, there is not enough manure to apply on the entire farm or it may be uneconomical to apply manure in remote areas because of transportation costs. Crop rotations with legumes usually improve grain yields as well as grain protein content. Deep-rooted perennial forages in crop rotations can help to improve soil fertility/quality and productivity through efficient nutrient cycling. Legume green manuring can replace summer fallowing and sustain crop production, mainly by addition of the most limiting nutrient, N, and by increasing the availability of some other nutrients to the succeeding crops. Intercropping of non-legume with legume annual crops and perennial grass–legume mixed stands are effective in maximizing economic yields of grain and forage and protecting soil.

Industrial byproducts/products such as thin stillage showed large beneficial effects in increasing crop yields. Alfalfa pellets, distiller grain, and fish food additives resulted in moderate increases in crop yields. *Penicillium bilaiae* and other microbial inoculants (e.g., AMF inoculant) were found effective in solubilizing native soil P and increasing P availability in some cases. But these products are not well adopted by organic producers, due to their lack of noticeable/consistent beneficial impact on crop yield.

Summer fallowing is a relatively widespread practice on organic farms, but the heavy reliance of organic production on tilled summer fallow to control weeds and increase soil fertility can be a serious concern for soil erosion/degradation and loss of organic matter and sustainable organic production in the long run. In addition to increasing crop yields, sustainable crop management practices and the use of amendments return more crop residues to soil, which could enhance soil quality and fertility.

Organic producers may apply integrated crop management practices, organic amendments, mineral amendments, industrial byproducts, or biofertilizers/microbial inoculants on their farms to improve sustainability of production systems, economic returns, produce quality, nutrient uptake, soil quality and fertility, and water and energy use efficiency while minimizing environmental damage.

REFERENCES

Arshad, M.A., Gill, K.S., 1996. Field pea response to liming of an acidic soil under two tillage systems. Can. J. Soil Sci. 76, 549–555.

Beauchamp, E.G., Voroney, R.P., 1994. Crop carbon contribution to the soils with different cropping and livestock systems. J. Soil Water Conserv. 49, 205–209.

Beauchamp, E.G., Kidd, G.E., Thurtell, G., 1982. Ammonia volatilization from liquid dairy cattle manure in the field. Can. J. Soil Sci. 62, 11–19.

Beckie, H.J., Brandt, S.A., 1997. Nitrogen contribution of field pea in annual cropping systems. 1. Nitrogen residual effect. Can J. Plant Sci. 77, 311–322.

Beckie, H.J., Schlechte, D., Moulin, A.P., Gleddie, S.C., Pulkinen, D.A., 1998. Response of alfalfa to inoculation with *Penicillium bilaii* (Provide). Can. J. Plant Sci. 78, 91–102.

Bolland, M., Gilkes, B., 1990. The poor performance of rock phosphate fertilizers in Western Australia: Part 1. The crop and pasture responses. Agric. Sci. 3, 43–48.

Chalk P.M., Souze R. de F., Urquiaga B.J.R., Boddey R.M., 2006. The role of arbuscular mychorrhiza in legume symbiotic performance. Soil Biol. Biochem. 38, 2944–2951.

Chang, C., Sommerfeldt, T.G., Entz, T., 1990. Rates of soil chemical changes with eleven annual applications of cattle feedlot manure. Can. J. Soil Sci. 70, 673–681.

Clapperton, M.J., Jansen, H.H., Johnston, A.M., 1997. Suppression of VAM fungi and micronutrient uptake by low-level P fertilization in long-term wheat rotations. Am. J. Alt. Agric. 12, 59–63.

Entz, M.H., Bullied, W.J., Katepa-Mupondwa, F., 1995. Rotational benefits of forage crops in Canadian prairie cropping systems. J. Prod. Agric. 8, 521–529.

Entz, M.H., Bullied, W.J., Foster, D.A., Gulden, R., Vessey, K., 2001a. Extraction of subsoil nitrogen by alfalfa, alfalfa-wheat and perennial grass systems. Agron. J. 93, 495–503.

Entz, M.H., Guilford, R., Gulden, R., 2001b. Crop yield and soil nutrient status on 14 organic farms in the eastern part of the northern Great Plains. Can. J. Plant. Sci. 81, 351–354.

Entz, M.H., Penner, R.R., Vessey, K.R., Zelmer, C.D., Thiessen Martens, J.R., 2004. Mycorrhizal colonization of flax under long-term organic and conventional management. Can. J. Plant Sci. 84, 1097–1099.

Freeze, B.S., Sommerfeldt, T.G., 1985. Breakeven hauling distance for beef feedlot manure in southern Alberta. Can. J. Soil Sci. 65, 687–693.

Getachew, A., Amare, G., Sinebo, W., 2006. Yield performance and land-use efficiency of barley and faba bean mixed cropping in Ethiopian highlands. Europ. J. Agron. 25, 202–207.

Goos, R.J., Johnson, B.E., Stack, R.W., 1994. *Penicillium billaie* and phosphorus fertilization effects on the growth, development, yield and common root rot severity of spring wheat. Fert. Res. 39, 97–103.

Haque, I., Lupwayi, N.Z., Ssali, H., 1999. Agronomic evaluation of unacidulated and partially acidulated Minjingu and Chilembwe phosphate rocks for clover production in Ethiopia. Europ. J. Agron. 10, 37–47.

Heitkamp, F., Raupp, J., Ludwig, B., 2011. Soil organic matter pools and crop yields as affected by rate of farm yard manure and use of biodynamic preparation in a sandy soil. Org. Agr. 1, 111–124.

Izaurralde, R.C., Juma, N.G., McGill, W.B., 1990. Plant and nitrogen yield of barley-field pea intercrop in cryoboreal-subhumid central Alberta. Agron. J. 82, 295–301.

Izaurralde, R.C., Juma, N.G., McGill, W.B., Chanasyk, D.S., Pawluk, S., Dudas, M.J., 1993. Performance of conventional and alternative cropping systems in cryoboreal subhumid central Alberta. J. Agric. Sci. 120, 33–41.

Izaurralde, R.C., Nyborg, M., Solberg, E.D., Janzen, H.H., Arshad, M.A., Malhi, S.S., et al., 1998. Carbon storage in eroded soils after five years of reclamation techniques. In: Lal, R., Kimble, J.M., Follett, R.F., Stewart, B.A. (Eds.), Soil Processes and the Carbon Cycle CRC Press, Boca Raton, FL, pp. 369–385.

Jeng, A.S., Haraldsen, T.K., Gronlund, R., Pedersen, P.A., 2009. Meat and bone meal as nitrogen and phosphorus fertilizer to cereals and ryegrass. Nutr. Cycl. Agroecosyst. 76, 183–191.

Joner, E.J., 2000. The effect of long-term fertilization with organic or inorganic fertilizers on mycorrhiza-mediated phosphorus uptake by subterranean clover. Biol. Fertil. Soils 32, 435–440.

Juma, N., Robertson, J.A., Izaurralde, R.C., McGill, W.B., 1997. Crop yield and soil organic matter over 60 years in a Typic Cryoboralf at Breton, Alberta. In: Paul, E.A., Paustian, K.H., Elliott, E.T., Cole, C.V. (Eds.), Soil Organic Matter in Temperate Agroecosystems – Long-Term Experiments in North America CRC Press, Boca Raton, FL, pp. 273–281.

Kirchmann, H., Bergstrom, L., 2001. Do organic farming practices reduce nitrate leaching? Commun. Soil Sci. Plant Anal. 32, 997–1028.

Kucey, R.M.N., 1987. Increased phosphorus uptake by wheat and field beans inoculated with a phosphorus-solubilizing *Penicillium bilaiae* strain and with vesicular-arbuscular mycorrhizal fungi. Appl. Environ. Microbiol. 53, 2699–2703.

Kutcher, H.R., Lafond, G., Johnston, A.M., Miller, P.R., Gill, K.S., Way, W.E., et al., 2002. Rhizobium inoculant and seed-applied fungicide effects on field pea production. Can. J. Plant Sci. 82, 645–651.

Malhi, S.S., 2012a. Relative effectiveness of various amendments in improving yield and nutrient uptake under organic crop production. Open J. Soil Sci. (In Press).

Malhi, S.S., 2012b. Short-term residual effects of various amendments on organic C and N, and available nutrients in soil under organic crop production. Agric. Sci. 3, 1–10.

Malhi, S.S., 2012c. Improving crop yield, N uptake and economic returns by intercropping bayley or canola and pea. Int. Res. J. Agric. Sci. (In Press).

Malhi, S.S., Heier, K., Solberg, E., 2000. Effectiveness of elemental S fertilizers on forage grass. Can. J. Plant Sci. 80, 105–112.

Malhi, S.S., Nyborg, M., Solberg, E.D., Dyck, M., Puurveen, D., 2011a. Improving crop yield and N uptake with long-term straw retention in two contrasting soil types. Field Crops Res. 124, 378–391. doi: 10.1016/j.fcr.2011.07.009.

Malhi, S.S., Nyborg, M., Solberg, E.D., McConkey, B., Dyck, M., Puurveen, D., 2011b. Long-term straw management and N fertilizer rate effects on quantity and quality of organic C and N, and some chemical properties in two contrasting soils in western Canada. Biol. Fertil. Soils 47, 785–800. doi: 10.1007/s00374-011-0587-8.

Malhi, S.S., Schoenau, J.J., and Leach, D. (2010). Maximizing N fertilizer use efficiency and minimizing the potential for nitrate-N accumulation and leaching in soil by balanced fertilization. *Proc. Great Plains Soil Fertility Conference*, 2-3 March, 2010, Denver, CO, U.S.A.

Malhi, S.S., Solberg, E.D., Nyborg, M., 2005. Influence of formulation of elemental S fertilizer on yield, quality and S uptake of canola seed. Can. J. Plant Sci. 85, 793–802.

Malhi, S.S., Zentner, R.P., Heier, K., 2002. Effectiveness of alfalfa in reducing fertilizer N input for optimum forage yield, protein concentration, returns and energy performance of bromegrass-alfalfa mixtures. Nutr. Cycl. Agroecosyst. 62, 219–227.

Manna, M.C., Ghosh, P.K., Ghosh, B.N., Singh, K.N., 2001. Comparative effectiveness of phosphate-enriched compost and single superphosphate on yield, uptake of nutrients and soil quality under soybean-wheat rotation. J. Agric. Sci. Camb. 137, 45–54.

McKenzie, R.H., Middleton, A.B., Solberg, E.D., DeMulder, J., Flore, N., Clayton, G.W., et al., 2001. Response of pea to rhizobia nitrogen and starter nitrogen in Alberta. Can. J. Soil Sci. 81, 637–643.

Miller, P.R., Buschena, D.E., Jones, C.A., Holmes, J.A., 2008. Transition from intensive tillage to no-tillage and organic diversified cropping systems. Agron. J. 100, 591–599.

Przednowek, D.W.A., Entz, M.H., Irvine, B., Flaten, D.N., Thiessen Martens, J.R., 2004. Rotational yield and apparent N benefits of grain legumes in southern Manitoba. Can. J. Plant Sci. 84, 1093–1096.

Rick, T.L., Jones, C.A., Engel, R.E., Miller, P.R., 2011. Green manure and phosphate rock effects on phosphorus availability in a northern Great Plains dryland organic cropping system. Org. Agr. 1, 81–90.

Scheller, E., Vogtmann, H., 1995. Case studies of nitrate leaching in arable fields of organic farms. Biol. Agr. Hort. 11, 91–102.

Schoenau, J.J., Bolton, K., Panchuk, K., 2000. Managing manure as a fertilizer. Farm Facts. Saskatchewan Agriculture and Food (ISSN 0840-9447).

Singh, P.K., 1981. Use of *Azolla* and blue-green algae in rice cultivation in India. In: Bose, P.V., Ruschel, A.P. (Eds.), Associative N_2-Fixation CRC Press, Boca Raton, FL, pp. 236–242.

Tomm, G.O., Walley, F.L., van Kessel, C., Slinkard, A., 1995. Nitrogen cycling in an alfalfa and bromegrass sward via litterfall and harvest losses. Agron. J. 87, 1078–1085.

Vessey, J.K., Heisinger, K.G., 2001. Effect of *Penicillium bilaiae* inoculation and phosphorus fertilisation on root and shoot parameters of field-grown pea. Can. J. Plant Sci. 81, 361–366.

Wani, S.P., McGill, W.B., Haugen-Kozyra, K.L., Juma, N.G., 1994a. Increased proportion of active soil N in Breton loam under cropping systems with forages and green manures. Can. J. Soil Sci. 74, 67–74.

Wani, S.P., McGill, W.B., Haugen-Kozyra, K.L., Robertson, J.A., Thurston, J.J., 1994b. Improved soil quality and barley yields with fababeans, manure, forages and crop rotation on a Gray Luvisol. Can. J. Soil Sci. 74, 75–84.

Warman, P.R., 1986. The effect of fertilizer, chicken manure and dairy manure on timothy yield, tissue composition and soil fertility. Agric. Wastes 18, 289–298.

Watson, C.A., Atkinson, P., Gosling, P., Jackson, L.R., Ryans, F.W., 2002. Managing soil fertility in organic farming systems. Soil Use Manage. 18, 239–247.

Welty, L.E., Prestbye, L.S., Engel, R.E., Larson, R.A., Lokerman, R.H., Speilman, R.S., et al., 1988. Nitrogen contribution of annual legumes to subsequent barley production. Appl. Agric. Res. 3, 98–104.

Whalen, J.K., Chang, C., Olson, B.M., 2001. Nitrogen and phosphorus mineralization potentials of soil receiving repeated annual cattle manure applications. Biol. Fertil. Soils 34, 334–341.

Yanan, T., Emteryd, O., Dianqing, L., Grip, H., 1997. Effect of organic manure and chemical fertilizer on N uptake and nitrate leaching in a Eum-orthic Anthrosol profile. Nutr. Cycl. Agroecosyst. 48, 225–229.

Yang, S., Malhi, S.S., Li, F., Suo, D., Jia, Y., Wang, J., 2007. Long-term effects of manure and fertilization on soil organic matter and quality parameters of a calcareous soil in northwestern China. J. Plant Nutr. Soil Sci. 170, 234–243.

Yaseen, M., Malhi, S.S., 2009. Differential growth response of 15 wheat genotypes for grain yield and P uptake on a low P soil without and with applied P fertilizer. J. Plant Nutr. 32, 1015–1043.

Yuan, X., Tong, Y., Yang, X., Li, X., Zhang, F., 2000. Effect of organic manure on soil nitrate accumulation. Soil Environ. Sci. 9, 197–200.

Zentner, R.P., Campbell, C.A., Biederbeck, V.O., Selles, F., Lemke, R., Jefferson, P.G., et al., 2004. Long-term assessment of management of an annual legume green manure crop for fallow replacement in the Brown soil zone. Can. J. Plant Sci. 84, 11–22.

Zentner, R.P., Basnyat, P., Brandt, S.A., Thomas, A.G., Ulrich, D., Campbell, C.A., et al., 2011. Input management and crop diversity: Effects on economic returns and riskiness of cropping systems in the semiarid Canadian Prairie. Renew. Agric. Food Syst., 1–16.

Zhang, T.Q., MacKenzie, A.F., Liang, B.C., Drury, C.F., 2004. Soil test phosphorus fractions with long-term phosphorus addition and depletion. Soil Sci. Soc. Am. J. 68, 519–528.

Chapter 6

Effective Management of Scarce Water Resources in North-West India

Sudhir-Yadav*, Balwinder-Singh*, Elizabeth Humphreys*, and Surinder Singh Kukal†

*International Rice Research Institute, Los Baños, Philippines, †Punjab Agricultural University, Ludhiana, India

1. INTRODUCTION

The Indo-Gangetic Plain (IGP) is one of the world's largest and most productive ecosystems. The region is named after the Indus and Ganges rivers which, together with their many tributaries, drain the region. The flat terrain with fertile soils, readily accessible surface and ground water, and favorable climate for year-round agriculture attracted human settlement. As a result, this is one of the most densely populated regions of the world, and now serves as home to almost one billion people. Being a rich source of ground and surface water, the region provides water for the economic base of agriculture as well as the urban and industrial sectors. The IGP covers about 700,000 km^2 across Pakistan, India, Nepal, and Bangladesh (Figure 6.1). The region is highly diverse in terms of cultural, social, and agro-ecological conditions. Based on physiography, climate, and vegetation, the IGP has been divided into five sub-regions (Narang and Virmani, 2001). The north-west (NW) IGP, which includes the northern portion of sub-regions 1 and 2 shown in Figure 6.1, has a semi-arid climate. The central IGP (sub-region 3) and eastern IGP (sub-regions 4 and 5) have a hot sub-humid climate. This chapter is mainly focused on NW India, which falls under sub-region 2 which comprises the two small Indian states of Punjab and Haryana. By area, these two states are less than 3% of the total geographical area of India but contribute about 69% of the total food procurement by the Government of India (about 54% of the rice and 84% of the wheat) (Yadvinder-Singh et al., 2003).

North-west India is endowed with rich water resources from snowmelt-fed rivers, monsoon rain, and groundwater. Rainfall mostly falls during the

Agricultural Sustainability. DOI: http://dx.doi.org/10.1016/B978-0-12-404560-6.00006-X

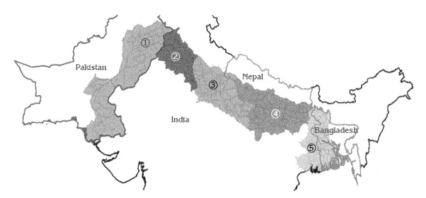

FIGURE 6.1   The location of the Indo-Gangetic Plain. *Source: Narang and Virmani (2001).*

warm monsoon months (June–September), and average rainfall varies in the range 235–998 mm across the region, with averages of 550 mm in Haryana and 650 mm in Punjab. Punjab (India) is traversed by four major rivers, namely the Sutlej, Beas, Ravi, and Ghaggar. In Haryana, surface water for irrigation comes from the Sutlej River via the Bhakra canal system and from the Yamuna River via the Western Yamuna system. The Ghaggar is the main seasonal river of both Haryana and Punjab. Both states have an extensive canal system which helps in the distribution of surface water to almost all parts of these states. The total extent of irrigation canals and distributaries in Punjab is approximately 14,500 km (Kaur et al., 2010). In addition to abundant surface water, the IGP has the world's largest underground aquifer system, containing 4,800 km^3 of water (Tanwar and Kruseman, 1985). There are currently around 1.3 million tubewells pumping groundwater for agriculture in Punjab, and 0.7 million tubewells in Haryana (Hira, 2009; Statistical abstract for Haryana). As a result, 97% and 83% of the cultivable area of Punjab and Haryana, respectively, is under irrigation, in comparison with an average of 39.5% for India as a whole.

2. THE DEVELOPMENT OF WATER SCARCITY FOR IRRIGATION

Traditionally, agriculture in the IGP was rainfed, with mixed cropping in the north-west, and predominantly rice in the east as a result of the higher monsoon rainfall and heavier soils. Surface irrigation schemes and canal networks were developed in the middle of the 19th century during the British era, to address the problem of frequent droughts and crop failure. This brought significant areas under irrigation; for example, 52% of the cultivable land area of Punjab was irrigated by the canal system in 1947 (Randhawa, 1977). In the 1960s, government policies and projects and large scale surface irrigation schemes played important roles in the success of the green revolution, which included a large increase in the area under cultivation. However, the performance of the surface irrigation schemes declined rapidly because of inappropriate design and lack of proper operation and maintenance. Between 1994 and 2003, India and Pakistan together lost more than 5.5 million ha of canal irrigated area, despite very large investment in rehabilitation of existing canals and construction of new ones (Mukherji et al., 2009). In the second half of the 20th century there was rapid expansion in groundwater pumping from tubewells because of the inability of the supply-driven canal system to meet the needs of farmers (Raina and Sangar, 2004). For example, the number of tubewells in Punjab, India increased more than 13-fold, from 98,000 in 1960–61 to about 1.3 million in 2011, even though the total geographical area of the state is only 5 Mha (Hira, 2009). In Punjab, groundwater now accounts for about 73% of the net irrigated area while in Haryana it is 58% (2009–10). As a result of assured water supply, higher-yielding rice varieties and various government subsidies, the area under rice increased from 0.39 to 2.80 Mha in Punjab and from 0.3 to 1.2 Mha in Haryana between 1970–71 and 2010–11 (GOP and GOH, several years).

Unfortunately, the rapid increase in groundwater extraction and cropping intensity resulted in a steady decline in the depth to the groundwater in NW India (Ambast et al., 2006; Hira, 2009; Rodell et al., 2009). The decline in the watertable has accelerated alarmingly in some areas in recent years; for example, in parts of Ludhiana district in central Punjab, the rate of ground-water decline increased from about $0.2\,m\ year^{-1}$ during 1973–2001 to about $1\,m\ year^{-1}$ during 2000–2006. The annual ground water budget estimated by the Central Ground Water Board, India showed that there is a negative annual budget for Punjab and Haryana (Figure 6.2).

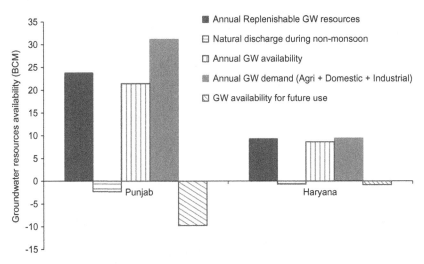

FIGURE 6.2 Ground water (GW) resources availability, demand, and deficit in Punjab and Haryana. *Adapted from Central Ground Water Report (2010).*

3. REASONS FOR GROUND WATER DEPLETION IN NW INDIA

3.1 Increase in the Area Under Cultivation

The Green revolution of the 1960s helped to make South Asian countries self-sufficient for food through large increases in the area under cultivation, cropping system intensity and yield. Traditionally, irrigation was only used as an insurance against crop failure in times of severe drought. The new, higher-yield potential cultivars, however, needed intensive irrigation to maximize yield and profitability for farmers. The net irrigated area in Punjab increased from 2.9 Mha (1970-71) to 4.1 Mha in 2009–10, of which 73% is irrigated through groundwater. In Haryana there was a two-fold increase in the irrigated area from 1.5 Mha to 3.1 Mha, with 58% irrigated from groundwater (Statistical Abstract of Punjab and Haryana).

The increase in groundwater pumping (Figure 6.3) was the result of the resourcefulness of the people, independent of government. Farmers with tube-wells have full control of water supply (timing and amount of water) in the absence of any solid government policy or control. It has been estimated that farmers and entrepreneurs invested around 12 billion USD on ground water pumping infrastructure in India in the last 20 years compared with investment of 20 billion USD of public money on surface water irrigation schemes over the last 50 years (International Water Management Institute, 2002).

3.2 Shift in Cropping Patterns

In addition to expansion of the cropped area and intensification, the green revolution in NW India also involved a shift in the types of crops grown. In particular there was a shift to a rice and wheat cropping system, replacing

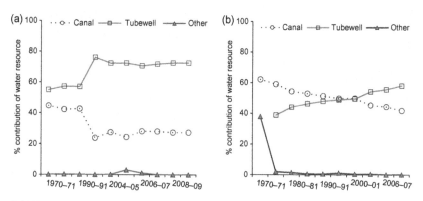

FIGURE 6.3 Trends in the use of different water resources for irrigation in (a) Punjab and (b) Haryana. *Source: Statistical Abstract of Punjab and Haryana.*

legumes and fodder crops. On an all-India basis, the area under rice increased from 30.8 Mha to 42.0 Mha and the wheat area increased from 9.75 Mha to 29.3 Mha between 1950–51 and 2010–11 (GOI, several years). The biggest shift took place in Punjab and Haryana. In Punjab, the increase in area under rice was mainly due to cultivation of fallow/grazing land and partially at the expense of maize, gram and other kharif pulses (Figure 6.4a). In Haryana, the shift to rice was at the expense of fodder crops like Bajra and Jowar, in addition to fallow/grazing land (Figure 6.4c). The increase in the area under wheat was partly at the expense of gram, especially in Haryana, and partly at the expense of fallow/grazing land (Figure 6.4b,d). The drivers of the shift to rice and wheat were assured minimum price for rice and wheat, good and efficient marketing, and subsidies on inputs, together with the availability of improved and high-yielding varieties. However, to achieve maximum yields and profitability of the modern rice and wheat varieties, irrigation on demand was needed, hence the rapid expansion in tubewell installations. A recent study (Anonymous, 2012) suggests that both increased groundwater extraction for paddy cultivation and lower irrigation return to the ground water system over the years are the main reasons for the steady decline of groundwater levels in central Punjab.

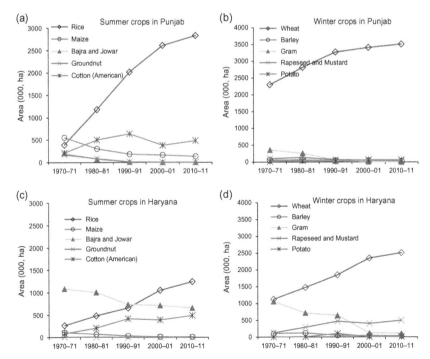

FIGURE 6.4 Shift in cropping pattern of summer and winter crops in Punjab and Haryana. *Source: Statistical Abstract of Punjab and Haryana.*

3.3   Injudicious Use of Surface and Ground Water

Rice, which is also called "water loving plant", is traditionally grown in tropical, high-rainfall areas on clayey soils in Asia. However, in sub-tropical NW India, rainfall during the monsoon varies approximately from 300 mm to 800 mm depending on location and season, and rice is often grown on soils varying from sandy loam to loam texture, and sometimes even on loamy sands. The majority of the agricultural soils in Punjab are sandy (sands, loamy sands, sandy loams, sandy clay loams), but 70% of the area is used for rice cultivation. The soils are highly permeable and require large amounts of irrigation water to keep the fields flooded, as per conventional farmer practice. While the percolation losses flow back to the groundwater, this is at the expense of considerable amounts of energy to pump the water back to the surface for irrigation.

The general recommendation for the time of transplanting rice is with onset of monsoon, which usually occurs in the last week of June in NW India. However, in practice the date of transplanting used to vary from early May to mid August. Early transplanting was often preferred to avoid peak demand for labor and to reduce pest and disease pressure which builds up as the season progresses. However, evaporative demand is extremely high in May and June (often in excess of 10 mm/d), meaning that excessively large amounts of irrigation water (and energy to pump it) are needed prior to the onset of the monsoon, and much of this water is lost from the system as evapotranspiration (ET).

The hydrostatic pressure of standing water results in high water losses in the forms of seepage and percolation, especially on permeable soils. It has been well known for over 30 years that alternate wetting and drying water management (also known as intermittent irrigation) greatly reduces irrigation input in comparison with continuous flooding (Sandhu et al., 1980; Bouman and Tuong, 2000; Kukal et al., 2005; Sudhir-Yadav et al., 2011a). Nonetheless, most famers in NW India still use continuous flooding for most of the season when growing rice, especially when sufficient electricity is available.

3.4   Degradation of Ground Water Quality

There are some regions in NW India where the groundwater is very shallow and abundant but unsuitable for irrigation because of salinity. Brackish groundwater underlies almost two-thirds of Haryana, an area characterized by poor natural drainage, rising watertables and secondary salinization (Hellegers et al., 2007). South west Punjab faces similar problems. The area where groundwater quality is very good is already overexploited, resulting in lowering of groundwater levels (pressure heads). The main risk of over-exploitation of the good quality groundwater in NW India arises from contamination by saline groundwater intrusion as a result of reversal of groundwater flows due to lowering of pressure heads in the fresh aquifers (Perveen et al., 2012).

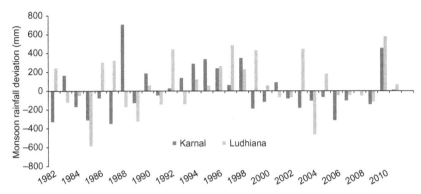

FIGURE 6.5   Deviation of total rainfall during the monsoon season (June–September) at Karnal (Haryana) and Ludhiana (Punjab) during 1982–2010 (seasonal rainfall minus long-term average).

3.5   Rainfall Distribution and Variability

In much of NW India, the total amount of rainfall during the south west monsoon is usually sufficient to meet ET requirements of summer crops, including rice. Rains are also important to recharge water resources. However, the total rainfall and monthly rainfall distribution during the monsoon (June to September) are highly variable from year to year (Figures 6.5 and 6.6). Rainfall distribution is generally very poor, most of the rain falling in a few large events, sometimes causing runoff and flooding in low-lying areas. There are usually many dry spells between significant rainfall events, hence the high dependence on irrigation to grow high-yielding summer crops.

Winters in the NW IGP are relatively dry, with average rainfall of around 150 mm during the wheat season, compared with crop water use requirement of about 350 mm. Furthermore, rainfall during the wheat season is extremely variable across years, ranging from almost nothing to around 300 mm. Hence, the heavy dependence on irrigation for high wheat yield. Using weather data at Ludhiana (Punjab), Jalota and Arora (2002) found that total rainfall during the rice season exceeded total ET by 141 mm (for rice transplanted in mid-June), while total rainfall during the wheat, long-fallow, and short-fallow periods was much less than ET. Net water depletion was greatest due to cultivation of wheat (198 mm) followed by the long fallow between wheat harvest and rice transplanting (67 mm).

3.6   Energy Subsidies for Farmers

In most of the IGP, water is considered as a free resource and its use is not limited by water pricing or allocation policies. In NW India, the state governments highly subsidize the provision of electricity to farmers, primarily to

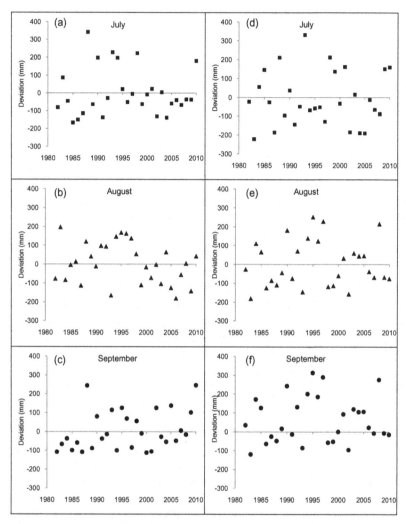

FIGURE 6.6　Deviation of the monthly rainfall from long-term monthly average in Karnal (a–c) and Ludhiana (d–f). *Sources: Data collected from Central Soil Salinity Research Institute (CSSRI), Karnal and Punjab Agricultural University (PAU), Ludhiana.*

pump the ground water. Furthermore, the small charge paid by farmers is not related to the amount of electricity used and thus the volume of water pumped. In Punjab, 60% of the annual electricity supply goes to tubewells for rice cultivation (Hira, 2009). The supply of electricity is limited and can be unreliable, therefore farmers tend to pump whenever electricity is available. When the supply of electricity is inadequate to meet irrigation needs, many farmers use diesel generators to power the tubewell pumps. Diesel is also highly subsidized for farmers, by up to 50%, by the Central Government. The high subsidies on electricity and diesel, and the lack of charging for the amount of electricity used, leave no incentive to the farmers to reduce irrigation input. Furthermore, the unreliable electricity supply is a disincentive to the adoption of alternate wetting and drying water management, which requires the ability to apply timely irrigations to avoid yield loss from water deficit stress.

4. "REAL" WATER SAVINGS

Water in the form of irrigation or rain that enters the field may be lost from the rootzone in a range of ways, namely deep drainage beyond the rootzone, surface runoff, evaporation, and transpiration. All losses except transpiration are non-productive losses, while transpiration is a productive loss which is directly related to biomass production and yield. Where deep drainage and runoff losses can be re-used, these are not real losses from the soil–groundwater system. This is often the case in groundwater-irrigated areas, such as most of the rice–wheat areas of NW India, where deep drainage flows back into the groundwater. Here, reducing deep drainage through practices such as alternate wetting and drying water management in rice reduces the water loss at the farmer field level, but from a regional groundwater resource perspective there is no net water saving (just an energy saving in terms of reduced pumping) or effect on groundwater depletion (Humphreys et al., 2010). Similarly, surface runoff from one field is often captured by downstream fields or other users, and reducing runoff at the field level does not necessarily mean that there is more water available for other users (Hafeez et al., 2007). However, reducing ET (evaporation + transpiration) is always a real water saving. There are many technologies which can reduce irrigation input at the farmer field scale by reducing drainage and runoff, but which are unlikely to have a significant effect on real water saving. Thus, from the perspective of regional water resources management in the rice–wheat systems of NW India, increasing water productivity based on ET losses needs to be the priority where the amount of available water is declining due to over-extraction.

5. IMPROVING MANAGEMENT OF WATER RESOURCES

Water for irrigation comes from harvesting of surface runoff from rainfall in catchments (dams) delivered to irrigation areas through canal networks, pumping from rivers and other surface water bodies, and pumping from groundwater. The groundwater is primarily derived from drainage of rainfall into the strata beyond the rootzone, and seepage from rivers. Discussion of the potential for increasing the supply of water in NW India through construction of additional large dams is beyond the scope of this chapter, except to note that dams are controversial. They are very expensive and whether there are net economic benefits to society is sometimes questionable. In addition to providing water for irrigation and various other uses, dams can be beneficial for flood control, and capture some of the water that would otherwise flow to the sea. While beneficial for irrigated agricultural production, there are often disadvantages for other land and water users, including those dispossessed of land and relocated as a result of dam construction, those downstream of the dam, and those downstream of the irrigated areas, including river and riparian ecosystems. These problems become more intractable when they involve rivers crossing political borders, as in NW India. Therefore, this section focuses on opportunities for making better use of the available water by reducing water losses at the canal network and farmer field scales.

5.1   Reducing Water Losses from Canal Networks

Water losses are inevitable during conveyance in canals/channels from dams to fields, and may be categorized as: (i) seepage through the floor and sides of the canal, (ii) evaporation from the water surface in the canals, (iii) leaks from structures in bad condition, and (iv) waste due to inefficient canal network operation; while waste (surface drainage out of the irrigation area) is a costly loss for a canal command area, the water may be used downstream, and is not necessarily a loss at higher spatial scales. The amount of water lost in all the above ways can be reduced by proper design, good maintenance, and optimum management of the system.

Optimum management involves delivering the right amount of water, at the right time, to where it is needed, with minimal amounts of water in the parts of the system where it is not needed. Modern irrigation management systems can greatly reduce all these losses through accurate, real time monitoring of flows and remote control of flow regulators on the supply side, together with good knowledge of historic land and water use patterns, and real time estimation of water demand (through real time monitoring of land use and weather across the command area).

The amount of water lost from canal networks by evaporation depends on the water surface area of the canal network, with reports of losses of 0.3% (Reid and Muller, 1983) of the amount of water conveyed. At this stage, there are no cost effective technologies for reducing evaporation losses from well designed canals.

Seepage is one of the main factors that reduce the efficiency of canal irrigation systems. In addition to loss of water, seepage causes water logging and loss of productivity in lands adjacent to the canals. A well maintained canal with 99% perfect lining reduces seepage by about 30-40% (Wachyan and Rushton, 1987), but it is extremely expensive. Seepage hotspots often account for large proportions of the total seepage loss, hence the importance of identifying the location of such hotspots. Electromagnetic induction (EM) survey and towed geo-electric arrays are techniques for rapidly identifying locations with high seepage losses (Khan et al., 2009; Akbar et al., 2011). While seepage from canals is a cost to canal network managers, it also recharges the groundwater and in many situations makes groundwater available to farmers who are poorly served by canal systems, such as those at the ends of some systems. Furthermore, the groundwater is not vulnerable to loss by evaporation, except where watertables are shallow (e.g., within about 1.5 m of the soil surface) (Khan et al., 2006; Young et al., 2007). However, in areas with shallow saline watertables, only a proportion of the seepage losses can be recovered, and this proportion varies greatly depending on the local hydrogeology.

5.2   Conjunctive Use of Surface and Ground Water

NW India faces the dual problems of rising groundwater levels in areas that are well served by canals, and falling groundwater levels in other areas which don't have sufficient access to surface water. Shallow groundwaters in canal-irrigated areas are the result of seepage from the canal network and deep drainage from irrigated fields, and have resulted in water logging, and salinity in areas where the groundwater is saline (such as South-west Punjab and central Haryana).

Conjunctive use of surface and groundwater can reduce the problems of rising groundwater in canal-irrigated areas, and falling groundwater in predominantly groundwater-irrigated areas. For example, shallow skimming wells can be used to tap the fresh water lens above the saline groundwater, providing water for irrigation and lowering the watertable (Saeed et al., 2002). Some of the poor quality water can also be mixed with the good quality canal water, increasing the overall availability of water for irrigation. On the other hand, the use of canal networks in predominantly groundwater-irrigated areas helps restore groundwater levels through recharge from the canals and fields irrigated from canal water, in addition to reducing extraction from the groundwater.

In canal-irrigated areas, one of the reasons for the increasing dependence on groundwater is the low capacity of the canal network to supply water at the peak time of water demand. Most of the canal irrigation systems in NW India don't have many storage/reservoir areas. Even if there is storage capacity, the delivery system has only a fraction of the capacity required to meet peak irrigation demands. Construction of small and medium storages within canal systems could help meet peak demand, and can also help to increase ground water recharge. Grewel and Dar (2004) estimated that 15-20% of the groundwater depletion in Punjab could be met by such measures.

5.3   Artificial Recharge of Groundwater

Several promising strategies to arrest the declining watertable by artificial recharge of ground water have been proposed at landscape, village, and urban levels (Hira et al., 2009). These methods include the use of recharge wells in areas where there is a thick sub-surface layer restricting downward percolation of water (Ambast et al., 2006). Desilting of village ponds will also increase recharge in the region. Village ponds are a sink for all of the runoff from the village, but the side and bottom surfaces are clogged by sediment, with the result that these ponds never dry up as they used to do in earlier times. The silt can also serve as a rich source of nutrients for field application. However, most village ponds are highly contaminated, as the runoff water includes dairy effluent and other contaminants. Therefore, increasing recharge from village ponds in the absence of water treatment is likely to exacerbate the current high levels of contamination of the shallower groundwater from where village drinking water is often drawn using hand pumps (Singh and Singh, 2007).

Adoption of soil conservation practices in the Shiwalik hills can help to reduce runoff and increase infiltration by increasing the residence time of runoff water at the surface: for example, contour trenches, vegetation barriers, and bunding of fields (Hira, 2009). The safe storage of the runoff water in dugouts or small dams for irrigation can also reduce pressure on other surface and ground water resources. Similarly, rainwater harvesting in cities through the use of grass saver tiles can help recharge the ground water for urban use.

5.4 Improved Crop Management Technologies

5.4.1 Laser Leveling

The staple food crops (rice, wheat, maize) in the IGP are irrigated by flooding. Large amounts (10–25%) of irrigation water are lost by deep drainage and evaporation during application because of poor management (inadequate flow rates) and uneven fields (Kahlown et al., 2002). Good land leveling is a prerequisite for achieving high efficiency of flood irrigation. Rice fields are puddled and leveled every year prior to transplanting; however, such fields are still uneven, and field slopes vary from 1° to 5° across the region (Jat et al., 2006). In a survey of 300 farmers' fields in Punjab, India, the difference between the highest and lowest parts of the fields ranged from 8 to 25 cm (H.S. Sidhu, unpublished data). In their review, Humphreys et al. (2010) found that laser leveling reduced irrigation input by about 100–200 mm in rice and by about 50–100 mm in wheat, and proposed that the reduction was due to shallower drainage, and that magnitude of the reduction would depend upon site conditions, water management, and irrigation system (flow rates relative to field size). On average, laser leveling had a small effect on rice yield, but a much larger effect on increasing wheat yield.

Another study in Punjab found that laser leveling in rice reduced irrigation water input by 362 mm with a yield increase of 0.78 t/ha (Sidhu and Vatta, 2010), much higher than the values in other studies. The reduction in irrigation input resulted in a reduction in electricity consumption of 213 kWh/ha, but a cost saving of only Rs. 610 per ha of rice over the season. Based on these data, but assuming average irrigation water saving of 150 mm, laser leveling of the entire rice growing area of 2.8 Mha in Punjab could reduce irrigation input by 0.42 Mha m (0.42×10^9 m^3) and save 247 GWh of electricity.

5.4.2   Crop Establishment Method

5.4.2.1   Rice

The traditional practice of growing rice in NW India, which involves puddling, transplanting, and continuous flooding until shortly before harvest, has many disadvantages including high tillage costs, high labor costs, impairment of soil structure for upland crops grown in rotation with rice, and high water requirement for puddling (100–250 mm). Therefore, replacement of the puddled transplanted rice (PTR) production system with a non-puddled system has many potential benefits. There are two main rice establishment methods which eliminate puddling and manual transplanting: dry seeding and mechanical transplanting. In both cases, crop establishment can take place following dry tillage, or in the absence of any tillage (zero till). Dry seeding may take place on raised beds or on the flat. Dry seeded rice (DSR) on the flat is a technology which does not require any specialized farm implements, and it can be sown using the same seed drill in more or less the same manner as other crops such as wheat. The duration of DSR in the main field is longer than for a transplanted crop, and DSR is sown at a time when evaporative demand is high and therefore needs several irrigations for establishment (Sudhir-Yadav et al., 2011b). However, both PTR and mechanically transplanted rice need continuous flooding for the first couple of weeks after transplanting, resulting in higher irrigation requirement for establishment than DSR.

5.4.2.2   Wheat

Zero till (ZT) wheat is the most successful resource-conserving technology to have been introduced to the IGP, with adoption on 0.7 Mha as of 2010–11 in NW India (0.2 Mha in Punjab and 0.5 Mha in Haryana) (Grover and Sharma, 2011). Adoption studies show irrigation water savings of about 20–35% with ZT (Erenstein and Laxmi, 2008), mainly because pre-sowing irrigation is not normally needed, and because the smooth soil surface results in faster movement of irrigation water across the field. The use of raised beds is another technology which reduces irrigation time in farmers' fields, while providing agronomic advantages such as reduced waterlogging. Raised beds have been successfully tested for a range of crops including wheat, maize, cotton, and legumes (Ram et al., 2005). The few published results from farmers' fields show reductions in irrigation amount of 35–50%. The saving in irrigation water is due to faster movement of the water along the furrows, reaching the other end of the field in less time than in a conventionally tilled field, resulting in reduced deep drainage losses during irrigation. The size of the reduction depends on soil type (soil permeability), irrigation flow rates relative to field size, and depth to the watertable.

5.4.3   Rice Transplanting Date

In the 1970s, rice transplanting in Haryana and Punjab used to start from June 30, after the onset of the monsoon rains, under conditions of low evaporative demand due to high humidity. Because of the low cost of pumping as a result of highly subsidized power, and the benefits of early transplanting (reduced disease pressure for earlier planted crops, greater labor availability), many farmers advanced the start of transplanting by early to mid-May. However, evaporative demand is very high at this time of year, with 40–45% of annual open-pan evaporation occurring during the two months from mid-April to mid-June (Minhas et al., 2010). Shifting transplanting back to the time when a crop can grow under reduced evaporative demand while maintaining yield, and without jeopardizing the ability to sow the next crop at the optimum time, can significantly reduce ET from rice. Many studies have shown a 70–189 mm reduction in ET by shifting the date of transplanting from early May to July 1 (Humphreys et al., 2010). In 2009, Punjab and Haryana passed the "Preservation of Sub-Soil Water Act" which prohibits transplanting before 10 June (Punjab) or 15 June (Haryana). Hira (2009) predicted that further delaying the transplanting date from 10 June to 15, 20, 25, and 30 June would reduce the average watertable decline rate from 64 cm/yr to 51, 39, 29, and 20 cm/yr, respectively.

5.4.4 Irrigation Scheduling

5.4.4.1 Rice

Rice in NW India is traditionally grown with continuous flooding, which requires large amounts of irrigation water. The main reasons for much higher water use in flooded rice than other crops are high seepage and percolation losses. Reduction of hydrostatic pressure is an important means for reducing such losses (Bouman et al., 1994). This is the principle behind alternate wetting and drying (AWD), also known as intermittent irrigation. This involves flooding the field to a shallow depth (e.g., 50 mm), allowing the water to dissipate, and re-irrigating some time after the soil surface has dried out. It has been well established that irrigation input can be reduced by the use of safe AWD, i.e., AWD managed to avoid yield loss. The reduction varies from 10–40% of the amount applied to a continuously flooded field, depending on soil hydraulic conductivity and depth to the watertable. In NW India the general rule, based on many experiments conducted 20 to 30 years ago, is to irrigate 2–3 days after the ponded water has disappeared (Humphreys et al., 2010). However, this practice does not take into account soil type and crop water requirement, which is determined by growth rate and evaporative demand. Better methods for determining when to irrigate, based on the soil water status of the rootzone, have now been developed in the form of the farmer-friendly PAU tensiometer (Kukal et al., 2007) and the field water tube or pani-pipe (Bouman et al., 2007). Both PTR and DSR can be irrigated using safe AWD using the same threshold (soil tension) of 15 kPa at a soil depth of 15–20 cm.

5.4.4.2 Wheat

In wheat, farmers often apply irrigation based on visual observations of the crop (signs of wilting) and/or soil appearance (dryness of the topsoil). However, these approaches do not take into account soil water availability, which can vary greatly depending on soil type, crop growth rate (which is influenced by management), root distribution, evaporative demand (which depends on the weather), and other management factors such as tillage and residue retention. As a result, crops may be stressed at some times and over-irrigated at others. In NW India, farmers tend to over-irrigate, applying 5–6 irrigations to wheat. With scheduling of irrigations according to plant-available water (e.g., based on soil tension), there is potential to save 1–3 irrigations. Reducing the number of irrigations by one (say 75 mm) over the wheat area of 6.0 Mha in NW India (3.5 Mha in Punjab, 2.5 Mha in Haryana) would reduce the volume of water applied by $4.5 \times 10^9 \, m^3$, resulting in a huge energy-saving equivalent to 265 GWh if all the water were pumped (based on the data of Sidhu and Vatta, 2010).

In more-developed countries, web-based decision-support systems are starting to become available to give farmers real-time information via SMS on the need to irrigate, and how much to apply. These systems are based on estimation of crop water use and requirement from remote sensing or crop modeling, together with reference ET (Singels and Smith, 2006; Hornbuckle et al., 2010).

5.4.5 Irrigation Method

Flood irrigation, the conventional method of irrigation in NW India, can be highly inefficient where flow rates are inadequate to complete the irrigation quickly (a couple of hours). The inefficiency is due to deep drainage below the rootzone. Flood irrigation also causes temporary waterlogging, with adverse effects on crops like wheat, maize, and legumes. Waterlogging is more prolonged and more severe on heavy textured soils, and on soils used for rice culture because of the well-developed, shallow, hard pan (slowly permeable) as a result of puddling. This leads to aeration stress in upland crops, especially in wheat (Kukal and Aggarwal, 2003). Modern, pressurized irrigation systems (center pivot and lateral move sprinkler, micro-sprinklers, surface drip and subsurface drip) have the potential to increase irrigation water use efficiency by providing water to match crop requirements, reducing runoff and deep drainage, and generally keeping the root zone drier. Drier soil also means less waterlogging, lower soil evaporation, and increased capacity to capture rainfall, further reducing runoff and deep drainage.

In developed countries, pressurized irrigation systems are more common for horticulture (fruit trees, grape vines, vegetables), and higher-value and more-waterlogging-sensitive staple food and fiber crops such as maize, soybeans, and cotton (Cornish, 1998). However, a wide range of crops, including rice and wheat, and pastures are also grown under sprinkler systems. Evaluation of such technologies for cropping systems in NW India is at an early stage. Surface drip and fixed sprinkler irrigation is currently under evaluation for rice–wheat systems at Modipuram, Western Uttar Pradesh. There was about 50% and 70% irrigation water saving in rice and wheat, respectively, under drip and sprinkler irrigation compared with flush/flood irrigation (Rajeev Kumar et al., unpublished data). In Australia, sprinkler irrigation of rice to replace evaporative loss reduced irrigation water use by 30–70% (Humphreys et al., 1989); however, even at frequencies of up to three times per week, yield declines of 35–70% occurred (Muirhead et al., 1989). The lack of floodwater to buffer against low night temperatures was probably part of the cause of lower yield, whereas this would not be an issue in NW India, where the weather is warm throughout the rice season. Also in Australia, irrigation water use on a heavy clay soil was reduced by about 200 mm (12%) in rice with subsurface drip commencing two weeks prior to panicle initiation compared with conventional flooded rice culture; however, yields with drip also decreased by 1 t/ha (12%), and there was no increase in irrigation water productivity (Beecher et al., 2006).

In contrast to the widely held belief that adoption of pressurized irrigation technologies leads to significant real water savings, Ward and Pulido-Velazquez (2008) have shown that these technologies reduce valuable return flows and limit aquifer recharge. Secondly, given that yields of the major crops in NW India can be maintained while greatly reducing irrigation input using flood irrigation in combination with the improved technologies above, there remains the question of economic feasibility of pressurized irrigation methods. Diversification into higher-value crops could assist, but on a limited scale.

5.4.6 Mulch

It is well established that surface retention of crop residues conserves water by suppressing soil evaporation (Bond and Willis, 1969). The rice–wheat system of NW India produces large amounts of crop residues. Wheat residues have high value and are normally used as animal feed. The rice residues are normally burnt prior to wheat establishment, as incorporation is costly and delays wheat sowing beyond the optimum time, and as it has not been possible until recently to sow wheat directly into the residues. The development of the Happy Seeder (Sidhu et al., 2007, 2008) brings the possibility of sowing wheat immediately after rice harvest, into the residual soil moisture, avoiding the need of pre-sowing irrigation. Many studies have shown that rice mulch significantly reduces the rate of soil drying and delays the need for irrigation of wheat (Balwinder-Singh et al., 2011a). However, whether or not mulch will reduce the number of irrigations during the wheat season will depend on the seasonal rainfall distribution in relation to the timing of irrigations with and without mulch, provided that irrigations are scheduled according to soil water status or crop demand. In a simulation study using the Agricultural Production System Simulator (APSIM) for well-irrigated wheat in central Punjab, India,

Balwinder-Singh et al. (2011b) found that mulch reduces the number of irrigations needed by one, or about 60 mm in 50% of years. Thus, with 6 Mha of irrigated wheat in NW India, and assuming 60 mm per irrigation, surface residue retention has the potential of reducing the amount of irrigation water applied by $3.6 \times 10^9 \, m^3$.

There is no scope for mulch to reduce evaporation from rice in the puddled transplanted system. However, the introduction of DSR brings the feasibility of ZT rice-based cropping systems and surface residue retention for the entire cropping system. With surface retention of rice residues in wheat, some residues (2–3 t/ha, Naveen Gupta, unpublished data) are still present on the soil surface at the time of wheat harvest. Whether the residual mulch would confer any soil evaporation (or other) benefits during establishment of ZTDSR in the following rice season is yet to be determined. In a ZT rice–wheat–mung system, surface retention of the mung residues may also assist.

5.4.7 Crop Diversification

Replacing rice with other summer crops such as maize, soybean, and cotton would greatly reduce the amount of irrigation water applied (e.g., Ram et al., 2005; Hira, 2009). Depending on the distribution and amount of the monsoon rains and soil type, these crops require from none to four to five irrigations after sowing, with total irrigation application to maize and soybeans of the order of one-fifth to one-tenth of that for rice (Humphreys et al., 2010). However, the irrigation water saving is primarily due to less drainage. The real losses (ET) from some non-rice cropping systems (sugarcane, cotton–wheat) may actually be higher than from the rice–wheat system (Jalota and Arora, 2002). Further studies are needed to quantify ET for a range of cropping systems under a range of site, seasonal, and management conditions.

To date, attempts to promote crop diversification have not been successful, because of the assured procurement and minimum support price for rice and wheat only, together with other government policies (see section 5.6). Hence, there is great importance of promotion of technologies for reducing irrigation water requirement and using the water more productively.

5.5 Rainfall Forecasting

Knowledge of the likelihood of rain (occurrence and amount) in the short (next few days), medium (next 2–3 weeks), and long (next season) term would be highly beneficial to both farmers and canal irrigation system managers. For farmers growing irrigated crops, accurate short-term forecasts would enable them to avoid unnecessary irrigation (and crop damage from waterlogging as a result of rain shortly after irrigation in the case of upland crops) and increase profitability (Wang and Cai, 2009). Such information could be delivered to farmers on an appropriate scale (e.g., district or sub-district) through services such as mobile phone SMS, radio, and TV. Accurate rainfall forecasting would also assist irrigation scheme managers to make decisions about the delivery of surface water to where it is most needed. Need-based delivery of water in canal systems would also help to reduce water losses through avoiding the situation of full canals in locations where the water is not needed.

5.6   Policies to Improve Water Management: Water Pricing and Allocation

In NW India, there are water-pricing policies related to consumption for urban domestic and industrial water use. On the other hand, farmers irrigating from groundwater are not charged for water use, and farmers in canal schemes are charged a flat rate on the basis of the area of land irrigated. Furthermore, many farmers use electricity to pump groundwater, especially in NW India. Out of the 1.3 million tubewells in Punjab, 1.0 million are powered by electricity and 0.3 million by diesel (Agriculture Statistic database, Punjab). Electricity is highly subsidized for agriculture. Therefore there is no incentive to reduce water use. In fact the incentive is the opposite—the supply of electricity to rural areas is restricted to a few hours a day, and can be unreliable, therefore farmers keep the pumps running whenever electricity is available to maximize the use of free power, and to minimize the need to use diesel-powered generators and the amount of money spent on diesel. However, even diesel is subsidized.

In recent years, some states in India started metering of electricity to tubewells used for agriculture and charging for electricity on a usage basis (e.g., West Bengal, Gujarat). The state of Punjab also needs to evolve such policies of charging for electricity for agriculture on a usage basis. Coordinated planning, policies, and regulations are also needed regarding the installation of tubewells, taking into account the existing groundwater conditions and sustainable yield.

Where the demand for water is greater than the supply, policies are needed to enable access based on social, economic, and environmental priorities. A range of policies and tools have been developed for attempting to achieve this in countries like Australia (Wei et al., 2011). These tools include water allocations which are based on the capacity of dams, or on the sustainable extraction limit of groundwater. The surface water allocations are adjusted proportionately on an annual basis depending on likely availability of water based on the amount of water in the dams and historic rainfall patterns.

6.   CONCLUSIONS

The sustainability of India's most productive agro-ecosystem, the rice–wheat cropping system of NW India, is threatened by water scarcity as a result of deficient surface irrigation systems and over-exploitation of the groundwater. About 60–75% of the total irrigation water supply in this region now comes from groundwater, and ground water levels are declining at an alarming rate, increasing extraction costs. However, the more serious threat to sustainability is saline intrusion into the fresh groundwater as a result of reversal of groundwater flows.

The primary reasons for the depletion of groundwater resources are the huge expansion of irrigated agriculture during the Green Revolution of the 1960s–1980s, together with lack of coordinated planning of surface and ground water resources development, and lack of policies to act as disincentives to over-use of water (with one notable, recent exception: the effectively implemented (2009) prohibition of rice transplanting before mid-June).

Many opportunities have been proposed for helping to arrest the decline in groundwater, requiring action at state government, community, and individual farmer levels. At government levels, these include rehabilitation and modernization of surface canal systems, development of coordinated plans for conjunctive use of surface and ground waters, and volumetric water pricing (charging for the amount of electricity used) and (gradual) removal of subsidies. At community levels, the opportunities include artificial recharge of the groundwater.

There are many opportunities at the farmer field level to greatly reduce irrigation inputs and thus the cost/energy consumption for irrigation. These include laser leveling, use of alternate wetting and drying water management in rice, changing from puddled transplanted rice to dry seeded rice, changing from conventional tillage to zero-till wheat and dry seeded rice, mulching, scheduling irrigations according to need using tensiometers or pani-pipes, and diversification to crops with lower irrigation requirement.

While all of these technologies reduce irrigation requirement to varying degrees, depending on site and seasonal conditions, their effects on real water savings are poorly understood at the farmer field scale, and have hardly been considered at higher spatial scales. The performance of the technologies needs to be assessed at a range of spatial scales to determine the extent to which widespread adoption will reduce groundwater depletion on a regional basis.

REFERENCES

Akbar, S., Abbas, A., Hanjra, M.A., Khan, S., 2011. Structured analysis of seepage losses in irrigation supply channels for cost-effective investments: case studies from the southern Murray-Darling basin of Australia. Irri. Sci. doi: 10.1007/s00271-011-0290-4.

Ambast, S.K., Tyagi, N.K., Raul, S.K., 2006. Management of declining groundwater in the trans Indo-Gangetic plain (India): some options. Agric. Water Manage. 82, 279–296.

Anonymous, 2012. Project Report on "Over-exploited Indo-Gangetic Alluvial Aquifer Characterization and Dynamic Mapping of Groundwater Resource Utilization: A Case Study of PAU Campus, Ludhiana, Punjab". National Geophysical Research Institute, Hyderabad.

Balwinder-Singh, Humphreys, E., Eberbach, P.L., Katupitiya, A., Yadvinder-Singh, Kukal, S.S., 2011a. Wheat growth, yield and water productivity of zero-till wheat as affected by rice straw mulch and irrigation schedule. Field Crops Res. 121, 209–225.

Balwinder-Singh, Humphreys, E., Eberbach, P.L. and Gaydon, D.S. (2011b). Evaluation of mulch for irrigated zero till wheat in north-west India. Paper presented at 5th World Congress of Conservation Agriculture, Brisbane, 26–29 September, 2011.

Beecher, H.G., Dunn, B.W., Thompson, J.A., Humphreys, E., Mathews, S.K., Timsina, J., 2006. Effect of raised beds, irrigation and nitrogen management on growth, water use and yield of rice in south-eastern Australia. Australian J. Exp. Agric. 46, 1363–1372.

Bond, J.J., Willis, W.O., 1969. Soil water evaporation – surface residue rate and placement effects. Soil Sci. Soc. Am. Pro. 33, 445–448.

Bouman, B.A.M., Tuong, T.P., 2000. Field water management to save water and increase its productivity in irrigated lowland rice. Agric. Water Manage. 49, 11–30.

Bouman, B.A.M., Wopereis, M.C.S., Kropff, M.J., Tenberge, H.F.M., Tuong, T.P., 1994. Water-use efficiency of flooded rice fields. 2. Percolation and seepage losses. Agric. Water Manage. 26, 291–304.

Bouman, B.A.M., Lampayan, R.M., Tuong, T.P., 2007. Water Management in Irrigated Rice: Coping with Water Scarcity. International Rice Research Institute, Los Baños (Philippines), (54 p).

Central Ground Water Report (2010). Ground water scenario of India 2009-10. Central ground water board, Ministry of water resources, Government of India.

Cornish, G.A., 1998. Pressurised irrigation technologies for smallholders in developing countries – a review. Irri. Drain. Syst. 12, 185–201.

Erenstein, O., Laxmi, V., 2008. Zero tillage impacts in India's rice-wheat systems: A review. Soil Till Res. 100, 1–14.

Grewel, S.S., Dar, S.K., 2004. Groundwater recharge by rainwater management in Shiwaliks of north India—An overview. In: Abrol, I.P., Sharma, B.R., Sekhon, G.S. (Eds.), Groundwater Use in North-West India—Workshop Paper Centre for Advancement of Sustainable Agriculture, New Delhi, India, pp. 91–98.

GOI, (several years). Statistical Abstract of India, Ministry of Agriculture, Government of India.

GOH, (several years). Statistical Abstract of Haryana, Department of Agriculture, Government of Haryana, India.

GOP, (several years). Statistical Abstract of Punjab, Department of Agriculture, Government of Punjab.

Grover, D.K., Sharma, T., 2011. Alternative resources conservative technologies in agriculture: impact analysis of zero-tillage technology in Punjab. Indian J. Agric. Res. 45, 283–290.

Hafeez, M.M., Bouman, B.A.M., van de Giesen, N., Vlek, P., 2007. Scale effects on water use and water productivity in a rice-based irrigation system (UPRIIS) in the Philippines. Agric. Water Manage. 92, 81–89.

Hellegers, P.J.G.J., Perry, C.J., Berkoff, J., 2007. Water Pricing in Haryana, India. In: Molle, F., Berkoff, J. (Eds.), Irrigation Water Pricing: The Gap Between Theory and Practice. CAB International, Wallingford, UK, pp. 192–207.

Hira, G.S., 2009. Water management in northern states and the food security of India. J. Crop Improv. 23, 136–157.

Hornbuckle, J.W., Car, N.J., Christen, E.W., Smith, D.J., 2010. Large scale, low cost irrigation scheduling – making use of satellite and ETo weather station information. Irrigation Australia. CRC Irrigation Future, Melbourne, VIC.

Humphreys, E., Muirhead, W.A., Melhuish, F.M., White, R.J.G., Blackwell, J., 1989. The growth and nitrogen economy of rice under sprinkler and flood irrigation in South East Australia. Irrigation Sci. 10, 201–213.

Humphreys, E., Kukal, S.S., Christen, E.W., Hira, G.S., Balwinder-Singh, Sudhir-Yadav, 2010. Halting the groundwater decline in North-West India – which crop technologies will be winners? Adv. Agron. 109, 155–217.

International Water Management Institute, IWMI-TATA Water Policy Program, 2002. The Socio-Ecology of Groundwater in India. International Water Management Institute (IWMI), Vallabh Vidyanagar, Gujarat, India, (6p. (IWMI Water Policy Briefing 004) <http://dx.doi.org/10.3910/2009.321>).

Jalota, S.K., Arora, V.K., 2002. Model-based assessment of water balance components under different cropping systems in north-west India. Agric. Water Manage. 57, 75–87.

Jat, M.L., Chandana, P., Sharma, S.K., Gill, M.A., Gupta, R.K., 2006. Laser land leveling: a precursor technology for resource conservation. Rice-wheat consortium technical bulletin series 7. Rice–Wheat Consortium for the Indo-Gangetic Plains, New Delhi, India, 48.

Kahlown, M.A., Gill, M.A., Ashraf, M., 2002. Evaluation of Resource Conservation Technologies in Rice-Wheat System of Pakistan. Pakistan Council of Research in Water resources (PCRWR), Research Report-I, Islamabad, Pakistan.

Kaur, B., Sidhu, R.S., Vatta, K., 2010. Optimal crop plans for sustainable water use in Punjab. Agric. Econo. Res. Review. 23, 273–284.

Khan, S., Tariq, R., Yuanlai, Cui, Blackwell, J., 2006. Can irrigation be sustainable? Agric. Water Manage. 80, 87–99.

Khan, S., Rana, T., Dassanayake, D., Abbas, A., Blackwell, J., Akbar, S., et al., 2009. Spatially distributed assessment of channel seepage using geophysics and artificial intelligence. Irrig. Drain. 58, 307–320.

Kukal, S.S., Aggarwal, G.C., 2003. Puddling depth and intensity effects in rice-wheat system on a sandy loam soil. II. Water use and crop performance. Soil Till Res. 74, 37–45.

Kukal, S.S., Hira, G.S., Sidhu, A.S., 2005. Soil matric potential-based irrigation scheduling to rice (Oryza sativa). Irri. Sci. 23, 153–159.

Kukal, S.S., Sidhu, A.S., Hira, G.S., 2007. Tensiometer-based irrigation scheduling to rice. Extension Bulletin 2007/02. Department of Soils, PAU, Ludhiana.

Minhas, P.S., Jalota, S.K., Arora, V.K., Jain, A.K., Vashist, K.K., Choudhary, O.P., et al., 2010. Managing water resources for ensuing sustainable agriculture: situational analysis and options for Punjab. Resarch Bulletin 2/2010, Directorate of Research, Punjab Agricultural University, Ludhiana-141004 (India), 40.

Muirhead, W.A., Blackwell, J., Humphreys, E., White, R.J.G., 1989. The growth and nitrogen economy of rice under sprinkler and flood irrigation in South East Australia. Irri. Sci. 10, 183–199.

Mukherji, A., Facon, T., Burke, J., de Fraiture, C., Faurès, J.-M., Füleki, B., et al., 2009. Revitalizing Asia's Irrigation: To Sustainably Meet Tomorrow's Food Needs. International Water Management Institute; Rome, Italy: Food and Agriculture Organization of the United Nations, Colombo, Sri Lanka.

Narang, R.S., Virmani, S.M., 2001. Rice–wheat cropping systems of the Indo-Gangetic Plain of India. In: Rice–Wheat Consortium Paper Series 11. Rice–Wheat Consortium for the Indo-Gangetic Plains, New Delhi, and International Crops Research Institute for the Semi-Arid Tropics (ICRISAT), Patancheru, India.

Perveen, S., Krishnamurthy, C.K., Sidhu, R.S., Vatta, K., Kaur, B., Modi, V., et al., 2012. Restoring groundwater in Punjab, India's breadbasket: finding agricultural solutions for water sustainability. Columbia Water Centre, 26.

Raina, R.S., Sangar, S., 2004. Institutional reform in knowledge for integrated water management – groundwater lessons from Haryana. In: Abrol, I.P., Sharma, B.R., Sekhon, G.S. (Eds.), Groundwater Use in North-west India – Workshop Paper Centre for Advancement of Sustainable Agriculture, New Delhi, India, pp. 178–193.

Ram, H., Yadvinder-Singh, Timsina, J., Humphreys, E., Dhillon, S.S., Kumar, K., et al., 2005. Performance of upland crops on beds in North-west India. In: evaluation and performance of permanent raised bed cropping systems in Asia, Australia and Mexico. In: Roth, C.H., Fischer, R.A., Meisner, C.A. (Eds.), ACIAR Proceedings No. 121 Australian Centre for International Agricultural Research, Canberra, Australia, pp. 41–58. (Retrieved August 14, 2012 from <http://www.aciar.gov.au/publication/term/18>).

Randhawa, M.S., 1977. Green revolution in Punjab. Agric. Hist. 51, 656–661.

Reid, P.C.M., Muller, H., 1983. Losses from Water Distribution Systems on Gravity Fed Irrigation Schemes. Paper Read at the Workshop on Hydrological Process and Water Supply Aspects of Irrigation Research. 28–30 November, 1983. Published under the auspices of the Water Research Commission, Republic of South Africa.

Rodell, M., Velicogna, I., Famiglietti, J.S., 2009. Satellite-based estimates of groundwater depletion in India. Nature 460, 999–1002.

Saeed, M.M., Ashraf, M., Asghar, M.N., Bruen, M., Shafique, M.S., 2002. Farmers' skimming well technologies: practices, problems, perceptions and prospects. Working Paper No. 40. International Water Management Institute, Lahore, Pakistan, (Retrieved August 14, 2012 from <http://www.iwmi.cgiar.org/Publications/Working_Papers/working/WOR40.pdf>).

Sandhu, B.S., Khera, K.L., Prihar, S.S., Singh, B., 1980. Irrigation needs and yield of rice on a sandy loam soil as affected by continuous and intermittent submergence. Indian J. Agric. Sci. 50, 492–496.

Sidhu, H.S, Manpreet-Singh, Humphreys, E., Yadvinder-Singh, Balwinder-Singh, Dhillon, S.S., et al., 2007. The happy seeder enables direct drilling of wheat into rice stubble. Aust. J. Exp. Agric. 47, 844–854.

Sidhu, H.S., Manpreet-Singh, Blackwell, J., Humphreys, E., Bector, V., Yadvinder-Singh, 2008. Development of the happy seeder for direct drilling into combine harvested rice. In "Permanent Beds and Rice-Residue Management for Rice-Wheat Systems in the Indo-Gangetic Plain". In: Humphreys, E., Roth, C.H. (Eds.), ACIAR Proceedings No. 127 Australian Centre for International Agricultural Research, Canberra, Australia, pp. 159–170. (Retrieved August 14, 2012 from <http://www.aciar.gov.au/publication/PR127>).

Sidhu, R.S., and Vatta, K. (2010). Economic Valuation of Innovative Agriculture Conservation Technologies and Practices in Indian Punjab. In: 12th Annual BIOECON Conference "From the Wealth of Nations to the Wealth of Nature: Rethinking Economic Growth" Centro Culturale Don Orione Artigianelli - Venice, Italy. September 27–28, 2010. Retrieved August 4, 2012 from <http://www.ucl.ac.uk/bioecon/12th_2010/>.

Singels, A., Smith, M.T., 2006. Provision of irrigation scheduling advice to small-scale sugarcane farmers using a web based crop model and cellular technology: A South African case study. Irrig. Drain. 55, 363–372.

Singh, G., Singh, S., 2007. Diaspora philanthropy in action: An evaluation of modernization in Punjab villages. J. Punjab. Stud. 14, 225–248.

Sudhir-Yadav, Gill, G., Humphreys, E., Kukal, S.S., Walia, U.S, 2011a. Effect of water management on dry seeded and puddled transplanted rice Part 1: crop performance. Field Crops Res. 120, 112–122.

Sudhir-Yadav, Humphreys, E., Kukal, S.S., Gill, G., Rangarajan, R., 2011b. Effect of water management on dry seeded and puddled transplanted rice Part 2: water balance and water productivity. Field Crops Res. 120, 123–132.

Tanwar, B.S., Kruseman, G.P., 1985. Saline groundwater management in Haryana State, India Hydrogeology in the Service of Man. Memoirs of the 18th Congress of the International Association of Hydrogeologists, Cambridge, (pp. 24–30. Retrieved August 12, 2012 from <http://iahs.info/redbooks/a154/iahs_154_03_0024.pdf>).

Wachyan, E., Rushton, K.R., 1987. Water losses from irrigation canals. J. Hydrol. 92, 275–288.

Wang, D., Cai, X., 2009. Irrigation scheduling—Role of weather forecasting and farmers' behavior. J. Water Resour. Plann. Manage. 135, 364–372.

Ward, F.A., Pulido-Velazquez, M., 2008. Water conservation in irrigation can increase water use. Proceedings of the National Academy of Sciences of the USA 105, 47. (Retrieved August 14, 2012 from <http://www.pnas.org/content/105/47/18215.full.pdf>).

Wei, Y., Langford, J., Willett, I.R., Barlow, S., Lyle, C., 2011. Is irrigated agriculture in the Murray Darling Basin well prepared to deal with reductions in water availability? Global. Environ. Change. 21, 906–916.

Yadvinder-Singh, Bijay-Singh, Nayyar, V.K., Jagmohan-Singh, 2003. Nutrient Management for Sustainable Rice-Wheat Cropping System. National Agricultural Technology Project, Indian Council of Agricultural Research, New Delhi, India, and Punjab Agricultural University, Ludhiana, Punjab, India.

Young, C., Wallender, W., Schoups, G., Fogg, G., Hanson, B., Harter, T., et al., 2007. Modeling shallow water table evaporation in irrigated regions. Irrig. Drain. Syst. 211, 119–132.

Chapter 7

Modeling for Agricultural Sustainability: A Review

Mukhtar Ahmed*, Muhammad Asif†, Arvind H. Hirani**,
Mustazar N. Akram*, and Aakash Goyal‡

*Department of Agronomy, PMAS Arid Agriculture University Rawalpindi, Pakistan,
†Agricultural Food and Nutritional Science, University of Alberta, Canada, **Department of
Plant Science, University of Manitoba, Winnipeg, Canada, ‡Bayer Crop Science, Saskatoon,
Saskatchewan, Canada

1. INTRODUCTION

Scientific developments are helping human beings to establish new horizons of life. Hundreds of new technologies are being patented each year, many of which are supposed to help society directly or indirectly. Simulation of crops is thought to be a new concept, particularly in developing countries; however, it started in the late 1970s. A model is a mathematical representation of a real system, and modeling is an efficient way to learn about complex biophysical systems. Crop simulation modeling is quantitative application of a crop-based model. Crop modeling holds a vital place in the development of innovative crop management strategies and agricultural sustainability under a continuously changing climate, as it expresses the response of crops to meteorological, edaphic, and biological factors. Crop modeling aids in decision-making, forecasting of crop growth and development, minimizing the yield gaps, and selecting suitable genotypes and appropriate sowing dates for sustainable crop production under changing climatic scenarios. It is becoming a valuable tool for increasing our understanding of crop physiology and ecology for sustainable agricultural production. Models are used for yield prediction (Tsuji et al., 1994), simulation of crop damage, making policies, finding out interaction effects (Ahmed and Hassan, 2011), such as genotype by environment (G × E) interaction (Heuvelink et al., 2007), soil moisture dynamics, nitrate losses (Liu et al., 2011), soil erosion, and other factors related to agriculture (Wang et al., 2003).

Models are designed in three classes based on end users' requirements. Models with detailed parameters are designed for researchers to study the complexity of fluctuating input parameters and their impact on final output.

Agricultural Sustainability. DOI: http://dx.doi.org/10.1016/B978-0-12-404560-6.00007-1

Relatively simple operational models are used for decision-making and in education, training, and technology transfer (Matthews and Stephens, 2002). Several static and dynamic models have been developed and are being used, such as Agricultural Production Systems Simulator (APSIM), Crop Environment Resource Synthesis (CERES), and Decision Support System for Agrotechnology Transfer (DSSAT).

In an agricultural system, crop productivity varies with varying climatic and edaphic conditions. Various models have been developed to best understand yield gaps and optimization of yield potential. CERES is a process-based, major, management-oriented model that can simulate crop growth, yield, and development by taking into account the effects of weather, soil, planting method, irrigation, and fertilizer management. Many later models have been developed, including DSSAT, APSIM, SPASS (Soil Plant Atmosphere System Simulation), RZWQM (Root Zone Water Quality Model), CROPGRO (Crop Growth), STICS crop model (Simulateur mulTIdisciplinaire pour les Cultures Standards), SWAP (Soil-Water-Atmosphere-Plant environment), WOFOST (World Food Study), SOYGRO (Soybean Crop Growth Simulation), and CropSyst (Cropping System Simulation).

An important task in experimenting with models is the testing of their performance in a wide range of circumstances to identify the scope of their validity and their limitations. As crop simulation models are site and crop specific in nature, they cannot be used in areas other than those for which they have been developed, until and unless validated under local conditions. Comparative studies of modeling approaches have been taken into account by few scientists. For simulation of yield and biomass of maize, Clemente et al. (2005) compared the CERES MAIZE and CropSyst modeling approaches and suggested that both models predicted very similar potential evapotranspiration results based on cultivar specific input. Singh et al. (2008) compared the CERES WHEAT and CropSyst models for water–nitrogen interaction in wheat and concluded that the CropSyst model is more efficient and suitable in predicting yield and growth of wheat under different water and nitrogen applications. In another study, CERES MAIZE and CropSyst models were compared and validated for the simulation of maize yield, and the investigators suggested that both models were accurate in wet and dry land for such simulation (Jara and Stockle, 1999). Comparison of different modeling approaches is beneficial to select a suitable crop model for a specific locality and climate so that it could be further used to predict and simulate agricultural productivity.

The kinetics of vertical and horizontal expansion of crop plants, known as crop canopy, is of vital importance and has been ignored in most crop models (Jones et al., 1998; Brisson et al., 2002). Mechanistic modeling of crop canopies has been carried out to simulate plant growth and development as a function of microclimate. Such models depend on the objectives for which they have been built. For example, sound prediction of the whole-crop canopy leaf area may be adequate for large-scale estimation of crop productivity, but does not allow an accurate estimation of distribution of light, photosynthesis, or of spatial distribution of materials applied to crops. Functional models of plant development aim to simulate physiological processes of crop growth and development as well as the physical architecture of crop plants in the field.

Atmospheric CO_2 concentration has increased to 385 ppmv from 280 ppmv over the last 150 years because of increased burning of fossil fuels, cement production, and use of modern agricultural techniques (Fan et al., 2007). Over the last century, temperatures have increased by 0.6–0.9°C on a global scale and are projected to further rise by 0.6–2.5°C in the upcoming fifty years and by 1.1–6.4°C during the next century (IPCC, 2007). Abrupt climatic changes such as variation in rainfall pattern, dry spells, intermittent droughts and floods threaten agricultural production and may result in severe reduction of crop yields and even total crop failure (Mishra et al., 2008).

Climate change, market infrastructure, and increasing population are major driving forces to change agricultural industry. New management options and methods of production are therefore necessary for sustainable agriculture and to meet these challenges (Dore et al., 2008). Finding best management options and mimicking climatic degradation are two key factors under consideration in agronomic research to enhance/sustain crop productivity. Crop simulation models offer an efficient substitute for agricultural systems under diverse climatic conditions and have been widely used to assess the impacts and adaptation to climate change in relation to agricultural production (Xiong et al., 2008). These models help as decision-making tools for better and sustainable agriculture (Amanullah et al., 2007).

Major assumptions have been made to upgrade results, as most crop models are designed and used to characterize plot scale, which made it difficult to forecast the impacts of climate change at a regional scale (Challinor et al., 2007). To obtain realistic results, all crop growth models need to be calibrated and validated for the environment of a particular region (Timsina and Humphreys, 2006). Simple static (Makowski et al., 2001) and dynamic crop models (Asseng and Herwaarden, 2003) have been extensively used to predict yield and grain protein content in wheat using soil and crop characteristics. These models were used to predict yield of different fields, which ultimately led to an optimization of the grading process (Le Bail and Makowski, 2004).

Crop modeling has been basically used to predict and forecast the yields of crops in relation to environmental changes. However, the predictive quality of models has often been very low due to the complexity of the systems (Barbottin et al., 2008). Many studies and experiments have been made to improve the predictive ability of models and to enhance their area of validity by inserting new parameters and equations and combining these equations with measured data (Naud et al., 2008). In the following section, we will discuss major simulation models that are being used to maintain and improve agricultural production on a sustainable basis.

2. MAJOR SIMULATION MODELS

2.1 APSIM (Agricultural Production System Simulator)

APSIM has been parameterized and validated in various agro-ecological zones to explain the impacts of climatic variation on wheat yield and to enhance climate forecast utilizations. It consists of a modular modeling framework that has been developed by the Agricultural Production Systems Research Unit in Australia. It was developed to simulate and predict biophysical processes in farming systems, to mimic climatic risks, and to sustain optimum yield potential. The modules included in the model deal with a diverse range of crops, pastures and trees, soil processes including water balance and N and P transformations, soil pH, erosion, and a full range of management controls. APSIM has vast applications: providing support for on-farm decision-making, designing farming systems for production, natural resource management, investigation of seasonal climate prediction, analysis of supply chain issues in agribusiness activities, waste management strategies, risk evaluation for policy makers, and as a guide for researchers and educationists (Keating et al., 2003). One of the studies has been conducted at three locations of the Pothwar region in Pakistan (Ahmed et al., 2010). The results lead to the conclusion that the APSIM model is appropriate for the simulation of wheat yield and can be further used for scenario analysis under changing climatic conditions (Ahmed et al., 2010). To develop a simple crop model, WHEATGROW for simulating wheat growth, an experiment has been conducted at Faisalabad, Pakistan to compare measured and simulated results, and the study revealed that the model performed well in simulating wheat growth (Wajid et al., 2007). The APSIM model has also been parameterized and calibrated to decide appropriate management options for wheat in a Mediterranean environment. The model was run with 40 years of weather and wheat crop data. Simulated results revealed that early sowing in November and late October gave maximum yield compared with December-sown wheat crop under particular field practices and in the absence of water logging and stress conditions (Bassu et al., 2009). The APSIM model was employed to simulate response of maize to N fertilizer, and the output was compared with the observed results from a long-term experiment conducted in Zimbabwe, leading to the conclusion that the model very accurately predicted grain yield and biomass at a different rate of N fertilizer within one standard error in all the seasons (Shamudzarira and Robertson, 2002). Similarly, the APSIM model was used to simulate N fertilizer response under rainfed conditions in wheat in loamy soils of three locations in Australia, and results suggested that predicted data were well supported by observed data for grain filling and yield (Asseng and van Herwaarden, 2003).

2.2 Ceres Wheat

Modeling efforts began in 1977 for wheat productivity. Models prior to that time were used for monthly weather data prediction and were mostly of a statistical nature. Initially the United States Department of Agriculture (USDA) developed three models, one of which is the CERES WHEAT model (Willis, 1985). To give predictions about leaf number and sizes by describing the interaction of genetics and climate for determining the duration of vegetative growth in cereals was the initial goal of the CERES model. This provided information to predict evaporation and interception of light by the plants. The model was developed by estimating the rate and duration of crop growth influenced by soil water status and environment, as well as by the linkage of nitrogen dynamics to the system (Godwin, 1987). Several research groups used different versions of the model for different purposes, including testing and improvement of CERES WHEAT. Interest in wheat modeling increased, and two more wheat models, named TAMW (Texas A&M Wheat) and SIMTAG (Simulation Model of Wheat Genotypes), were developed. The development of several other cereal models similar to CERES WHEAT was made possible by the financial support of IBSNAT (International Benchmark Sites Network for Agrotechnology Transfer). The CERES WHEAT model was designed to simulate the effects of weather, soil water, planting density, cultivar, and nitrogen on crop growth, development, and yield. It was developed for useful decision-making at farm level by predicting the effect of alternative management strategies on crop yield (Wilkerson, et al., 1983). It is also useful for making decision and risk analysis and can be run on microcomputers, as it is written in computer language and thus requires minimum time for computation and uses readily available water, genetics, and soil inputs (Boote et al., 1994). To accurately simulate growth and yield, the model must take into consideration processes such as soil nitrogen transformation, uptake and partitioning among plant parts, water usage by the crop and water balance, biomass accumulation and partitioning, extension in growth of leaf, stem, and roots, and phenological development.

Realistic model predictions depend on accurate information. This model uses specific weather, soil, genetics, and management data to evaluate strategies to attain farmers' goals of maximum profit, minimum variance of yield, minimum environmental degradation, and maximum stability of income. This model uses a system of inputs and outputs (IBSNAT, 1986) and can be run in three modes: (i) single-treatment simulation, in which the model works through a day simulation and inputs are displayed on screen; (ii) multiple-treatment simulation, in which treatments can be simulated without further inputs and, at the end, all outputs are displayed on the screen for closer examination; (iii) multiple-year run, which is the fastest form of model operation and examines the year-to-year variability of a strategy.

Weather and climatic variability have considerable impact on agricultural production. Various research studies have been carried out to determine the appropriate sowing date as a management option, to mimic the yield loss under varying climatic conditions. One such study conducted by Sultana et al. (2009) demonstrated that a late-sown genotype boosted production compared with early-sown, which clearly indicated that the wheat-sowing date has shifted towards colder months in arid, humid, and sub-humid regions, due to climatic variation.

Wheat yield can be enhanced by selecting appropriate genotypes and altering planting dates, mostly in the regions where water logging and water stress is a huge problem under changing climatic perspectives. Results have revealed that wheat and barley respond differently under changing climate scenarios. About 10–20% reduction in wheat yield and 4–5% in barley was recorded consequent to 10–20% reduction in rainfall (Al-Bakri et al., 2010). Similarly, increase in temperature has also affected the yield (Al-Bakri et al., 2010). More recently, the CERES WHEAT model was calibrated and validated using a data set of the last thirty years to assess durum wheat genotypes under changing environmental perspectives in Italy. The model performed well in simulating and forecasting yield of wheat and might be further used to better understand crop performance under changing climatic scenarios (Dettori et al., 2011). The CERES WHEAT model was used with satisfactory prediction of simulated results of grain yield, N uptake, and nitrate accumulation in the soil through several years of variable weather and soil conditions in Hungary (Kovacs et al., 1995). Sustainable crop production can be achieved through implementation of the CERES WHEAT model in different cereal crops with long-term data sets to simulate and predict scenarios of changing climate and its impact on production in specific regions.

2.3 DSSAT (Decision Support System for Agrotechnology Transfer)

DSSAT v2.1 was developed by the International Benchmark Sites Network for Agrotechnology Transfer (IBSNAT) Project (IBSNAT, 1989) with the aim of making it available for global use (Jones et al., 2003). DSSAT became popular among researchers immediately after its advent and its initial version had the capability of simulating 16 crops (Jones et al., 2003). Soil, crop, weather, and management components have been added in DSSAT to make the parameterization of the model easier (Jones et al., 2003). Crop production is dependent upon a multitude of variables, which include uncertain climate and socioeconomic factors (Ahmed and Hassan, 2011; Sanfo and Gerard, 2012). Lack of information in the models mostly gives limited results. To solve this problem, soil, climate, crop, and management factors have been improved in DSSAT and it includes most of the climatic parameters such as sunlight incidence, rainfall, surrounding temperature, soil water-holding capacity, and crop execution factors (White et al., 2006). DSSAT is now available for 28 crops and is being used by researchers, students, educators, and policy makers in over one hundred countries (http://www.dssat.net/).

The minimum data set (MDS) for DSSAT is local weather conditions, including two extremes of temperature in a day, sunlight hours, rainfall, physicochemical properties of the local soil, crop factors like cultivar, and management activities such as sowing time, plant to plant distance, and fertilizer and irrigation applications. All the data is encoded in the model.

The DSSAT crop simulation model has been calibrated to assess the potential impacts of climatic variability on wheat and barley crop under rainfed conditions. DSSAT performed well in capturing yield trend over years. For example, simulated yield of wheat ($1176\,kg\,ha^{-1}$) was very close to observed ($1173\,kg\,ha^{-1}$). Similarly, in barley, simulated and observed average yield were recorded as 927 and $922\,kg\,ha^{-1}$, respectively. DSSAT has been employed in various aspects of crop production, including crop management (Wafula, 1995), fertilizer management (Jagtap et al., 1999), irrigation budgeting and scheduling (MacRobert and Savage, 1998; Yang et al., 2010; Liu et al., 2011), climate change (Muchena and Iglesias, 1995), pest management (Luo et al., 1997), tillage management (Castrignano et al., 1997), soil nitrogen dynamics and effect of excessive moisture in several soils (Lizaso and Ritchie, 1997), extent of leaching of chemicals (Pang et al., 1998), soil moisture dynamics, nitrate losses (Liu et al., 2011), multi-environmental trials for varietal performance (Mavromatis et al., 2001), and estimating crop biological and economical yields (Singh et al., 1998). DSSAT has also been used for decision-making, as it saves the time and labor needed to explore complex systems (Tsuji et al., 1998). This model is being used in many developed and developing countries as a research and risk management tool. Several research groups have used the DSSAT model to simulate different parameters which are directly or indirectly involved for sustainable crop production in agricultural systems. Based on various sources of experimental evidence, it can be concluded that DSSAT could be more effectively used in enhancing sustainable production of various crops in various geographical locations, helping to increase production and productivity with minimum input resources.

2.4   SALUS (System Approach to Land Use Sustainability)

The SALUS model is designed for multi-year prediction of continuous crop, soil, water and nutrient conditions under different management practices. Three main input parameters can be used for simulation: (i) biotic interaction; (ii) weather conditions; (iii) management practices, including crop rotations, planting dates, plant population, irrigation, fertilizer applications, and tillage regimes. The SALUS model can be deployed to predict yield, atmospheric fluxes, runoff, soil erosion, drainage, and leaching of nutrients. SALUS can also be used to simulate plant growth and soil conditions every day over a longer period. In each management strategy, all the major components of the crop–soil–water model can be executed. Recently, a web-based version of SALUS has become available in English and Chinese languages with a simple interface of sustainability index (http://140.134.48.19/salus/). The SALUS crop simulation model has been used to measure nitrogen uptake by wheat crop in Italy.

The simulation was performed by using two N levels (0 and 90kg ha^{-1}) for a wheat crop grown long term in a monoculture cropping system. A further simulation was performed for different N levels. Results revealed that grain yield was more responsive for 120kg-N ha^{-1}. The simulation approach proved to be beneficial in decision-making for appropriate nitrogen management strategies (Basso et al., 2010). Agro-economic models provide information regarding economics of farming practices and use of inputs for crop growth. These models could be used in conjunction with SALUS to simulate agricultural yield and to determine agricultural land use planning for the future (Apfelbeck et al., 2007).

2.5   NDICEA (Nitrogen Dynamics in Crop Rotation in Ecological Agriculture)

Nitrogen use efficiency is an important parameter in organic farming. The NDICEA model (www.ndicea.nl) was developed by the Louis Bolk Institute to determine the assessment of organic fertilization strategies and crop rotations using easily obtainable input parameters (Koopmans and Van der Burgt, 2005). NDICEA is a useful model to optimize nitrogen efficiency and reduce loses. The NDICEA model describes soil water dynamics, nitrogen mineralization, and inorganic nitrogen dynamics in relation to crop and weather demand. An organic farming system completely relies on soil manure application in order to meet the nitrogen demand of the crop. Prediction of the nitrogen requirement of the specific crop and losses by soil can be simulated by the NDICEA model for sustainable production in organic as well as in conventional farming systems. The NDICEA model can be used to make two standards: one is a rotation standard in which one can manage the whole rotation schedule by using previous data; the second applies to the field situation, in which future requirement of nitrogen application can be calculated by entering present data of soil N (van der Burgt and Timermans, 2009). The NDICEA model uses soil as well as agronomic information for its interpretation. Modeling nitrogen dynamics and agronomic parameters can help in decision-making at the farm level and to improve nitrogen use efficiency of the crop. This is particularly relevant in developing countries, where farmers generally apply nitrogen to soil without having information about the requirement for it. Therefore, this model can help to measure the nitrogen availability in the soil by estimating residues from the crop and from fertilizer application in the previous years/ seasons. The NDICEA model offers an easy way to obtain a consistent sign of nitrogen and organic dynamics of the farm for productive crop rotations. It is a simple model and can easily be operated by the farmer (van der Burgt and Timermans, 2009).

Farmers of arable land are challenged to regulate fertilization rates in order to obtain good crop yields with a minimal release of nitrogen to the environment. Nutrient management for crops, especially availability of nitrogen, is difficult to determine, because of losses by mineralization, leaching, percolation, and evaporation. Prediction of nitrogen requirement of a crop can be simply performed by the NDICEA model with easily available input parameters to simulate and predict total nitrogen requirement and application times depending on soil and weather conditions. The NDICEA model offers the advantage of a dynamic calculation in which several climate and soil parameters are integrated. Synchronization of crop nitrogen demand and soil nitrogen accessibility through the NDICEA model can be used to optimize efficiency of nitrogen for crop production in various soil–water dynamics. Nitrogen use efficiency can be improved by the NDICEA model at various crop rotations, soil profiles, and water availability levels with minimum input parameters that are easily generated or collected by farmers. However, addition of input parameters such as soil mineral nitrogen and plant nitrogen content will improve the accuracy of predictions. The NDICEA model could be deployed for budgeting for nitrogen fertilizer cost of a crop, based on available soil nitrogen, weather condition, and soil–water–weather interactions.

Koopmans and Bokhorst (2002) evaluated the NDICEA model on eight organic farms and research sites to estimate the effect of crop rotations and manure applications on the mineral nitrogen availability in different phases of a crop rotation. They suggested that the model accurately predicted nitrogen level for the top 30 cm of the soil, with modeling efficiency of 0.4 and coefficient of determination of 0.65. They concluded that the NDICEA model can be used to simulate sustainable farming systems at the farm level, based on easily available farm data and readily available climate data as source input.

2.6   Rhizome: A Model of Clonal Growth

The Rhizome model is basically designed to simulate the growth of plant roots. This model has the capability to check many processes which can take place in the roots. These processes include the downward movement of the roots and their further division into branches and fragments; it can also check the process in which primary roots and secondary roots are formed (Herben and Suzuki, 2002). The primary roots are normally long - having no branches - are quick-growing, and have no ramets attached on their surface. The ramets are plant parts in which photosynthesis is very active, and these parts are helpful in providing the resources for root growth (Wolfer et al., 2006). The ramets may be attached to the plants, which are established by the seeds and remain fixed to their original place, or they are found attached to the nodes. The Rhizome model can check the formation of secondary roots, which are normally short growing, having many branches on their surface, and the ramets are also attached to their surface. The Rhizome model can predict the formation of the ramets at any part of the roots.

The Rhizome model has also the capability to check the various competitions which can take place between the roots for the utilization of resources, or the competition for resources with the surrounding environment (Jackson and

Caldwell, 1993). These competitions are responsible for the downward growth of roots, formation of side branches, establishment of ramets, and also the mortality of roots if they are unable to compete. The roots act as one of the major sinks that utilize the food synthesized in the leaves. Rhizome has the capability to check the movement of this food in different parts of the roots. The Rhizome model has the capacity to capture different sizes of the ramets and roots. The ramets have a limited life and die after a specific period of time. The model has the capacity to check the mortality of these ramets. In single or multiple plant species cropping systems (crop–weed or crop–crop systems), numbers of ramets and position of ramets on roots determine competition for nutrients between plants of the same species or different species. Planting distance, within rows and between rows, can be determined to minimize nutrient competition in heterogeneous soils (Pacala and Silander, 1990). Robinson et al. (1994) reported capture of nitrate from soil by wheat in relation to root growth and availability of nitrogen in the soil by the Rhizome model, and suggested that inflow of nutrients impacts plant growth and development more than availability of nutrients.

The Rhizome model has the capacity to check in which way the ramets and roots have been established from the roots, and it can also track the origin of all ramets and roots. It can also check the mortality of ramets resulting from disturbance by any external factor, such as fire. The rhizome model works with four different classes of objects: (i) the plants which are established from the seeds; (ii) roots or root fragments which are formed from the fragmentation process of larger root systems; (iii) the nodes which can compose the rhizomes; (iv) the ramets which provide the resources for root growth.

Different variables are defined by each of the above objects. In the plants, which are established by the seeds, variables include the genetic makeup, code of the particular species, the time during which the plant is established from the seed, and the ramet, which is the important part of such plants. The absence of ramets in these plants can lead to the death of the whole plant. The variables which are defined for the fragments of roots include the genetic makeup of the fragment, the species code, the time at which the fragment was established from the seed, and the list of all the nodes and ramets which are attached at the surface of each root fragment.

2.7 LINTUL (Light Interception and Utilization Simulator)

The LINTUL model was designed by Wageningen University, and extensively used to predict crop potential production based on light use efficiency, under optimal growth conditions in different macro- and microenvironments. LINTUL is a simple crop model that simulates the triggering of dry matter formation according to light use efficiency and light interception during the exponential phase. Appropriate temperature and light distribution are crucial parameters for expansion of the leaf area of a plant. Overall temperature and light intensity during the photoperiod determines apportion of dry matter in plants.

Senescence of the plant leaves depends on environmental temperature; however, increase in leaf area index is an important factor for its initiation. Area of leaf canopy is one of the most important variables affecting the efficiency of photosynthesis, and hence production and productivity, for a crop.

The LINTUL model can be used to simulate biotic and abiotic stresses in potatoes, such as drought, late blight, and potato cyst nematodes. An important use of LINTUL is to measure the gap between actual and potential crop. In other words, with input of specific weather parameters, the model can estimate the yields that might be produced in a specific environment compared with crop production under current environmental conditions. Farre et al. (2000) tested and validated LINTUL for the prediction of flowering time, leaf area index, and yield, and suggested that the model satisfactorily predicted results for all the parameters in maize. They suggested that LINTUL could be used as a tool for exploring the consequences, for yields, of different irrigation strategies in various locations for sustainable crop production.

LINTUL can be helpful in measuring biotic and abiotic stress. The model can also help researchers in finding the CO_2 fixation during photosynthesis. LINTUL can determine light absorption rates and processing during photosynthesis. It can determine the accumulation of dry matter under sufficient supply of water and nutrients, pest, disease, and weed free conditions, and under existing weather conditions. Dry matter is largely the product of light interception. The growth rate of both monocot and dicot can be found with the help of LINTUL. The original version of LINTUL is a potential crop growth model in which storage organ yield can be simulated by multiplying the harvest index by biomass. Different types of LINTUL models are LINTUL 1, LINTUL 2, LINTUL 3 and LINgra, which are used for different simulations and predictions. The LINTUL model was used in potato, grasses, maize, oilseed, and rice in optimal water and water stress conditions for potential yield prediction (Spitters and Schapendonk, 1990; Habekotté, 1997; Schapendonk et al., 1998; Farre et al., 2000; Shibu et al., 2010). A modified version of the model, LINTUL 2 is also used to classify the effect of nitrogen on biomass production, especially in rice crop (Shibu et al., 2010). The LINTUL model is also very helpful in finding or checking soil water processes: drainage, runoff, and evapotranspiration. LINTUL 2 is especially used for this purpose to check the condition of soil and water availability for vegetation, which can serve as information for crop sowing. The cropping pattern and management practices can be adapted according to predicted availability of soil water for the crop. The LINTUL model produces outputs for a range of variables, using the following equations in its simulation.

2.7.1   Light Use Efficiency and Biomass Production

Biomass production is directly regulated by light use efficiency of the crop canopy. Daily growth rate can be determined as the product of intercepted PAR (photosynthetically active radiation) and LUE (light use efficiency).

According to Lambert-Beer's law, photosynthesis activity (0.5) of canopy depends on intercepted PAR and LAI (leaf area index) (Shibu et al., 2010).

$$Q = 0.5Q_0[1 - e^{(-kLAI)}]$$

where, Q is intercepted PAR ($MJ\,m^{-2}d^{-1}$), Q_0 is daily global radiation ($MJ\,m^{-2}d^{-1}$), and k is the attenuation coefficient for PAR in the canopy.

2.7.2 Biomass Partition

Growth and development of crop plant is governed by biomass partition into different organs such as root, stem, leaf, flower, etc. The LINTUL model simulates biomass partition of crop to predict yield and dry matter production and also provide information about possible input options to enhance sustainability of the process.

$$\left(\frac{dW}{dt}\right)_i = P_{ci}\left(\frac{dW}{dt}\right)$$

where (dW/dt) is the rate of biomass growth ($g\,m^{-2}d^{-1}$); $(dW/dt)_i$ and Pc_i are the rate of growth ($g\,m^{-2}d^{-1}$) of, and the biomass partitioning factor to (g-organ-$i\,g^{-1}$-biomass), organ i, respectively.

2.7.3 Leaf Area Development

Leaf area development is the net rate of change in leaf area, which is the difference between its growth rate during juvenile growth phase and its death rate as a result of senescence.

$$\left(\frac{dLAI}{dt}\right) = \left(\frac{dGLAI}{dt}\right) - \left(\frac{dDLAI}{dt}\right)$$

where $(dGLAI/dt)$ is the leaf area growth rate and $(dDLAI/dt)$ is the leaf area death rate.

2.7.4 Soil–Crop Nitrogen Balance

Soil–crop nitrogen balance is also referred to as mineral nitrogen balance of the soil, which is the difference between nitrogen applied through mineralization and/or fertilizer, and nitrogen removed by crop uptake or losses due to leaching. The net rate of change of nitrogen in soil, $(dN/dt)(g^{-2}d^{-1})$, is simulated by:

$$\left(\frac{dN}{dt}\right)_{Soil} = N_{min} + (QN_{fer}NRF) - \left(\frac{dNU}{dt}\right)$$

where, N_{min} is the nitrogen supply through mineralization and biological fixation, QN_{fer} is the nitrogen fertilizer rate, NRF is the nitrogen fertilizer recovery fraction, and dNU/dt is the rate of nitrogen uptake by the crop, which is calculated as the minimum of nitrogen supply from the soil and the demand from the crop.

2.7.5   Nitrogen Demand

The total nitrogen demand of the crop equals the sum of the nitrogen demands of its individual organs, excluding storage organs). Nitrogen demand of individual organs can be calculated based on the difference between maximum and actual organ nitrogen contents.

$$TN_{dem} = \sum_{i=1}^{n} \left(\frac{N_{max,i} - AN_i}{\Delta t} \right)$$

where $N_{max,i}$ is the maximum nitrogen concentration of organ i (g-N g^{-1}-biomass, with i referring to leaves, stems, and roots), and AN_i is the actual nitrogen content of organ i (g-N m^{-2}).

2.7.6   Nitrogen Uptake

Nitrogen uptake by crop is determined by available soil nitrogen content by mineralization and fertilizer applications. Demand of nitrogen by individual organs cumulatively impacts on total nitrogen uptake during plant growth and development.

$$\left(\frac{dNU}{dt} \right)i = \left(\frac{N_{dem,i}}{TN_{dem}} \right) - \left(\frac{dNU}{dt} \right)$$

where $(dNU/dt)_i$, and $N_{dem,i}$ are the rate of nitrogen uptake (g m^{-2} d^{-1}) and nitrogen demand (g m^{-2} d^{-1}) of organ i (i refers to leaves, stems, and roots), respectively.

2.7.7   Nitrogen Stress

The crop is assumed to be affected by variation of available nitrogen fertilizer under water stress and optimal water conditions. The LINTUL model simulates nitrogen stress on a crop with the following formula to predict nitrogen requirement in a given soil–water condition.

$$NNI = \frac{actual\,crop[N] - residual[N]}{critical[N] - residual[N]}$$

Numerous studies reported that the LINTUL model performed relatively well in different crop species under various climatic conditions to simulate socioeconomic, environmental, and productivity parameters (Spitters and Schapendonk, 1990; Habekotté, 1997; Schapendonk et al., 1998; Farre et al., 2000; Shibu et al., 2010). The model could further be employed in broader areas of crop production to effectively utilize nitrogen to minimize input cost for sustainable production in wet and dry land areas.

2.8   WaTEM (Water and Tillage Erosion Model)

WaTEM is a model used to imitate soil erosion and sedimentation by water as well as by tillage activities in a farm. It is a spatially distributed model, meaning it focuses more on geography than on temporal variability. Therefore, it allows incorporation of parameters related to landscape or geographical organization units and connectivity between these units.

WaTEM is a simple model and it works on the basis of topography of land to avoid major problems related to variability in spatial parameters and their uncertainties. The potential uses of WaTEM include calculation of soil erosion and deposition that takes place due to flow of water, to estimate the rates and the patterns of soil erosion and its deposition due to tillage practices, to find out the combined effect of water and tillage on rate and pattern of soil erosion and its deposition, to calculate effects of erosion due to tillage and water on the land structure, and to indicate the exact position of erosion-prone areas in an agricultural landscape.

WaTEM has different parameters related to water and tillage, which can be manipulated for simulation for suitable output, as follows.

2.8.1   Water Parameters

The water component of WaTEM uses the Universal Soil Loss Equation (USLE), as parameter (approximate) standards are easily obtainable for different areas:

$$A = R\,K\,LS\,C\,P$$

where A denotes the potential long-term average annual soil loss in tons per acre per year; R denotes rainfall and runoff factor by geographic location; K denotes soil erodibility factor; LS denotes the slope length-gradient factor; C denotes the crop/vegetation and management factor; and P denotes the support practice factor.

In the options, we can incorporate R, K, C, and P factors. The LS parameters are automatically calculated from the Digital Elevation Model (DEM).

2.8.2   Tillage Parameters

Recent research has recognized that direct soil movement by different tillage operations has some relationship with soil erosion. The tillage components of

WaTEM use an equation in which the strength of the tillage process is represented by one factor: tillage transport coefficient or k_{til} rate.

Tillage transport coefficients are available in, or calculated from, the literature, for different implements, tillage directions, tillage speed, and tillage depth. In addition, bulk density of soil can also be incorporated.

2.9   SPAC (Soil Plant Atmosphere Continuum)

This model is used to analyze crop water consumption in various topographic conditions of the soil. Many factors interact to determine crop yield, and with the help of this model limited resources can be managed according to requirements. SPAC analyses the radiation energy balance on each topographic condition and is used to determine the irrigation demand of different crops in different regions. Models that are used in xylem conductance can also predict transpiration, soil moisture, soil type, and rate of fertilization within the species (Ewers et al., 2000; Estrella and Gonzalez, 2003). This precise analysis has prime importance because it is a reliable tool for computation of water consumption. The SPAC model consists of three sub-models to perform the soil water consumption analysis in various topographic conditions for different crops. These three sub-models are: (i) radiation energy sub-model, (ii) soil moisture and heat flow analysis sub-model; (iii) soil water uptake evaluation sub-model.

For efficient modeling results, the minimum data set (MDS) required by the SPAC model includes appropriate information on:

1. Climatic data—average weather data is necessary for good results by SPAC simulation;
2. Soil data—the type of soil, to a depth of 0–120 cm;
3. Albedo—this is the ratio between incident and reflected radiation. Being high in black soils, it increases the temperature, resulting in higher water consumption than other soils.

The SPAC model can be a very useful tool for researchers, policy makers, and scientists to help with the evaluation of alternative management practices. Evapotranspiration is an important parameter for analyzing water consumption of crops. SPAC can be used for analyzing crop water consumption on various topographic conditions and therefore to contribute to irrigation scheduling of the area.

Being based on different parameters related to crop growth, inputs, and environment, different models vary in their predictions. Comparative studies may be useful in identifying discrepancies and selecting the best suited model for the available set of conditions. There have been a few comparative studies of crop simulation. CERES WHEAT and CropSyst models have been compared to validate and check the performance of models for a wheat crop. Singh et al. (2008) reported that the CropSyst model performed well. Celemente et al. (2005) compared the performance of CERES MAIZE and CropSyst in simulating a maize crop and found that these models satisfactorily predicted evapotranspiration; however, cultivar-specific input data are required to achieve reliable output. Three modeling approaches—WOFOST, SUCROS2-97, and PS123—were compared and validated for calculating water uptake by roots (Berg et al., 2002). Another comparative study was carried out to evaluate the performance of SWATER-SUCROS and CERES MAIZE crop simulation models in predicting biomass yield, leaf area index, and soil water content (Xevi et al., 1996). Phenological development of crops might be the most important attribute in crop adaptation to changing climate, as the increasing temperature has a huge impact on it. Calibration and testing of APSIM and CERES WHEAT models led to the conclusion that temperature has a significant impact on the rate of development of the crop, as increase in temperature shortened the flowering period. However, the duration of the post-flowering phase was unchanged, because there is less change in temperature at that stage (Sadras and Monzon, 2006).

Experiments on two contrasting soils were conducted in Italy to compare the performance of simulation approaches. Three models—SWAP, CropSyst, and MACRO—were parameterized and compared to simulate and predict soil water content on the basis of their consistency. It was concluded that the SWAP model performed well in simulating surface infiltration, followed by CropSyst, and MACRO proved to be inefficient (Bonfante et al., 2010). A simulation study comparing CERES, WOFOST, and SWAP models for simulation of soil water content during the growing season of wheat and barley in Austria found that the CERES and SWAP models simulated wheat and barley yield more precisely than WOFOST (Eitzinger et al., 2004).

Agriculture is the most important industry for the population on the planet. In recent years, increasing population and climate change have appeared as the biggest threat for food security and agriculture. The world population is increasing by over 70 million per year, and it will be 8 billion by 2030. To meet the demand for food for this increased population we have to increase our production by at least 40% by 2030. Moreover, the rapidly changing environment is going to affect agriculture and decrease production in certain areas. Global climatic changes such as rising temperature, uncertain heavy regional rains, flooding, draught, pest and disease, together with modernization and mechanization of agriculture have major effects on food production and productivity. In order to meet the current and upcoming demand for food and to secure healthy human life, food production needs to be enhanced sustainably. Sustainable crop production can be achieved with integrated approaches in combination with predictive models for efficient use of input resources to achieve maximum yield in changing macro- and microclimatic conditions. We need to use these models more effectively to provide advance information as a basis for decisions on sustainable agriculture, and implement new technologies to enhance crop production and sustainability.

Comparative studies of different models under similar experimental conditions in different crop species, to test, to validate, and to predict results with reliability are important for the use of these models in the quest for sustainable crop production. The performance of the models discussed above generally agreed well with the respective long- or short-term experimental data. It is therefore suggested that models could be implemented further in wider areas at different geographical regions to reduce risk from climatic fluctuation caused by global warming and natural calamities. Predictions about basic parameters such as irrigation and fertilizer application, sowing time, and overall weather impact on biotic and abiotic stresses could help farmers to minimize yield losses and to increase production potential of field crops.

3. OUTLOOK

Modeling in agriculture is an important tool to predict the results of crop production systems. Based on the predictions, management actions can be taken to minimize losses in production due to biotic and abiotic stresses, for example by adapting crop cultivation practices and rotation systems. In order to improve the prediction accuracy of each model, additional input parameters and simulation cofactors need to be edited in the equations used for calculation. According to comparative studies on different models, the CERES WHEAT, CERES MAIZE, and CropSyst models are more accurate than others in their predictions; these can be further tested and validated at different geographical locations to enhance their application in real agricultural production systems.

Furthermore, interdisciplinary research needs to be conducted to develop new models which can be implemented to tackle the current pressing problems of farmers in various crop production systems, including grain, vegetable, fruit, and fiber crops. Establishment of interdisciplinary and inter-organizational collaborations for developing, testing, validating, and providing appropriate training to end users for existing and new models may help in enhancing crop production to meet the current and future food demands of the increasing global population.

REFERENCES

Ahmed, M., Hassan, F.U., 2011. APSIM and DSSAT models as decision support tools. 19th International Congress on Modelling and Simulation, Perth, Australia, 12–16 December 2011. Available at: <http://mssanz.org.au/modsim2011>.

Ahmed, M., Hassen, F., Aslam, M., 2010. Climatic resilience of wheat comparing modeled and observed crop yields in Pothwar/East Pakistan. Int. J. Clim. Chan. 2, 32–48.

Al-Bakri, J., Suleiman, A., Abdulla, F., Ayad, J., 2010. Potential impact of climate change on rainfed agriculture of a semi-arid basin in Jordan. Phy. Chem. Earth 36, 125–134.

Amanullah, M., Mohamed, C.K., Safiullah, A., Selvam, S., Siva-kumar, K., 2007. Crop simulation growth model in Cassava. Res. J. Agri. Biol. Sci 3, 255–259.

Apfelbeck, J., Huigen, M., Krimly, T., 2007. Simulating the effects of climatic change on district- and farm-level decision-making in the Danube catchment area. In: "Agricultural Economics Society's 81st Annual Conference", April 2–4, 2007. Reading University, UK.

Asseng, S., van Herwaarden, A.F., 2003. Analysis of the benefits to wheat yield from assimilates stored prior to grain-filling in a range of environments. Plant Soil 256, 217–229.

Barbottin, A., Makowski, D., Le Bail, M., Jeuffroy, M-H., Bouchard, C., Barrier, C., 2008. Comparison of models and indicators for categorizing soft wheat fields according to their grain protein contents. Eur. J. Agron. 29, 175–183.

Basso, B., De Simone, L., Ferrara, A., Cammarano, D., Cafiero, G., Yeh, M.L., et al., 2010. Analysis of contributing factors to desertification and mitigation measures in Basilicata region. Ital. J. Agron. 3, 33–44.

Bassu, S., Asseng, S., Motzo, R., Giunta., F., 2009. Optimising sowing date of durum wheat in a variable Mediterranean environment. Field Crops Res. 111, 109–118.

Berg, M.V.D., Driessen, P.M., Rabbinge, R., 2002. Water uptake in crop growth models for land use systems analysis: II. Comparison of three simple approaches. Eco. Modell. 148, 233–250.

Bonfante, A., Basilea, A., Acutisb, M., De Mascellisa, R., Mannac, P., Peregob, A., et al., 2010. SWAP, CropSyst and MACRO comparison in two contrasting soils cropped with maize in Northern Italy. Agric. Water Mgt. 97, 1051–1062.

Boote, K.J., Bennett, J.M., Sinclair, T.R., Paulsen, G.M. (Eds.), 1994. Physicology and Determination of Crop Yield ASA-CSSA-SSSA, Madison, WI.

Brisson, N., Ruget, F., Gate, P., Lorgeau, J., Nicoullaud, B., Tayot, X., et al., 2002. STICS: a generic model for simulating crops and their water and nitrogen balances. 2. Model validation for wheat and maize. Agronomie 22, 69–92.

Castrignano, A., Colucci, R., Degiorgio, D., Rizzo, V., Stelluti, M., 1997. Tillage effects on plant extractable soil water in a silty clay vertisol in southern Italy. Soil Tillage Res. 40, 227–237.

Challinor, A.J., Wheeler, T.R., Craufurd, P.Q., Ferro, C.A.T., Stephenson, D.B., 2007. Adaptation of crops to climate change through genotypic responses to mean and extreme temperatures. Agri. Ecosyst. Environ. 119, 190–204.

Clemente, R.S., Asadi, M.E., Dixit, P.N., 2005. Assessment and comparison of three crop growth models under tropical climate conditions. J. Food Agri. Environ. 3, 254–261.

Dettori, M., Cesaraccio, C., Motroni, A., Spano, D., Duce, P., 2011. Using CERES-Wheat to simulate durum wheat production and phenology in Southern Sardinia, Italy. Field Crops Res. 120, 179–188.

Dore, T., Clermont-Dauphin, C., Crozat, Y., David, C., Jeuffroy, M.H., Loyce, C., et al., 2008. Methodological progress in regional agronomic diagnosis. A review. Agron. Sustain. Dev. 28, 151–161.

Eitzinger, J., Trnka, M., Hosch, J., Zalud, Z., Dubrovsky, M., 2004. Comparison of CERES, WOFOST and SWAP models in simulating soil water content during growing season under different soil conditions. Eco. Modell. 171, 223–246.

Estrella, H.F., Gonzalez, J.A., 2003. SPAC: an alternative method to estimate earthquake site effects in Mexico City. Geo. Intern. 42, 227–236.

Ewers, B.E., Oren, R., Sperry, J.S., 2000. Influence of nutrient versus water supply on hydraulic architecture and water balance in *Pinus taeda*. Plant Cell Environ. 23, 1065–1066.

Fan, Y., Liang, Q.-M., Wei, Y.-M., Okada, N., 2007. A model for China's energy requirements and CO_2 emissions analysis. Environ. Model. Softw. 22, 378–393.

Farre, I., Van Oojen, M., Leffelaar, P.A., Faci, J.M., 2000. Analysis of maize growth for different irrigation strategies in northeastern Spain. Eur. J. Agrono. 12, 225–238.

Godwin, D.C. (1987) Simulation of nitrogen dynamics in wheat cropping system. PhD. Thesis, University of New England, NSW, Australia.

Habekotté, B. (1997) Description, parameterization and user guide of LINTULBRASNAP 1. A crop growth model of winter oilseed rape (Brassica napus L.). In: Quantitative Approaches in Systems Analysis No. 9. Wageningen Agricultural University, Wageningen, The Netherlands, 40.

Herben, T., Suzuki, J., 2002. A simulation study of the effects of architectural constraints and resource translocation on population structure and competition in clonal plants. Evol. Ecol. 15, 403–423.

Heuvelink, E., Marcelis, L., Bakker, M., Van der Ploeg, A., 2007. Use of crop growth models to evaluate physiological traits in genotypes of horticultural crops. Frontis 21, 221–231.

IBSNAT (1986) International benchmark site network for agrotechnology transfer. Experimental design and data collection procedures for IBSNAT: the minimum data set for system analysis and crop simulation. second ed. Technical Report 1. Department of Agronomy and Soil Science, College of Tropical Agriculture and Human Resources. University of Hawaii, Honolulu, Hawaii, USA.

IBSNAT (1989) International benchmark site network for agrotechnology transfer. The decision support system for agrotechnology transfer version 2.1 (DSSAT): user's guide. Department of Agronomy and Soil Science, College of Tropical Agriculture and Human Resources. University of Hawaii, Honolulu, Hawaii, USA.

IPCC. (2007). Summary for policymakers of climate change: the physical science basis. Contribution of Working Group I to the Fourth Assessment Report of the Intergovernmental Panel on Climate Change (IPCC). Cambridge University Press, Cambridge.

Jackson, R.B., Caldwell, M.M., 1993. Geostatistical patterns of soil heterogeneity around individual perennial plants. J. Ecol. 81, 683–692.

Jagtap, S.S., Abamu, F.J., Kling, J.G., 1999. Long-term assessment of nitrogen and variety technologies on attainable maize yields in Nigeria using CERES Maize. Agri. Syst. 60, 77–86.

Jara, J., Stockle, C.O., 1999. Simulation of water uptake in maize, using different levels of process detail. Agron. J. 91, 256–265.

Jones, J.W., Tsuji, G.Y., Hoogenboom, G., Hunt, L.A., Thornton, P.K., Wilkens, P.W., et al., 1998. Decision support system for agrotechnology transfer; DSSAT v3. In: Tsuji, G.Y., Hoogenboom, G., Thornton, P.K. (Eds.), Understanding Options for Agricultural Production Kluwer Academic, Dordrecht, pp. 57–177.

Jones, J.W., Hoogenboom, G., Porter, C.H., Boote, K.J., Batchelor, W.D., Hunt, L.A., et al., 2003. The DSSAT cropping system model. Eur. J. Agron. 18, 235–265.

Keating, B.A., Carberry, P.S., Hammer, G.L., Probert, M.E., Robertson, M.J., Holzworth, D., et al., 2003. An overview of APSIM, a model designed for farming systems simulation. Eur. J. Agron. 18, 267–288.

Koopmans, C.J., Bokhorst, J., 2002. Nitrogen mineralization in organic farming systems: a test of the NDICEA model. Agronomie 22, 855–862.

Koopmans, C.J., Van der Burgt, G. (2005) NDICEA as a user friendly model tool for crop rotation planning in organic farming. Poster Presented at Researching Sustainable Systems – International Scientific Conference on Organic Agriculture, Adelaide, Australia, 21–23.

Kovács, G.J., Németh, T., Ritchie, J.T., 1995. Testing simulation models for the assessment of crop production and nitrate leaching in Hungary. Agric. Syst. 49, 85–397.

Le Bail, M., Makowski, D., 2004. A model-based approach for optimizing segregation of soft wheat in country elevator. Eur. J. Agron. 21, 171–180.

Liu, H., Yang, J., Tan, C., Drury, C., Reynolds, W., Zhang, T.Q., et al., 2011. Simulating water content, crop yield and nitrate-N loss under free and controlled tile drainage with subsurface irrigation using the DSSAT model. Agri. Water Manage. 98, 1105–1111.

Lizaso, J., Ritchie, J., 1997. A modified version of CERES to predict the impact of soil water excess on maize crop growth and development. Syst. Appr. Sustain. Agri. Devel. 5, 153–168.

Luo, Y., Teng, P.S., Fabellar, N.G., Tebeest, D.O., 1997. A rice-leaf blast combined model for simulation of epidemics and yield loss. Agri. Syst. 53, 27–39.

MacRobert, J.F., Savage, M.J., 1998. The use of a crop simulation model for planning wheat irrigation in Zimbabwe. In: Tsuji, G.Y., Hoogenboom, G., Thornton, P.K. (Eds.), Understanding Options for Agricultural Production. Kluwer Academic Publishers, Dordrecht, the Netherlands, pp. 205–207.

Makowski, D., Wallach, D., Meynard, J.M., 2001. Statistical methods for predicting responses to applied nitrogen for calculating optimal nitrogen rates. Agron. J. 93, 531–539.

Matthews, R.B., Stephens, W., 2002. Crop-Soil Simulation Models: Applications in Developing Countries. CAB International., (2002).

Mavromatis, T., Boote, K.J., Jones, J.W., Irmak, A., Shinde, D., Hoogenboom, G., 2001. Developing genetic coefficients for crop simulation models with data from crop performance trials. Crop Sci. 41, 40–51.

Mishra, A., Hansen, J.W., Dingkuhn, M., Baron, C., Traore, S.B., Ndiaye, O., et al., 2008. Sorghum yield prediction from seasonal rainfall forecasts in Burkina Faso. Agri. Forest Meteorol. 148, 1798–1814.

Muchena, P., Iglesias, A., 1995. Vulnerability of maize yields to climate change in different farming sectors in Zimbabwe. In: Rosenzweig, C., Allen, L.H., Harper, L.A., Hollinger, S.E., Jones, J.W. (Eds.), Climate Change and Agriculture: Analysis of Potential International Impacts (ASA Special Publication 59) American Society of Agronomy, Madison, Wisconsin, pp. 229–232.

Naud, C., Makowski, D., Jeuffroy, M.H., 2008. Is it useful to combine measurements taken during growing season with a dynamic model to predict the nitrogen status of winter wheat? Eur. J. Agron. 291, 300.

Pacala, S.W., Silander, J.A., 1990. Field tests of neighborhood population dynamic models of two annual weed species. Ecol. Monogr. 60, 113–134.

Pang, X.P., Gupta, S.C., Moncrief, J.F., Rosen, C.J., Cheng, H.H., 1998. Evaluation of nitrate leaching potential in Minnesota glacial outwash soils using the CERES Maize model. J. Environ. Qual. 27, 75–85.

Robinson, D., Linehan, D.J., Gordon, D.C., 1994. Capture of nitrate from soil by wheat in relation to root length, nitrogen inflow and availability. New Phytol. 128, 297–305.

Sadras, V.O., Monzon, J.P., 2006. Modelled wheat phenology captures rising temperature trends: Shortened time to flowering and maturity in Australia and Argentina. Field Crops Res. 99, 136–146.

Sanfo, S., Gerard, F., 2012. Public policies for rural poverty alleviation; the case of agriculture households in the Plateau Central area of Burkina Faso. Agri. Syst. 110, 1–9.

Schapendonk, A.H.C.M., Stol, W., Van Kraalingen, D.W.G., Bouman, B.A.M., 1998. LINGRA, a sink/source model to simulate grassland productivity in Europe. Eur. J. Agron. 9, 87–100.

Shamudzarira, Z., Robertson, M.J., 2002. Simulating response of maize to nitrogen fertilizer in semi-arid Zimbabwe. Exp. Agri. 38, 79–96.

Shibu, M.E., Leffelaar, P.A., van Keulena, H., Aggarwal, P.K., 2010. LINTUL3, a simulation model for nitrogen-limited situations: application to rice. Eur. J. Agron. doi: 10.1016/j.eja.2010.01.003.

Singh, A.K., Tripathy, R., Chopra, U.K., 2008. Evaluation of CERES-wheat and Crop System models for water–nitrogen interactions in wheat crop. Agri.Water Manage. 95, 776–786.

Singh, U., Matthews, R.B., Griffin, T.S., Ritchie, J.T., Hunt, L.A., Goenaga, R., 1998. Modelling growth and development of root and tuber crops. In: Tsuji, G.Y., Hoogenboom, G., Thornton, P.K. (Eds.), Understanding Options for Agricultural Production. Kluwer Academic Publishers, Dordrecht, The Netherlands, pp. 129–133.

Spitters, C.J.T., Schapendonk, A.H.C.M., 1990. Evaluation of breeding strategies for drought tolerance in potato by means of crop growth simulation. Plant Soil 123, 193–203.

Sultana, H.G., Hoogenboom, G., Iqbal, M.M., 2009. Planting window as an adaptation option to offset the likely impacts of climate change with emphasis on rain fed areas. Earth Environ. Sci. 6, 34–41.

Timsina, J., Humphreys, E., 2006. Performance of CERES rice and CERES wheat models in rice-wheat system; a review. Agri. Syst. 90, 5–31.

Tsuji, G.Y., Uehara, G., Balas, S. (Eds.), 1994. Decision Support System for Agrotechnology Transfer (DSSAT) Version 3. University of Hawaii, Honolulu, Hawaii.

Tsuji, G.Y., Hoogenboom, G., Thornton, P.K. (Eds.), 1998. Understanding options for agricultural production. Systems Approaches for Sustainable Agricultural Development, Kluwer Academic Publishers, Dordrecht, The Netherlands, pp. 400.

Van der Burgt, G.J.H.M., Timmermans, B.G.H., 2009. The NDICEA model: a supporting tool for nitrogen management in arable farming. In: 2nd Scientific Conference within Framework of the 9th European Summer Academy on Organic Farming. Lednice na Morave, Czech Republic, June 24–26 2009.

Wafula, B.M., 1995. Applications of crop simulation in agricultural extension and research in Kenya. Agri. Syst. 49, 399–412.

Wajid, A., Hussain, K., Maqsiid, M., Khaliq, T., Ghaffar, A., 2007. Simulation modelling of growth, development and grain yield of wheat under semi-arid conditions in Pakistan. Pak. J. Agri. Sci. 44, 194–199.

Wang, F., Fraisseb, C.W., Kitchenc, N.R., Sudduth, K.A., 2003. A site specific evaluation of cropgro soybean model on Missouri claypan soils. Agri. Syst. 76, 985–1005.

White, E.M., Wilson, F.E.A., 2006. Responses of grain yield, biomass and harvest index and their rates of genetic progress to nitrogen availability in ten winter wheat varieties. Irish J. Agric. Food Res. 45, 85–101.

Wilkerson, G.G., Jones, J.W., Boote, K.J., Ingram, K.T., Mishoe, J.W., 1983. Modelling soybean growth for crop management. Trans. ASAE 26, 63–73.

Willis, W.O., 1985. ARS wheat yield project. Agriculture Research Services Publication No. 38. U.S. Department of Agriculture, Washington, D.C., USA.

Wolfer, S.R., van Nes, E.H., Straile, D., 2006. Modelling the clonal growth of the rhizomatous macrophyte *Potamogeton perfoliatus*. Eco. Modell. 192, 67–82.

Xevi, E., Gilley, J., Feyen, J., 1996. Comparative study of two crop yield simulation models. Agri. Water Mgt. 30, 155–173.

Xiong, W., Holman, I., Conway, D., Lin, E., Li, Y., 2008. A crop model cross calibration for use in regional climate impacts studies. Ecol. Model. 213, 365–380.

Yang, Y.Y., Moiwo, J.P., Hu, Y., 2010. Estimation of irrigation requirement for sustainable water resources reallocation in North China. Agri. Water Manage. 97, 1711–1721.

Molecular, Biotechnological, and Industrial Approaches for Enhancement of Crop Production and Quality

Biotechnological Approaches for Increasing Productivity and Sustainability of Rice Production

D.S. Brar* and G.S. Khush[†]

*School of Agricultural Biotechnology, Punjab Agricultural University, Ludhiana, Punjab, India,
[†]University of California, Davis, CA, USA

Rice (*Oryza sativa* L.) is the primary food source for nearly half of the world's population and accounts for 35–60% of the calories consumed by 3 billion people in Asia. It is planted on 159 million ha the world over, with annual production of 690 million tonnes (Table 8.1). More than 90% of rice is grown and consumed in Asia, under a wide range of agroclimatic conditions. China is the largest producer of rice (197.2 Mt) and India has the largest area under rice but ranks second in production (143.9 Mt). Rice growing area is classified into four major ecosystems: irrigated (55%), rainfed lowland (25%), rainfed upland (12%), and deep water (8.0%).

During the last few decades, significant advances have been made in increasing rice productivity. World rice production has more than doubled from 257 Mt in 1966 to 696 Mt in 2010. This was achieved primarily through the application of principles of Mendelian genetics and conventional plant breeding. The current world population of 7.1 billion is expected to reach 8.0 billion by 2030, and rice production must increase by 25% to meet the growing demand. However, rice productivity is continually threatened by diseases (bacterial blight, blast, tungro virus, rice yellow mottle virus, sheath blight, etc.), insects (plant hoppers, stem borers, and gall midge), and abiotic stresses (drought, submergence, salinity, cold, heat, soil toxicities, etc.). The major challenge, at present, is to overcome these constraints particularly in the context of global climatic changes and newly emerging pests and unpredictable weather regimes. The other concern is to produce more rice with less land, less water, fewer chemicals, and less labor. To overcome these constraints and meet challenges to increase and sustain rice production, there is a need to develop rice varieties with higher yield potential, durable resistance to diseases and insects, and tolerance to abiotic stresses.

Agricultural Sustainability. DOI: http://dx.doi.org/10.1016/B978-0-12-404560-6.00008-3

TABLE 8.1 Area, Production, and Yield of Rice Paddy by Country in 2010

Country/Region	Area (10^3 ha)	Production (10^3 tonne)	Yield (t/ha)
World	159,416	696,324	4.37
Asia	142,065	631,842	4.37
Bangladesh	11,700	50,061	4.28
Cambodia	2,776	8,245	2.97
China	30,117	197,212	6.55
India	42,560	143,963	3.38
Indonesia	13,253	66,469	5.02
Iran (Islamic Republic)	563	2,288	4.06
Japan	1,628	10,600	6.51
Republic of Korea	892	6,136	6.88
Laos	855	3,070	3.59
Malaysia	673	2,548	3.78
Myanmar	8,051	33,204	4.12
Nepal	1,481	4,023	2.72
Pakistan	2,365	7,235	3.06
Philippines	4,354	15,771	3.62
Sri Lanka	1,060	4,300	4.06
Thailand	10,990	31,597	2.88
Vietnam	7,513	39,988	5.32
Egypt	459	4,329	9.42
Brazil	2,722	11,236	4.13
Africa	9,341	22,977	2.46
Australia	19	196	10.41
USA	1,462	11,027	7.54

1.   ADVANCES IN RICE BIOTECHNOLOGY

Recent advances in biotechnology, particularly in cellular and molecular biology and genomics, have provided new opportunities to develop improved germplasm with new genetic properties and to understand the function of rice genes (Khush and Brar, 2001). Some of the developments in biotechnology include:

- a dense molecular map consisting of more than 4,000 DNA markers and availability of a new generation of markers such as single nucleotide polymorphism (SNPs);
- many important genes/quantitative traits loci (QTLs) governing abiotic and biotic stresses tagged with molecular markers;
- marker-assisted selection (MAS) practiced for resistance to bacterial blight (BB), blast, submergence, gall-midge, and brown plant hopper (BPH), etc.;
- map-based cloning of several agronomically important genes such as yield related traits, blast, BB, etc. and gene-based markers for MAS;
- transfer of alien genes for resistance to BPH, BB, blast tungro, etc. from wild species across crossability barriers;
- extensive synteny with genomes of other cereals;
- availability of new genetic resources for functional genomics: a large set of T-DNA insertion lines (>100,000), 60,000 deletion mutants, and yeast artificial chromosomes (YAC), expressed sequence tag (EST), and bacterial artificial chromosome (BAC) libraries of cultivated and wild species;
- high-throughput methods/materials (SNP chips) for MAS and gene chips for gene discovery and gene expression analysis;
- high-throughput methods for *Agrobacterium* mediated genetic transformation and production of transgenics with new genetic properties for resistance to biotic and abiotic stresses with improved nutritional quality;
- availability of whole-genome sequence data of both indica and japonica rice for future research in forward and reverse genetics.

In addition, genome-wide selection, sequence-based MAS, and next-generation sequencing offer new opportunities to support plant breeding research for genetic enhancement of rice.

2. APPROACHES FOR INCREASING YIELD POTENTIAL OF RICE

During the last few decades, major progress has been made in rice production. The adoption of semi-dwarf high-yielding varieties coupled with improved production technologies has ushered in the Green Revolution. Several reviews are available on the development of such high-yielding varieties through conventional breeding and the way forward for increasing yield potential (Khush, 1995, 1999, 2001, 2005). Plant physiologists proposed increased photosynthetic efficiency and sink size as possible approaches to increase yield potential (Yoshida, 1972). Some strategies to improve yield potential include: (i) breeding new ideotypes for higher yield, (ii) developing hybrids with higher yield potential, (iii) introducing yield-enhancing loci/QTLs (wild species alleles), (iv) pyramiding cloned genes/QTLs governing yield-related traits, and (iv) developing C_4 rice.

The concept of breeding for ideotypes, based on plant architecture, was given as early as 1968 for increasing yield of cereals (Donald, 1968). Since then, several scientists have suggested breeding for high-yield-based ideotypes. Since the release of IR8 in 1966, major progress has been made in developing short-growth-duration varieties possessing multiple resistances to diseases and insects and with improved grain quality. However, the yield potential of rice did not increase significantly in tropical environments beyond the level of IR8 (10 t/ha). Modern high-yielding varieties (HYVs) have a harvest index of 0.5 and a total biomass of about 20 t/ha under optimum conditions. Khush (1995) suggested that by raising biomass to about 22 t/ha and harvest index to 0.55, it should be possible to increase yield potential to 12 t/ha. Breeding a new ideotype dubbed as the new plant type (NPT) began in 1989 at the International Rice Research Institute (IRRI, 1989). Several donors for target traits were identified in "bulu" or japonica germplasm, commonly referred to as "tropical japonica", mainly from Indonesia. Intensive breeding efforts were continued at IRRI and in other rice-growing countries such as China. Crosses were made between diverse germplasm: indica × tropical japonica, indica × temperate japonica, indica × indica. The second generation NPTs were derived from crosses of NPT × indica to improve grain quality. As a result, some promising high-yielding varieties have been released in the Philippines and China.

2.1 Developing Hybrids with Higher Yield Potential

The phenomenon of heterosis, or hybrid vigor, has been successfully exploited since the 1950s for increasing the productivity of cross-pollinated crops such as maize and sorghum. Major progress has been made to exploit hybrid vigor in self-pollinated crops such as rice. China has taken the lead in developing hybrid rice technology; about 50% of the rice area in China is under hybrid rice cultivation. However, the global area under hybrid rice, excluding China, is only 3.4 million hectares. In general, hybrids have 10–15% higher yield than inbreds, with 1 t/ha yield advantage over inbred cultivars. With new advances in genetics, and better understanding of heterosis, hybrid rice offers potential to further increase rice productivity. Traditional breeding methods combined with new molecular technologies are being employed in hybrid rice breeding to speed up product development. The yield of hybrid seed has increased significantly, thus making it easier for farmers to purchase seed at a much lower cost. Molecular marker technology has helped in the determination of the purity of parental lines and hybrid seed and further accelerated efforts to transfer genes for fertility restoration and resistances into parental lines. Heterotic gene pools have been identified to obtain higher heterosis. QTLs/genes for hybrid vigor and heterotic gene blocks are being identified, and parental lines of hybrid rice with new yield potential are being developed.

Biotechnological approaches can be used to: (i) identify highly heterotic gene pools of parental lines based on molecular marker data, (ii) pyramid genes/QTLs and to develop parental lines resistant to multiple biotic stresses, e.g., BB, blast, and BPH, and abiotic stresses such as drought, (iii) enhance outcrossing using wild species such as *O. longistaminata*, (iv) identify heterotic gene blocks/regions on rice chromosomes and QTLs for yield heterosis, and (v) explore research for a one-line system—apomixis—using comparative genomics and genetic engineering techniques.

2.2   Introgression of Yield-Enhancing Loci/Wild-Species Alleles

Crop gene pools are widened through hybridization of crop cultivars with wild species, weedy races as well as intra-subspecific crosses. Such gene pools are exploited for improving many traits, including yield. So far, in most plant breeding programs, intraspecific hybridization involving diverse crosses between indica × indica and indica × japonica (both tropical and temperate japonica) has been used in rice improvement. Plant breeders are now tapping new genes to improve yield using wild species of *Oryza*. These wild species, representing 10 genomic types, although inferior in phenotype and having poor plant type and grain characteristics, are an important source of new useful genes for rice improvement. Furthermore, with the new molecular markers available, it has become easy to track the introgression of wild species alleles into segregating populations and in elite breeding lines, and thus to tag the introgressed genes for use in marker-assisted breeding.

Several reports are available on transgressive segregation for yield-related traits from crosses of cultivated and wild species, suggesting that wild species contain genes that can improve quantitative traits such as yield (McCouch et al., 2007). Tanksley and Nelson (1996) proposed advanced-backcross (AB) QTL analysis to discover and transfer valuable genes from unadapted germplasm into the elite breeding lines. Xiao et al. (1996) reported increased yield in rice using wild species (*O. rufipogon*), where alleles at two marker loci, RM5 (*yld1-1*) on chromosome 1 and RG 256 on chromosome 2 (*yld2-1*), were associated with enhanced yield (18%). Later, Xiao et al. (1998) also reported 68 QTLs, 35 (51%) of which had trait-improving alleles derived from phenotypically inferior wild species.

Septiningsih et al. (2003) found that 33% of the QTL alleles originating from *O. rufipogon* had beneficial effect for yield and yield components in the IR64 background. Twenty-two QTLs (53.4%) were located in similar regions as previously reported rice QTLs, suggesting the existence of stable QTLs across genetic backgrounds and environments. Advance-backcross progenies derived from the crosses of an elite breeding line of new plant type (NPT) rice with *O. longistaminata* showed transgressive segregation for yield and yield components. Three introgression lines carried *O. longistaminata* alleles as detected through RM530, RM8255, and RM122 in the long arm of chromosome 2 where *yld 2-1* QTLs for yield are located (Reintar, 2007).

Cheema et al. (2008a) also reported genomic regions in advanced lines derived from crosses of rice and *O. rufipogon* that are associated with yield increase. Imai et al. (2011) evaluated advanced backcross lines (BC_3, BC_4) derived from wide crosses in field trials in Arkansas, and showed an average yield enhancement of 23% compared with the recurrent parent "Jefferson". McCouch et al. (2007) summarized the results of various studies on transgressive segregation and QTLs responsible for increased yield from crosses of *O. sativa* × *O. rufipogon* supporting that yield-enhancing loci from wild species could increase yield potential of both inbred and hybrid rice varieties. In addition, identification and introgression of QTLs from different AA genome wild species, including from *O. glaberrima*, offer opportunities to broaden the gene pool to further increase yield potential of rice inbreds and hybrid varieties. Emphasis should be placed on identifying and fine mapping of QTLs from wild species and introgression into high-yielding elite breeding lines.

2.3 Pyramiding of Cloned Genes/QTLs for Yield-Related Traits

So far, most genetic research has focused on identifying and pyramiding genes/QTLs for resistance to biotic and abiotic stresses. However, with new tools of genomics, it has now become easier to clone genes/QTLs even for yield-related traits. Map-based strategy and specialized genetic stocks such as near-isogenic lines (NILs) and chromosome segmental substitution lines (CSSLs) have been used to clone genes/QTLs for key yield components such as (i) number of grains per panicle, (ii) grain weight, and (iii) number of tillers per plant including grain filling. Completion of the rice genomic sequence has further facilitated the identification and cloning of genes/QTLs for yield-related traits. There are several reviews on approaches to identification and cloning of QTLs for agronomic traits, including yield, and research reports with information on cloned genes/QTLs governing yield-related traits (Li et al., 2003; Takai et al., 2005; Ashikari and Matsuoka, 2006; Fan et al., 2006; Song et al., 2007; Sakamoto and Matsuoka, 2008).

A set of 334 introgression lines (ILs) derived from crosses between IR64 and ten donor parents was developed through backcrossing at IRRI. Variations in agronomic characters were characterized, and introgressed segments were determined by SSR markers in each IL. Fifty-four regions associated with agronomic traits such as days to heading, culm length, panicle number per plant, grain weight, etc. in the genetic background of elite variety IR64 were identified. Fujita et al. (2011) identified QTLs for high total spikelet number per panicle in the long arm of chromosome 4. Jiao et al. (2010) reported on the regulation of *OsSPL14*, where point mutation in this gene results in modification of plant architecture with reduced tiller number, increased lodging resistance, and enhanced grain yield. The number of grains per panicle under the control of the *Gn1a* locus (a QTL for increased grain number; Ashikari et al., 2005) along with *OsSPL14* could allow fine-tuning of panicle architecture favorable to the breeding of high-yielding varieties. The genes *GS3*, *GW2*, *qSW5*, and *tgw6*, governing grain size, have also been reported to increase grain yield in rice. Pyramiding of genes/QTLs for yield components would provide new evidence if the individual yield-related QTLs, when combined, were found to increase yield potential and overcome the compensation effect resulting from interaction of agronomic traits. This could be achieved by developing NILs for each of the target genes through marker-assisted backcrossing. Such isogenic lines with well-defined genes could be used in pyramiding to ultimately analyze whether the compensation or interaction among yield-relative traits could be overcome to break the yield ceiling in rice, not achieved so far.

2.4 C_4 Rice—Modifying Photosynthetic Systems to Raise Yield

Improving radiation use efficiency (RUE) through regulation of rubisco, introduction of C_4-like traits such as CO_2 concentrating mechanisms, improvement of light interception, and improvement of photosynthesis at the whole-canopy level, is being explored as an option to increase the yield potential in rice. Efforts must focus on identification and utilization of photosynthetically efficient germplasm, and the genes/QTLs for efficient mobilization and loading of photosynthates from source to sink should be identified. High-throughput protocols for selection of photosynthetically efficient germplasm in the breeding program are needed.

C_3 plants such as rice, during the first steps in CO_2 assimilation, form a pair of three-carbon atoms, whereas C_4 plants (maize) initially form four-carbon-atom molecules. There are consistent and large differences in radiation-use efficiency (RUE) between C_3 and C_4 plants. Ultimately, RUE depends on the surplus of photosynthesis over respiration. Little can be done to decrease respiration, while higher RUE, biomass, and yields require increased photosynthesis. Sheehy et al. (2000, 2007) have discussed various strategies for conversion of C_3 to C_4 rice. Converting C_3 plants to C_4 would be a long-term (15–20 years) option, but C_4 rice could yield 25% more than the existing C_3 rice and in addition could have better nutrient- and water-use efficiency. The C_4 photosynthesis system is more efficient and offers several advantages over C_3, such as: (i) faster and more complete translocation of assimilates from leaves; (ii) reduced photorespiration; (iii) almost twice the efficiency in dry matter production per unit of water transpired, and (iv) greater photosynthetic efficiency at high temperature; and it requires less rubisco and hence less nitrogen and less water.

The most outstanding feature of high-capacity photosynthetic systems is the presence of two cell types—mesophyll cells and bundle sheath cells—which cooperate in carbon fixation. One of the major challenges to convert C_3 plants to C_4 plants is to understand: (i) what mechanisms are used to coordinate the activities of the two cell types; (ii) which of the enzymes of C_4

photosynthesis are already expressed in the bundle sheath and how metabolism has been altered compared to the bundle sheath of C_3 plants; and (iii) what processes in the C_3 bundle sheath might be disrupted by introducing C_4 photosynthesis.

Some approaches for converting C_3 to C_4 involve: (i) searching for genetic variability in the wild-species germplasm/relatives of C_3 for C_4 photosynthesis; (ii) incorporating such variability into C_3 plants through wide hybridization; (iii) identifying genes responsible for compartmentalization; (iv) identifying candidate genes for key components in C_4 plants; (v) using comparative genomics for the identification of chromosomal regions carrying key traits/genes for C_4-ness; and (vi) using genetic engineering approaches to transfer C_4 characteristics/key enzymes from different species into rice.

The main challenge is to make the photosynthetic pathway of C_3 resemble that of C_4, by eliminating photorespiration. In C_4, photorespiration is rarely greater than 5% of the rate of photosynthesis, whereas in C_3 it can exceed 30% of the rate of photosynthesis above 30 °C. In drought-prone ecosystems, yield could be maintained or increased with less water and less fertilizer, especially when coupled with the rising atmospheric concentration of CO_2. There is need to understand the complex interactions between canopy architecture and the biochemistry of the photosynthetic apparatus. C_4 is a long-term project requiring extensive experimentation and exploratory research on several basic components before C_3 plants could be successfully converted to C_4 for commercial use. IRRI has a collaborative project involving several laboratories of Europe and the USA to alter the photosynthesis of rice from C_3 to C_4 pathway by introducing cloned genes from maize and other systems that regulate the production of enzymes responsible for C_4 synthesis. If successful, the yield potential of rice may increase by 30–35%.

3. MAPPING GENES/QTLS AND MARKER-ASSISTED BREEDING IN RICE

Since the development of the first molecular map of rice in 1988 (McCouch et al., 1988), many reports have become available on the construction of high-density molecular maps with a diversity of molecular markers (Harushima et al., 1998; McCouch et al., 2002, 2010). This has led to the major breakthrough in mapping genes/QTLs for various agronomic traits such as tolerance to biotic and abiotic stresses and different grain quality and productivity traits.

Khush and Brar (2001) and Xu and Croch (2008) summarized the list of major genes/QTLs tagged with molecular markers in rice. Collard et al. (2005) have described details on marker analysis for QTL mapping. MAS (marker-assisted selection) protocols have become available for many traits, and pyramiding of genes/QTLs for tolerance to biotic and abiotic stresses has become priority research in many laboratories. MAS products resistant to BB, BPH, and submergence tolerance have been released as varieties for commercial cultivation. Successes on respective traits are discussed in the following subsections.

3.1 Breeding Against Biotic Stresses

A number of donors have been identified for resistance to major diseases and insects, and incorporated to develop varieties resistant to biotic stresses. Classical genetic analysis has revealed 30 genes for resistance to BB, 40 for blast, 21 for BPH, 6 for white-backed plant hopper (WBPH), 10 for green leaf hopper (GLH), 3 for zigzag leaf hopper (ZLH), and 10 for gall midge. A large number of rice varieties with multiple resistances to diseases and insects have been developed through conventional breeding. Cultivation of these varieties has reduced the use of chemicals and saved millions of dollars. However, there is a paucity of donors for resistance to sheath blight, virus diseases, and stem borer. Also, no resistance source is available for the newly emerging disease: false smut.

Molecular markers have made it possible to tag genes/QTLs governing agronomic traits such as tolerance to biotic and abiotic stresses (Khush and Brar, 2001; Nino-Liu et al., 2006). Huang et al. (1997) used PCR-based markers in pyramiding genes for resistance to BB, and several lines with different gene combinations have been developed. *Xa4*, *xa5*, *xa13*, and *Xa21* were also pyramided into one breeding line. The pyramided lines showed a wider spectrum and higher level of resistance than lines with only a single gene for resistance. MAS has also been employed for moving genes from pyramided lines into new plant types (Sanchez et al., 2000), as well as into improved varieties grown in India (Singh et al., 2001). Several institutes in rice-growing countries have practiced MAS for transfer of BB resistance. As many as 14 varieties resistant to BB have been released in India, the Philippines, China, and Indonesia through MAS (Table 8.2). Recently, BB resistance has been transferred through MAS into parental lines (Pusa 1601, Pusa 1605) of one of the high-yielding hybrid RH10 (Personal Communication, A.K. Singh). Besides, MAS has also been used to transfer BB resistance into parental lines of hybrid rice by different breeding programs in rice-growing countries. Perez et al. (2008) reported introgression of *Xa4*, *Xa7*, and *Xa21* into TGMS lines. Advances in molecular technology, involving a new generation of markers such as SNPs, would further accelerate the breeding of rice varieties with multiple resistance to different pathogens.

Some of the disease resistance genes have been cloned using map-based strategy (13 for blast resistance and 6 for BB resistance: *Xa1*, *xa5*, *xa13*, *Xa21*, *Xa26*, and *Xa27*). This has facilitated the use of gene-based MAS for developing resistant varieties. There is urgent need to identify new genes/QTLs for resistance to sheath blight, brown spot, and the newly emerging false smut. Similarly, pyramiding of genes for BPH and WBPH resistance is emphasized to develop varieties with a wider spectrum of resistance to these pests.

TABLE 8.2 MAS Products Tolerant to Biotic and Abiotic Stresses Released as Varieties in Rice

Inbreds/Hybrids	Year	Gene(s)	Institute/Country
Bacterial blight resistance[a]			
NSICRc 142 (Tubigan 7)	2006	*Xa4 + Xa21*	PhilRice, Philippines
NSICRc 154 (Tubigan 11)	2007	*Xa4 + Xa21*	PhilRice, Philippines
Improved Sambha Mahsuri	2007	*Xa5 + xa13 + Xa21*	India
Improved Pusa Basmati 1	2007	*Xa5 + xa13 + Xa21*	India
Angke	2002	*Xa4 + xa5*	Indonesia
Conde	2002	*Xa4 + Xa7*	Indonesia
Xieyou 218	2002	*Xa21*	China
Zhongyou 218	2002	*Xa21*	China
Guodao 1	2002	*Xa4 + xa5 + xa13 + Xa21*	China
Guodao 3	2004	*Xa4 + xa5 + xa13 + Xa21*	China
Neizyou	2002	*Xa4 + xa5 + xa13 + Xa21*	China
Ilyou 8006	2005	*Xa4 + xa5 + xa13 + Xa21*	China
Ilyou 218	2005	*Xa21*	China
ZhongbaiYou 1	2006	*Xa21*	China
BPH resistance			
Suweon 523	2011	*Bph18*	Korea
Submergence tolerance[b]			
IR64-Sub1(Submarino1)	2009	*Sub1*	Philippines
IR64-Sub1	2009	*Sub1*	Indonesia
Swarna-Sub1	2009	*Sub1*	India
Swarna-Sub1(INPARA-4)	2009	*Sub1*	Indonesia
Swarna-Sub1 (BRRIdhan51)	2010	*Sub1*	Bangladesh
Swarna-Sub1	2011	*Sub1*	Myanmar
Swarna-Sub1	2011	*Sub1*	Nepal
Sambha Mahsuri-Sub1	2011	*Sub1*	Nepal

[a] *Xa21 gene has also been transferred into many elite inbreds and parental lines of hybrids by several institutes in India, the Philippines, and Thailand.*
[b] *CR1009-Sub1, TDK-Sub1, Ciherang-Sub1, PSBRC18-Sub1 are in the pipeline for release.*
Modified from Brar and Singh (2011).

Enhancing resistance to stem borers through accumulation of QTLs with different mechanisms of resistance is recommended. MAS is becoming popular to pyramid genes for gall-midge resistance. One japonica rice variety with *Bph-18* has been developed through MAS (Table 8.2).

3.2 Breeding for Tolerance to Abiotic Stresses in Rice

Rice production in unfavorable environments is mostly constrained by abiotic stresses. The most serious abiotic stresses currently affecting rice production in Asia are drought, submergence, and salt stress, which annually affect about 23, 20, and 15 million ha, respectively. Low temperature adversely affects rice at high elevations as well as where rice is grown during the winter season in the subtropics; heat stress is also emerging as a serious threat to rice production as a consequence of climate change. Availability of markers provides new opportunities to develop stress-tolerant varieties.

3.2.1 Submergence Tolerance

Rice is sensitive to flooding during germination, which hinders direct seeding in rainfed areas, and also during the vegetative stage when completely submerged. Stagnant partial flooding of 0–50 cm for most of the season also affects considerable areas, estimated at more than 5 million ha in India and Bangladesh alone. Major progress has been made to identify a submergence-tolerance (*Sub1*) gene (Xu et al., 2006), and this gene has been transferred through MAS into seven mega-varieties: IR64, Swarna, TDK, BR11, PSBVRC18, Ciherang, and Sambha Mahsuri. MAS products tolerant to submergence have been released as varieties in the Philippines, India, Bangladesh, Indonesia, Myanmar, and Nepal (Table 8.2). The *Sub1* gene confers an advantage of 1–3 t of grain yield following flooding for 10–15 days. *Sub1* is now introduced into a wide range of genetic backgrounds to develop more-tolerant varieties that are also adapted to longer-term stagnant flooding. The gene is also being transferred into African mega-varieties, using IRRI donors. Further attempts are being made through gene discovery approaches to identify alleles/genes other than *Sub1* conferring submergence tolerance. Pyramiding of *Sub1* with the newly identified genes will enhance tolerance of submergence over a longer period (3–5 weeks of submergence). IRRI is pyramiding QTLs for tolerance to multiple abiotic stresses such as drought, submergence, and anaerobic germination.

3.2.2 Drought Tolerance

Conventional breeding has been practiced for the last several years, and many drought-tolerant varieties have been released. However, the level of tolerance of drought, particularly at flowering stage, is very low. There is thus an urgent need to enhance tolerance of drought at various stages of plant growth.

Marker-assisted mapping and the introgression of major-effect QTLs could be an efficient and fast-track approach for breeding drought-tolerant rice varieties. Several QTLs offering advantage to retain grain yield under water stress have been identified by different researchers. Four of these QTLs (qtl 12.1, qtl 3.1, qtl 1.1, and qtl 9.1) are being pyramided with the hope that the drought tolerance of rice varieties will increase significantly.

Vikram et al. (2011) undertook studies to discover QTLs for grain yield (GY) under reproductive stress (RS) with a large and consistent effect in the background of Swarna, IR64, and MTU1010 (mega varieties). A major QTL for GY under RS, $qDTY_{1.1}$, was identified on chromosome 1 flanked by RM11943 and RM431 in three populations. In combined analysis over 2 years, it showed an additive effect of 29.3%, 24.3%, and 16.1% of mean yield in N22/Swarna, N22/IR64, and N22/MTU 1010, respectively. It is interesting to note that $qDTY_{1.1}$ has shown effect in multiple elite genetic backgrounds under both stress and non-stress conditions.

Swamy et al. (2011) integrated 15 maps, resulting in a consensus map with 531 markers and a total map length of 1821 cM. Fifty-three yield QTLs reported in 15 studies were projected on a consensus map, and meta-analysis was performed which identified 14 meta-QTLs on 7 chromosomes. Three groups of genes such as stress-inducible genes, growth and development-related genes, and sugar transport-related genes were found in clusters in most of the meta-QTLs. Validation of major-effect QTLs on a panel of random drought-tolerant lines revealed the presence of at least one major QTL in each line. $DTY_{12.1}$ was present in 85% of the lines, followed by $DTY_{4.1}$ in 79%, and $DTY_{1.1}$ in 64% of the lines. Comparative genomics of meta-QTLs with other cereals also revealed homologous regions with QTLs for grain yield under drought in the case of maize, wheat, and barley.

In addition to the above-stated approaches, the availability of saturated transposon insertion-mediated knockout rice libraries may provide another route to gene discovery for drought tolerance. Finally, *in silico* neural network-mediated analysis of the promoter sequences of various known drought-responsive genes will help discover additional and novel genes tightly co-regulated under drought stress.

3.2.3 Salt Stress, Heat Stress, and Phosphorus Deficiency Tolerance

Poor soils with excess salt and deficiency in certain plant nutrients limit rice productivity in most rainfed rice areas, and several million ha of land suited to rice production in Asia and Africa are currently unexploited because of salinity and other related soil problems.

Singh and Flowers (2010) have summarized the progress made in identification of QTLs for tolerance to salinity with different mechanisms of tolerance. A number of QTLs have been identified for different components (Na, K absorption, Na/K ratio, shoot/root length, weight, and other plant traits) related to salinity tolerance. A few genes/QTLs with different modes of tolerance have been identified (Ismail et al., 2007). *Saltol* and other QTLs for seedling-stage tolerance and one for the reproductive stage are being tagged to develop varieties tolerant at both stages. MAS have been practiced to transfer the *Saltol* gene into a mega-variety of Bangladesh (BR11). Map-based cloning of genes responsible for salt tolerance led to the identification of *qSKcl* controlling K^+/Na^- homeostasis under salt stress.

Similarly, rising temperatures due to climate change are becoming an alarming problem. Heat stress causes high sterility, leaf yellowing, and accelerated development, leading to low yield potential in sensitive rice varieties. Germplasm screening (both improved as well as traditional rice varieties) has been done to identify donors tolerant to high temperature. Advanced breeding lines involving crosses between elite rice lines and heat-tolerant donors (N22, IR6) have been developed and are being evaluated at hot spots in different countries to identify promising heat-tolerant lines. Mapping populations have been developed to map genes/QTLs for heat tolerance. One of the wild species, *O. officinalis*, has become an important genetic resource for early-morning flowering, thus avoiding heat stress (Ishimaru et al., 2010). Available introgression lines from crosses of wild species (*O. officinalis and O. minuta*) are being screened for early day flowering (EDF). QTL mapping is in progress to facilitate the use of MAS in developing improved heat-tolerant cultivars.

Also, nutrient deficiencies are found as companion stresses together with other abiotic stresses, for example, phosphorus deficiency stress and drought are often interdependent (Sigrid Heuer, personal communication). It is, therefore, imperative to study these stresses systematically together, since a focus on either one of them might only have limited impact on yield improvement. A major QTL for phosphorus deficiency tolerance, *Pup1*, was identified in the traditional rice variety Kasalath. The *Pup1* locus was sequenced, and a protein kinase gene named phosphorus-starvation tolerance, *PSTOL1*, was identified (Gamuyao et al., 2012). The gene is absent in sensitive varieties, and over-expression of *PSTOL1* significantly enhanced grain yield on phosphorus-deficient soils. Introgression of this gene into rice would enhance productivity under low phosphorus conditions, and pyramiding with genes/QTLs for tolerance to other stresses will perhaps be a promising approach to sustain rice production under multiple stresses.

3.2.4 Increased Elongation Ability under Deep-Water Conditions

Rice productivity is quite low under deep-water conditions. Recently, QTLs have been identified for elongation ability under such conditions. Hattori et al. (2009) identified three QTLs on chromosomes 1, 3, and 12 and developed pyramided lines containing these QTLs (1 + 3, 1 + 12, 3 + 12, and 1+ 3 + 12). Pyramid lines with three QTLs showed additive effect on elongation. Positional cloning and gain of function analysis resulted in the identification of two genes, *SNORKEL1* and *SNORKEL2*, that trigger deep-water response by encoding an ethylene response factor involved in ethylene signaling. Under deep-water conditions, ethylene accumulates in the plant and induces expression of these two genes. Transfer of these QTLs into high-yielding genotypes would prove beneficial for rice agro-systems that are affected by deep-water conditions.

3.3 Other Approaches to Facilitate MAS

New opportunities exist to enhance selection efficiency for agronomic traits. One such approach involves genome selection (GS) that comprises methods that use genotypic data across the whole genome to predict any trait with accuracy sufficient to allow selection on that prediction alone. The requirements range from at least 200 to at most 10,000 molecular markers and observations. With the baseline, it can greatly accelerate the breeding cycle while also using marker information to maintain genetic diversity and potentially prolong gain beyond what is possible with phenotypic selection. With the cost of marker technologies continuing to decline and the statistical methods becoming more routine, the results indicate that GS will play a large role in plant breeding. Heffner et al. (2009) and Lorenz et al. (2011) have described the usefulness and limitations of genomic selection to accurately estimate breeding values, accelerate the breeding cycle, and introduce greater flexibility in phenotypic evaluation and selection. Genomic selection has great potential but needs to be tested in actual breeding programs.

Recent advances in large-scale genome-sequencing projects have also opened up new possibilities for the application of conventional mutation techniques in not only forward but also reverse genetic strategies. TILLING (Targeting Induced Local Lesions IN Genomes) protocols provide a high frequency of point mutations distributed randomly in the genome. High-throughput TILLING allows rapid and low-cost recovery of new alleles that could be useful in rice breeding. Sasaki (2011) highlighted the usefulness of sequencing data to support breeding activities. The next-generation sequencing can help in finding new alleles for useful genes. A genome-wide collection of SNPs among land races and advanced breeding lines/varieties could be useful in designing DNA markers for map-based cloning or in MAS. Genome-sequencing data would constitute a valuable resource for understanding allelic variations to facilitate genomics-assisted breeding.

4. BROADENING THE GENE POOL OF RICE THROUGH WIDE HYBRIDIZATION

The genus *Oryza* to which cultivated rice (*Oryza sativa* L.) belongs has 22 wild species (2n = 24,48) representing 10 genomic types: AA, BB, CC, BBCC, CCDD, EE, FF, GG, HHJJ, and HHKK. These wild species constitute an important reservoir of useful genes for resistance to diseases and insects and tolerance to abiotic stresses, besides being a source of productivity traits (Brar and Khush, 1997, 2006; Brar and Singh, 2011). Genes can be transferred through direct crosses of rice and AA genome wild species; however, embryo rescue is required for production of interspecific hybrids and progenies with all the wild species other than AA genomes. Molecular markers have facilitated fast-track introgression of segments from wild species and in mapping genes/QTLs introgressed from wild species. A number of genes from wild species for tolerance to biotic and abiotic stresses have been transferred into rice, and some varieties have been released (Tables 8.3 and 8.4). Genes for resistance to BPH were successfully transferred from *O. officinalis* and other wild species of rice to elite breeding lines (Jena and Khush, 1990; Multani et al., 1994; Jena et al., 2006). Similarly, genes for resistance to BB have been transferred from *O. minuta* and *O. rufipogon* to improved rice germplasm (Brar et al., 1996; Cheema et al., 2008b).

Several agronomically important genes have been tagged with molecular markers (Ishii et al., 1994; Jena et al., 2006; Rahman et al., 2008; Brar and Khush 2006; Brar and Singh, 2011), and one of these genes, *Xa21* for BB resistance, has been transferred from *O. longistaminata* through MAS into many breeding lines (Table 8.2), and 12 varieties have been released. QTLs introgressed for aluminum toxicity and drought tolerance have also been identified (Nguyen et al., 2003; Bimpong et al., 2011).

Wild-species germplasm is being investigated for novel traits such as C_4-ness and biological nitrogen fixation (BNF). In preliminary studies, some wild-species germplasm has shown endophytes to support BNF. Improving the endophytic associations between rice and nitrogen-fixing bacteria could prove useful for developing rice capable of nitrogen fixation.

5.   ROLE OF ANTHER CULTURE IN RICE BREEDING

Haploids are important to reduce the breeding cycle of varieties through fixation of recombinants in the immediate generation from segregating populations, unlike conventional breeding that needs 3–4 generations of selfing to get homozygous lines from diverse crosses. The production of haploids was reported as early as 1968 from anther culture of rice (Niizeki and Oono, 1968). Since then, anther culture has been greatly refined. More than 100 breeding lines and varieties have been developed through anther culture in China. Anther culture-derived lines have also been released as varieties in the Philippines (PSBRc50) and the Republic of Korea (Hwacheogbyeo, Joryeongbyeo, Hwajinbyeo). However, most of the anther culture-derived varieties are japonicas, as indica varieties are generally recalcitrant. Recently a large number of doubled haploid (DH) lines have been generated from crosses of indica with tropical japonicas (Grewal et al., 2011). Attempts are being made to identify QTLs for high anther-culturability for transfer into indica

TABLE 8.3 Introgression of Genes from Wild *Oryza* Species into Rice

Trait	Donor *Oryza* Species		
	Wild Species	**Gene**	**Genome**
Grassy stunt resistance	*O. nivara*	GS	AA
Bacterial blight resistance	*O. rufipogon*	Xa23	AA
	O. longistaminata	Xa21	AA
	O. nivara	Xa38	AA
	O. officinalis	Xa29(t)	CC
	O. minuta	Xa27	BBCC
	O. latifolia	Unknown	CCDD
	O. australiensis	Unknown	EE
	O. brachyantha	Unknown	FF
Blast resistance	*O. glaberrima*[1]	Unknown	AA
	O. rufipogon	Unknown	AA
	O. minuta	Pi9	BBCC
	O. australiensis	Pi40	EE
Brown plant hopper resistance	*O. officinalis*	bph11, bph12	CC
	O. eichingeri	Bph14, Bph15	CC
	O. minuta	Bph20, Bph21	BBCC
	O. latifolia	Unknown	CCDD
	O. australiensis	Bph10, Bph18	EE
White-backed plant hopper resistance	*O. officinalis*	Wbph7(t), Wbph8(t)	CC
	O. latifolia	Unknown	CCDD
Cytoplasmic male sterility	*O. sativa f. spontanea*	Unknown	AA
	O. perennis	Unknown	AA
	O. glumaepatula	Unknown	AA
	O. rufipogon	Unknown	AA
Tungro tolerance	*O. rufipogon*	Unknown	AA
Tolerance to iron toxicity	*O. rufipogon*	Unknown	AA
	O. glaberimma[1]	Unknown	AA
Drought-related traits	*O. glaberimma*[1]	QTLs	AA
Tolerance to aluminum toxicity	*O. rufipogon*	QTL	AA
Tolerance to acidic conditions	*O. glaberrima*[1]	Unknown	AA
	O. rufipogon	Unknown	AA
Tolerance to phosphorus deficiency	*O. rufipogon*	Unknown	AA
	O. glaberimma[1]	Unknown	AA
Yield-enhancing loci	*O. rufipogon*	QTL, yld1, yld2	AA
	O. grandiglumis	QTL	CCDD
Yellow stem borer (larval mortality)	*O. longistaminata*	QTL	AA
Increased elongation ability	*O. rufipogon*	Unknown	AA

[1] O. glaberrima – *African rice species*
Modified from Brar and Khush (2006) and Brar and Singh (2011).

TABLE 8.4 Rice Varieties Developed through Wide Hybridization

Key Trait	Wild Species	Varieties Released	Country
Grassy stunt resistance	O. nivara	Many rice varieties	Rice-growing countries in Asia
BPH resistance	O. officinalis	MTL 98, MTL 103, MTL 105, MTL114	Vietnam
Acid sulfate tolerance	O. rufipogon	AS 996	Vietnam
Salinity tolerance	O. rufipogon	BRRIdhan55 (As996)	Bangladesh
Tungro resistance	O. rufipogon	Matatag 9	Philippines
Bacterial blight resistance	O. longistaminata	NSICRc 112	Philippines
Blast resistance	O. rufipogon	Dhanrasi	India
	O. glaberrima[a]	Yun Dao	YAAS, China
High yield, earliness, weed competitive ability, and tolerance to abiotic stresses	O. glaberrima[a]	Many Nerica lines/ varieties	African countries

[a]African rice species.
Modified from Brar and Singh (2011).

cultivar, which could lead to successful use of anther culture in accelerating rice breeding.

Just as the large-scale production of haploids through chromosome elimination in wide-crosses of barley and wheat has been demonstrated, the system needs to be established for rice as well. In addition, exploratory research to identify haploid inducer gene/stocks in rice, similar to maize, could be pursued. These approaches might help to achieve high-throughput production of DH in recalcitrant indica rice and thus accelerate breeding.

6. GENETIC ENGINEERING APPROACHES FOR SUSTAINABLE RICE PRODUCTION AND ENHANCED NUTRITIONAL QUALITIES

The introduction of alien genes from bacteria, viruses, fungi, animals, and unrelated plants into crop species allows plant breeders to achieve breeding objectives that were not considered possible. So-called genetically modified (GM) or transgenic rice has been produced in many laboratories. Earlier polyethylene glycol (PEG), electroporation, and biolistic methods were used.

TABLE 8.5 Some Examples of GM Rice Carrying Agronomically Important Genes

Trait	Transgene	Transformation Method(s)
Herbicide resistance	*Bar*	PEG, biolistic
Tolerance to virus	*coat protein*	Electroporation, biolistic
Sheath blight resistance	*chitinase*	PEG
Stem borer resistance	(Bt genes) *cryI(A)*, *cryI(B)*, *cryI(C)*, *cry(2A)*,	Electroporation, biolistic, *Agrobacterium*
	CpTI	PEG
Tolerance to BPH	*Gna*	Biolistic, *Agrobacterium*
BB resistance	*Xa21*	Biolistic
Blast resistance	*Afp*	*Agrobacterium*
Salt tolerance	*Cod A*	Electroporation
Drought tolerance	*OtsA*, *OtsB*, *DREBIA*, *HRD*	*Agrobacterium*
Increased iron content	*ferritin*	*Agrobacterium*
Pro-vitamin A (Golden rice-1)	*psy*, *crtl*, *lcy*	*Agrobacterium*
Pro-vitamin A (Golden rice-2)	*psy*(maize), *crtl*	*Agrobacterium*
Increased photosynthetic activity	PEP, PEPC, PPDK, PEPC + PPDK, NADP-ME	*Agrobacterium*

Modified from Brar and Khush (2006).

However, now high-throughput methods using *Agrobacterium*-mediated transformation have become available both for indica and japonica rice (Hei et al., 1994). Transgenic rice lines were produced as early as 1988 (Toriyama et al., 1988). Since then, several laboratories all over the world have produced transgenic rice lines with new genetic properties/traits such as resistance to herbicides, diseases, and insects, tolerance to abiotic stresses, and improved nutritional quality, such as Golden rice with pro-vitamin A (Table 8.5). Confined-field trials on transgenics have also been conducted; however, so far no commercial release of transgenic rice has been made for cultivation in farmer's fields.

Conventional breeding approaches to develop stem borer resistance have not met with much success in rice. A number of *Bt* genes—*cryI(A)*, *cryI(B)*, *cryI(C)*, *cryI(2A)*—conferring resistance to stemborers have been transferred through transgenic approaches in many laboratories. These transgenes (*Bt* genes) have also been pyramided to develop a wide spectrum of resistance to stem borers (Zhang, 2009). Intensive research has also been carried out on development and deployment of transgenics for increased tolerance to drought. Garg et al. (2002) produced transgenic plants of indica rice variety PB-1 using *E. coli* trehalose biosynthetic genes (*OtsA* and *OtsB*). The transgenic plants accumulated trehalose at levels 3–10 times that of non-transgenic plants and were found to be tolerant to salt and drought stress. These results demonstrated the feasibility of engineering rice for increased tolerance to abiotic stress and enhanced productivity through tissue-specific or stress-dependent over-production of trehalose. Using another approach, Karaba et al. (2007) introduced the *HRD* gene, an AP_2/ERF-like transcription factor from *Arabidopsis*, into rice. The transgenic rice plants showed increase in leaf biomass and bundle sheath cells that contributed to enhanced photosynthesis assimilation and efficiency. Over-expression of the *HARDY* gene resulted in improvement in drought tolerance and water use efficiency. Approaches using transcription factors such as dehydration-responsive element binding (DREB) and HARDY, offer new potential to further increase the tolerance of drought beyond what has been achieved already. Many laboratories, including IRRI, are using the *DREB1* gene to enhance drought tolerance in rice. At IRRI, transgenic lines of IR64 are under field evaluation to determine their performance under drought stress.

Besides tolerance to biotic and abiotic stresses, improvement of nutritional quality of rice is being pursued as a priority. Screening of rice germplasm has not shown any accession with pro-vitamin A in the rice grain. Hence, to develop pro-vitamin-A-enriched rice, a genetic engineering approach was undertaken. Transgenic rice that produces carotenoids in the endosperm gives a characteristic yellow or golden color, hence dubbed as golden rice. Ye et al. (2000) first established a proof of concept and developed rice with pro-vitamin A properties. Since then, several attempts have been made to enhance pro-vitamin A. The daffodil gene encoding phytoene synthase (*psy*) was found to be one of the limiting factors in β-carotene accumulation of Golden rice-1. In Golden rice-2, *psy* from maize has been used in combination with *Erwinia uredovera* carotene desaturase (*crt l*) (Paine et al., 2005), which led to as high as 25 µg/g carotenoid content accumulation in the rice grains. The carotenoid locus from the leading GR2 event has been introgressed at IRRI into three mega-varieties of rice (IR64, PSBRc18, and BR 29) using marker-aided backcrossing. Evaluations of transgenic golden rice introgression lines in screenhouse and confined-field tests have shown that these mega-varieties are similar in performance and in agronomic characteristics to the recurrent parents, and have higher carotenoid content. The Golden rice introgression lines have been shared with the National Agricultural Research and Extension Systems (NARES) in India, the Philippines, Bangladesh, and Indonesia for evaluation, following biosafety protocols of the respective countries. Investigations on bioavailability and suitability for human consumption are in progress. It is expected that, by 2014, golden rice will become available for commercial release. However, regulatory and biosafety requirements need to be met before that occurs.

In a recent review, Bhullar and Gruissem (2012) discussed various strategies to enhance the nutritional properties of rice. Similar to pro-vitamin A, genetic diversity for rice endosperm iron has been found to be low as well, and it has not been possible so far to achieve the target iron levels through conventional breeding approaches. One IRRI breeding line, IR68144-2B-2-2-3-2, was released as a high-iron rice variety (NSICRc172); however, polished rice has only 2.5 µg/g iron in this line. Thus, genetic engineering approaches are being used to achieve the target of 14 µg/g iron in the polished grain. So far, transgenic rice has been produced with as high as 8 µg/g iron. To further enhance the iron content, new gene constructs including iron transporter genes coupled with an appropriate choice of promoters are being used to raise the iron content in polished rice grain.

High-throughput transformation protocols, inducible expression of transgenes, and marker-free transgenics are important to develop superior transgenic products. RNAi also offers new potential to develop improved germplasm for resistance to diseases and insects and with altered quality characteristics (Kusaba 2004; Small 2007). RNAi transgenes are dominant and can be used in different genetic backgrounds for any known gene in the genome.

7. ADVANCES IN FUNCTIONAL GENOMICS SUPPORT RICE BREEDING

Since the first report on sequencing of the rice genome (IRGSP, 2005), many exciting developments have occurred, particularly in functional genomics of rice. The availability of sequence data in indica and japonica rice has ushered in the era of exploring reverse genetics and functional genomics. Furthermore, many resources for genomic research—such as specialized genetic stocks, isogenic lines, CSSL, T-DNA insertion lines, mutant resources, BAC, EST libraries, molecular genetic and physical maps, and syntenic maps—have become important for future research on structural and functional genomics of rice. Sequence data in rice and wild species are providing valuable information on gene expression, DNA replication, chromosome organization, recombination, specialization, and evolution. The activator-dissociator (AC-DS) maize transposable elements, retrotransposons, and miniature inverted repeat transposable elements (MITEs) also provide a wealth of genetic resources for functional genomics. Other resources for functional genomics include: 100,000 T-DNA-tagged insertion lines carrying large number of well characterized T-DNA inserts; Tos17 retrotransposon insertional mutants with 30,000 lines carrying more than 250,000 Tos17s; and more than 50,000 deletion mutants produced by fast neutron, gamma radiation, and chemical mutagenesis. Maize transposon constructs have been used in the transformation of japonica and indica rice cultivars for knockout and gene deletion–insertion studies.

Availability of high-throughput genotyping and other "omics" technologies using microarrays or gene chips have led to the identification of several candidate genes for biotic and abiotic stress pathways, including for grain quality and C_4 traits and to assess gene functions at genomic level. Genome-wide selection, association mapping, sequence-based MAS, next-generation sequencing, and new advances in proteomics, transcriptomics, and metabolomics are becoming important to support rice improvement programs. The combination of forward genetics and reverse genetics offers new potential to apply modern tools of biotechnology for precise understanding of the complexity of the genetic architecture of the rice genome. The biological functions encoded by the genomic sequence through genetic and phenotypic analysis are important components of functional genomics. Identification of genes for various biotic and abiotic stresses, grain quality and productivity traits through functional genomics, and their manipulation would be another major breakthrough in rice genetics and breeding.

Integration of new tools of biotechnology and conventional breeding would be important in food security and for increased productivity and sustainability of rice, particularly in the context of global climate change.

REFERENCES

Ashikari, M., Matsuoka, M., 2006. Identification, isolation and pyramiding of quantitative trait loci for rice breeding. Trends Plant Sci. 11, 344–350.

Ashikari, M., Sakakibara, H., Lin, S., Yamamoto, T., Takashi, T., Nishimura, A., et al., 2005. Cytokinin oxidase regulates rice grain production. Science 309, 741–745.

Bhullar, N.K., Gruissem, W., 2012. Nutritional enhancement of rice for human health: the contribution of biotechnology. Biotechnol. Adv. doi: 10.1016/j.biotechadv.2012.02.001.

Bimpong, K., Serraj, R., Chin, C.J., Ramos, J., Mendoza, E.M., Hernandez, J.E., et al., 2011. Identification of QTL's for drought related traits in alien introgression lines derived from crosses of rice (*Oryza sativa* cv.IR 64) x *O.glaberrima* under low land moisture stress. J. Plant Biol. 54, 237–250.

Brar, D.S., Khush, G.S., 1997. Alien introgression in rice. Plant Mol. Biol. 35, 35–47.

Brar, D.S., Khush, G.S., 2006. Cytogenetic manipulation and germplasm enhancement of rice (*Oryza sativa* L.). In: Singh, R.J., Jauhar, P.P. (Eds.), Genetic Resources, Chromosome Engineering and Crop Improvement CRC Press, Boca Raton, FL, pp. 115–158.

Brar, D.S., Singh, K., 2011. Oryza. In: Kole, C. (Ed.), Wild Crop Relatives: Genomic and Breeding Resources: Cereals Springer-Verlag, Berlin, pp. 321–365.

Brar, D.S., Dalmacio, R., Elloran, R., Aggarwal, R., Angeles, R., Khush, G.S., 1996. Gene transfer and molecular characterization of introgression from wild *Oryza* species into rice. In: Khush, G.S. (Ed.), Rice genetics III International Rice Research Institute, Manila, Philippines, pp. 477–486.

Cheema, K.K., Bains, N.S., Mangat, G.S., Das, A., Brar, D.S., Khush, G.S., 2008a. Introgression of quantitative trait loci for improved productivity from *Oryza rufipogon* into *O. sativa*. Euphytica 160, 401–409.

Cheema, K.K., Grewal, N.K., Vikal, Y., Das, A., Sharma, R., Lore, J.S., et al., 2008b. A bacterial blight resistance gene from *Oryza nivara* mapped to 38 kbp region on chromosome 4L and transferred to *O. sativa*. L. Genet. Res. 90, 397–407.

Collard, B.C.V., Jahufer, M.Z.Z., Brouwer, J.B., Pang, E.C.K., 2005. An introduction to markers, quantitative trait loci (QTL) mapping and marker-assisted selection for crop improvement. Euphytica 142, 169–196.

Donald, C.M., 1968. The breeding of crop ideotypes. Euphytica 17, 385–403.

Fan, C., Xing, Y., Mao, H., Lu, T., Han, B., Xu, C., et al., 2006. GS3. A major QTL for grain length and weight and minor QTL for grain width and thickness in rice, encodes a putative transmembrane protein. Theor. Appl. Genet. 112, 1164–1171.

Fujita, D., Tagle, A.G., Ebron, L.A., Fukuta, Y., Kobayashi, N., 2011. Characterization of near-isogenic lines carrying QTL's for high spikelet number in the genetic background of an indica rice variety IR 64 (Oryza sativa L.). Breed. Sci. 62, 18–26.

Gamuyao, R., Chin, C.J., Tanaka, J.P., Pesaresi, P., Catusan, S., Dalid, C., et al., 2012. The protein kinase Pstol 1 from traditional rice confers tolerance of phosphorus deficiency. Nature 488, 535–539.

Garg, A.K., Kim, J.K., Thomas, G.O., Ranwala, A.P., Choi, Y.D., Kochian, L.V., et al., 2002. Trehalose accumulation in rice plants confers high tolerance levels to different abiotic stresses. Proc. Natl. Acad. Sci. USA 99, 15898–15903.

Grewal, D., Manito, C., Bartolome, V., 2011. Doubled haploids generated from crosses of elite indica and japonica cultivars and/or lines of rice: large scale production, agronomic performance and molecular characterization. Crop Sci. 51, 2544–2553.

Harushima, Y., Yano, M., Shomura, A., Sato, M., Shimano, T., Kuboki, Y., et al., 1998. A high-density rice genetic linkage map with 2275 markers using a single F2 population. Genetics 148, 479–494.

Hattori, Y., Nagai, K., Furukawa, S., Song, S.J., Kawano, R., Sakakibara, H., et al., 2009. The ethylene response factors SNORKEL1 and SNORKEL2 allow rice to adapt to deep water. Nature 460, 1026–1030.

Heffner, E.L., Sorrells, M.E., Jannink, J.L., 2009. Genomic selection for crop improvement. Crop Sci. 49, 1–12.

Hei, Y., Ohta, S., Komari, T., Kumashiro, T., 1994. Efficient transformation of rice (Oryza sativa L.) mediated by Agrobacterium and sequence analysis of the boundaries of T-DNA. Plant J. 6, 271–282.

Huang, N., Angeles, E.R., Domingo, J., Magpantay, G., Singh, G., Zhang, G., et al., 1997. Pyramiding of bacterial blight resistance genes in rice: marker assisted selection using RFLP and PCR. Theor. Appl. Genet. 95, 313–320.

Imai, I., McCouch, S.R., Mcclung, A.M., 2011. NILs associated with yield enhancement of Oryza sativa x O. rufipogon cross. Plant & Animals Genomes XIX Conference. Town & Country Convention Center, San Diego, California.

IRGSP (International Rice Genome Sequencing Project), 2005. The map-based sequence of the rice genome. Nature 436, 793–800.

IRRI, 1989. IRRI towards 2000 and Beyond. International Rice Research Institute, Manila, Philippines, (P.O. Box 933).

Ishii, T., Brar, D.S., Multani, D.S., Khush, G.S., 1994. Molecular tagging of genes for brown planthopper resistance and earliness introgressed from Oryza australiensis into cultivated rice, O. sativa. Genome 37, 217–221.

Ishimaru, T., Hirabayashi, H., Ida, M., Takai, T., Yumiko, A.S., Yoshinaga, S., et al., 2010. A genetic resource for early-morning flowering trait of wild rice Oryza officinalis to mitigate high temperature-induced spikelet sterility at anthesis. Ann. Bot. 106, 515–520.

Ismail, A.M., Heuer, S., Thomson, M.J., Wissuwa, M., 2007. Genetic and genomic approaches to develop rice germplasm for problem soils. Plant Mol. Biol. 65, 545–570.

Jena, K.K., Khush, G.S., 1990. Introgression of genes from *Oryza Officinalis* Well ex Watt to cultivated rice, *O. sativa* L. Theor. Appl. Genet. 80, 737–745.

Jena, K.K, Jeung, J.U., Lee, J.H., Choi, H.C., Brar, D.S., 2006. High resolution mapping of a new brown planthopper (BPH) resistance gene, *Bph18*(t), and marker-assisted selection for BPH resistance in rice (*Oryza sativa L.*). Theor. Appl. Genet. 112, 288–297.

Jiao, Y., Wang, Y., Xue, D., Wang, J., Yan, M., Liu, G., et al., 2010. Regulation of *OsSPL14* by OsmiR156 defines ideal plant architecture in rice. Nat. Genet. 42, 541–544.

Karaba, A., Dixit, S., Greco, R., Aharoni, A., Trijatmiko, K.R., Marsch-Martinez, N., et al., 2007. Improvement of water use efficiency in rice by expression of *HARDY*, an *Arabidopsis* drought and salt tolerance gene. Proc. Natl. Acad. Sci. USA 104, 15270–15275.

Khush, G.S., 1995. Breaking the yield frontier of rice. Geo J. 35, 329–332.

Khush, G.S., 1999. Green revolution: preparing for the 21st century. Genome 42, 646–655.

Khush, G.S., 2001. Green revolution: the way forward. Nat. Rev. Genet. 2, 815–822.

Khush, G.S., 2005. What it will take to feed 5.0 billion rice consumers in 2030. Plant Mol. Biol. 59, 1–6.

Khush, G.S., Brar, D.S., 2001. Rice genetics from Mendel to functional genomics. In: Khush, G.S., Brar, D.S., Hardy, B. (Eds.), Rice Genetics Science Publishers, Inc. and International Rice Research Institute, Manila, Philippines, pp. 3–25.

Kusaba, M., 2004. RNA interference in crop plants. Curr. Opin. Biotechnol. 15, 139–143.

Li, X., Qian, Q., Fu, Z., Wang, Y., Xiong, G., Zeng, D., et al., 2003. Control of tillering in rice. Nature 422, 618–621.

Lorenz, A.J., Chao, S., Asoro, F.G., Heffner, E.L., Hayashi, T., Iwata, H., et al., 2011. Genomic selection in plant breeding: knowledge and prospects. Adv. Agron. 110, 77–123.

McCouch, S.R., Kochert, G., Yu, Z.H., Wang, Z.Y., Khush, G.S., Coffman, W.R., et al., 1988. Molecular mapping of rice chromosomes. Theor. Appl. Genet. 76, 815–829.

McCouch, S.R., Teytelman, L., Xu, Y., Lobos, K.B., Clare, K., Walton, M., et al., 2002. Development and mapping of 2240 new SSR markers for rice (*Oryza sativa L.*). DNA Res. 9, 199–207.

McCouch, S.R., Sweeney, M., Li, J., Jiang, H., Thomson, M., Septiningsih, E., et al., 2007. Through the genetic bottleneck: *O. rufipogon* as a source of trait-enhancing alleles for *O. sativa*. Euphytica 154, 317–339.

McCouch, S.R., Zhao, K., Wright, M., Tung, C.W., Ebana, K., Thomson, M., et al., 2010. Development of genome-wide SNP assays for rice. Breed. Sci. 60, 524–535.

Multani, D.S., Jena, K.K., Brar, D.S., Delos Reyes, B.G., Angeles, E.R., Khush, G.S., 1994. Development of monosomic alien addition lines and introgressions of genes from *Oryza australiensis* Domin to cultivated rice, *O. sativa* L. Theor. Appl. Genet. 88, 102–109.

Nguyen, B.D., Brar, D.S., Bui, B.C., Nguyen, T.V., Pham, L.N., Nguyen, N.T., 2003. Identification and mapping of the QTL for aluminum tolerance introgressed from new source, *Oryza rufipogon* Griff into indica rice (*Oryza sativaL.*). Theor. Appl. Genet. 106, 581–593.

Niizeki, H., Oono, K., 1968. Induction of haploid rice plants from anther culture. Proc. Jpn. Acad. Sci. 44, 544–557.

Nino-Liu, D.O., Ronald, P.C, Bogdanove, A.J., 2006. *Xanthomonas orzae* pathovars: model pathogens of a model crop. Mol. Plant Path. 7 (5), 303–324.

Paine, J., Shipton, C., Chaggar, S., Howells, R., Kennedy, M., Vernon, G., et al., 2005. Improving the nutritional value of golden rice through increased pro-vitamin A content. Nat. Biotechnol. 23, 482–487.

Perez, L.M, Redoña, E.D., Mendioro, M.S., Vera Cruz, C.M., Leung, H., 2008. Introgression of *Xa4*, *Xa7* and *Xa21* for resistance to bacterial blight in thermosensitive genetic male sterile rice (*Oryza sativa L.*) for the development of two line hybrids. Euphytica 164, 627–636.

Rahman, M.D., Jiang, W., Chu, S.H., Qiao, Y., Ham, T.H., Woo, M.O., et al., 2008. High resolution mapping of two brown planthopper resistance genes, *Bph20(t)* and *Bph21(t)* originating from *Oryza minuta*. Theor. Appl. Genet. 119, 1237–1246.

Reintar, R.S., 2007. Molecular Characterization of Introgression from *Oryza Longistaminata* A.Chev.et.Roehr into Rice (*Oryza sativa L.*). M.Sc.thesis. University of the Philippines, Los Baños (UPLB), Laguna, Philippines.

Sakamoto, T., Matsuoka, M., 2008. Identifying and exploiting grain yield genes in rice. Curr. Opin. Plant Biol. 11, 209–214.

Sanchez, A.C., Brar, D.S., Huang, N., Li, Z., Khush, G.S., 2000. STS marker-assisted selection for three bacterial blight resistance genes in rice. Crop Sci. 40, 792–797.

Sasaki, T., 2011. Next generation sequencing: gold mine or tsunami for breeding science. Breed. Sci. 61, 1.

Sheehy, J.E., Mitchell, P.L., Hardy, B. (Eds.), 2000. Redesigning Rice Photosynthesis to Increase Yield. International Rice Research Institute, Manila, Philippines.

Sheehy, J.E., Mitchell, P.L., Hardy, B. (Eds.), 2007. Charting new Pathways to C4 Rice. International Rice Research Institute, Manila, Philippines.

Septiningsih, E.M., Prasetiyano, J., Lubis, E., Tai, T.H., Tjurbayat, T., Moeljopawiro, S., et al., 2003. Identification of quantitative trait loci for yield and yield components in an advanced back cross population derived from the *Oryza sativa* variety IR64 and the wild relative, *O. rufipogon*. Theor. Appl. Genet. 107, 1409–1432.

Singh, R.K., Flowers, T.J., 2010. The physiology and molecular biology of the effects of salinity on rice. In: Pessarakli, M. (Ed.), Third Edition of Handbook of Plant and Crop Stress. Taylor and Francis, Boca Raton, FL, pp. 901–942.

Singh, S., Sidhu, J.S., Huang, N., Vikal, Y., Li, Z., Brar, D.S., et al., 2001. Pyramiding three bacterial blight resistance genes (*xa5, xa13* and *Xa21*) using marker-assisted selection into indica rice cultivar PR106. Theor. Appl. Genet. 102, 1011–1015.

Small, I., 2007. RNAi for revealing and engineering plant gene function. Curr. Opin. Biotechnol. 18, 148–153.

Song, X.J., Huang, W., Shi, M., Zhu, M.Z., Lin, H.X., 2007. A QTL for rice grain width and weight encodes a previously unknown RING-type E3 ubiquitin ligase. Nat. Genet. 39, 623–630.

Swamy, B.P.M., Vikram, P., Dixit, S., Ahmed, H., Kumar, A., 2011. Meta-analysis of grain yield QTL identified during agricultural drought in grasses showed consensus. BMC Genomics 12, 319–336.

Takai, T., Fukuta, Y., Shiraiwa, T., Horie, T., 2005. Time-related mapping of quantitative trait loci controlling grain-filling in rice (*Ozyza sativa L.*). J. Exp. Bot. 56, 2107–2118.

Tanksley, S.D., Nelson, J.C., 1996. Advanced backcross QTL analysis: a method for the simultaneous discovery and transfer of valuable QTLs from unadapted germplasm into elite breeding lines. Theor. Appl. Genet. 92, 191–203.

Toriyama, K., Arimoto, Y., Uchimiya, H., Hinata, K., 1988. Transgenic rice plants after direct gene transfer into protoplasts. Nat. Biotechnol. 6, 1072–1074.

Vikram, P., Swamy, B.P., Dixit, S., Ahmed, H.U., Cruz, M.T.S., Singh, A.K., et al., 2011. *qDTy 1.1*, a major QTL for rice grain yield under reproductive-stage drought stress with a consistent effect in multiple elite genetic backgrounds. BMC Genet. 12, 89–103.

Xiao, J., Grandillo, S., Ahn, S.N., McCouch, S.R., Yuan, L.P., Tanksley, S.D., 1996. Genes from wild rice improve yield. Nature 384, 223–224.

Xiao, J., Li, J., Gradillo, S., Ahn, S.N., Yuan, L., Tanksley, S.D., et al., 1998. Identification of trait improving quantitative trait loci alleles from wild rice relative, *Oryza rufipogon*. Genetics 155, 899–909.

Xu, K., Xu, X., Fukao, T., Canlas, P., Rodriguez, R.M., Heuer, S., et al., 2006. *Sub1A* is an ethylene-responsive factor-like gene that confers submergence tolerance to rice. Nature 442, 705–708.

Xu, Y., Croch, J.H., 2008. Marker-assisted selection in plant breeding: from publication to practice. Crop Sci. 48, 391–407.

Ye, X., Al-Babili, S., Kloti, A., Zhang, J., Lucca, P., Beyer, P., et al., 2000. Engineering the Provitamin A (-B carotene) biosynthetic pathway into (carotenoid-free) rice endosperm. Science 287, 303–305.

Yoshida, S., 1972. Physiological aspects of grain yield. Ann. Rev. Plant Physiol. 23, 437–464.

Zhang, Q., 2009. Strategies for developing green super rice. Proc. Natl. Acad. Sci. USA 104, 16402–16409.

Biofortification of Staple Crops

Vishal Chugh and Harcharan S. Dhaliwal

Akal School of Biotechnology, Eternal University, Sirmour, Himachal Pradesh, India

1. INTRODUCTION

Micronutrient malnutrition, also called "hidden hunger", caused by the deficiency of micronutrients in the diet, afflicts more than two billion people worldwide, especially women and preschool age children (Welch and Graham 2002; White and Broadley, 2009) in the developing countries who are largely dependent on staple food crops. The consequences, in terms of malnutrition and health, are devastating and can result in blindness, stunting, diseases, and even death. The major reason for micronutrient deficiency in the populations of third world countries is the predominance of non-diversified cereal and plant-based diets, which are poor in micronutrients, as compared with the meat-rich diets of people in developed countries (FAO, 2004; Grotz and Geurinot, 2006; Gomez-Galera et al., 2010). Moreover, processes like polishing, milling, and pearling of cereals make them even poorer in micronutrients (Welch and Graham, 2004; Borg et al., 2009). Anti-nutritional factors such as phytic acid, fibers, and tannins further reduce the bioavailability of these minerals from dietary intakes by preventing their absorption in the intestine (White and Broadley, 2005; Brinch-Pederson et al., 2007; Pfeiffer and McClafferty, 2007).

1.1 Magnitude and Causes of Micronutrient Malnutrition

Over two billion people in the developing countries suffer from micronutrient malnutrition (Table 9.1). The costs of these deficiencies in terms of lives lost and poor quality of life are staggering. Iron and zinc deficiencies are the most common and widespread, afflicting more than half of the human population (World Health Organization, 2002; White and Broadley, 2009) among which developing countries of Asia and Africa are the most affected (Hotz & Brown, 2004; Gomez-Galera et al., 2010). More than two billion people are iron deficient, and the estimates of zinc deficiency are also near this proportion (ACC/SCN, 2000; Prasad, 2003; Welch and Graham 2002; Gibson, 2006;

Agricultural Sustainability. DOI: http://dx.doi.org/10.1016/B978-0-12-404560-6.00009-5

TABLE 9.1 Level and Effects of Micronutrient Malnutrition

Deficiency	Prevalence in Developing Countries	Groups Most Affected	Consequences
Iron	2 billion people	All, but especially women and children	Reduced cognitive ability; childbirth complications; reduced physical capacity and productivity
Zinc	May be as widespread as iron deficiency	Women and children	Illness from infectious diseases; poor child growth; pregnancy and childbirth complications; reduced birth weight
Vitamin A	250 million children	Children and pregnant women	Increased child and maternal mortality; blindness

Source: ACC/SCN (2000).

Thacher et al., 2006). Iron takes part in most of the redox reactions in the body and also acts as a cofactor in numerous vital enzymatic reactions (Kim and Geurinot, 2007). Likewise, zinc is an essential micronutrient for regulating gene expression and maintaining structural integrity of proteins. It acts as a cofactor in more than three hundred enzymatic reactions (King and Keen, 1999). Deficiencies of iron and zinc lead to poor growth and compromised psychomotor development of children, weakness, fatigue, irritability, hair loss, reduced immunity, wasting of muscles, sterility, morbidity, and even death in acute cases (Prasad et al., 1961; Pfeiffer and McClafferty 2007; Wintergerst et al., 2007; Stein, 2010).

Similarly, approximately three million preschool age children suffer with visible eye damage owing to vitamin A deficiency globally, and an estimated quarter-million to half-million preschool children go blind from this deficiency annually, and about 66% of them die within months (Mason et al., 2001). Even more importantly, the last two decades have brought an awareness that vitamin A is essential for immune functions.

Cereal grains have inherently low concentrations of micronutrients such as Zn and Fe, particularly when grown on micronutrient-deficient soil. In wheat and rice, only a small fraction of iron is transported to the grains from the senescing leaves whereas more than 70% of zinc is mobilized into grains (Grusak and DellaPenna, 1999; Grusak et al., 1999). In cereals these micronutrients are primarily stored in husks, the aleurone layers, and embryos which are lost during milling and polishing processes (Welch and Graham, 2004; Borg et al., 2009; Lombi et al., 2011). For better nutrition of human beings, Zn or Fe content of wheat grains should be improved to around 40–60 mg/kg, from the existing available amount of 10–30 mg/kg (Cakmak et al., 2000).

1.2 Strategies for Alleviating Micronutrient Malnutrition

To alleviate micronutrient malnutrition, a comprehensive strategy involving various interventions—such as supplementation, dietary diversification, fortification, and biofortification—adapted to conditions in different countries and regions is required (Zimmerman and Hurrel, 2007; Stein, 2010). Supplementation involving the oral delivery of micronutrients in the form of pills and syrups has been used in chronic deficiencies. Fortification is the addition of the desired minerals to food and stuffs like iodine in salts. Recurring expenditure and lack of a robust distribution system and careful implementation are some of the problems associated with these approaches. The most economical and feasible approach to alleviate hidden hunger is biofortification, a strategy for producing staple food cultivars whose edible portions have a higher concentration of bioavailable minerals and vitamins.

2. BIOFORTIFICATION: A NEW TOOL TO REDUCE MICRONUTRIENT MALNUTRITION

Biofortification refers to genetically increasing the bioavailable mineral content of food crops (Bouis, 2000; Brinch-Pederson et al., 2007). Developing biofortified crops also improves their growth and production in soils with depleted or unavailable minerals (Cakmak et al., 2008; Borg et al., 2009). Conventional and molecular breeding and genetic engineering techniques are the approaches that may be used to biofortify the crops with iron, zinc, and vitamin A (DellaPenna, 1999; Johns and Eyzaguirre, 2007; Pfeiffer and McClafferty, 2007; Tiwari et al., 2010).

This approach offers multiple advantages over other approaches. First, it capitalizes on the regular daily intake of a consistent and large amount of food staples by all family members. Second, after one-time investment to develop seeds that fortify themselves, recurrent costs are low, and the germplasm can be shared internationally and will be cost effective. The third advantage is that the biofortified crop system is highly sustainable. Fourth, biofortification provides a feasible means of reaching undernourished populations in remote rural areas, delivering naturally fortified foods to people with limited access to commercially marketed fortified foods. Fifth, breeding for higher trace mineral density in seeds will not incur a yield penalty. With mineral-dense seeds, more seedlings survive and initial growth is more rapid. Ultimately, yields are higher, particularly in trace mineral "deficient" soils of arid and calcareous regions (Monasterio et al., 2007).

HarvestPlus leads the global effort to make the familiar staple foods more nutritious and available to those suffering from hidden hunger. Founded in 2003 it is now a part of the CGIAR Research Program on Agriculture for Nutrition and Health. HarvestPlus uses biofortification to breed higher amounts of vitamins and minerals in staple foods, including bean, cassava, orange sweet potato, rice, maize, pearl millet, and wheat. Its activities are coordinated by the International Center for Tropical Agriculture (CIAT) and the International Food Policy Research Institute (IFPRI). The HarvestPlus biofortification program mainly focuses on three micronutrients—iron, zinc, and vitamin A—that are widely recognized by the World Health Organization (WHO) as limiting.

2.1 Conventional and Molecular Breeding Approaches for Biofortification

During the "green revolution", plant breeding efforts led to enhanced grain productivity of staple cereal crops. Such breeding efforts, along with improved agriculture technologies, succeeded in providing enough calories and protein to prevent the threatened massive starvation and famines predicted in the early 1960s in many regions. Breeding for micronutrient-enriched staple plant foods can be pursued (Bouis, 1996; Graham et al., 1998, 1999; Graham and Welch, 1996) for combating hidden hunger. The predominant cultivars of cereal crops have limited variability for grain iron and zinc concentrations/contents (Cakmak et al., 2000; Graham et al., 2001; Bouis, 2003; Rawat et al., 2009). A wide range of wheat germplasm has been screened at CIMMYT, Mexico for concentration of Fe and Zn in the whole grain over different environments. Graham et al., (1999) reported a range of $28.8–56.5\,\mu g\,g^{-1}$ (mean $= 37.2\,\mu g g^{-1}$) for Fe and $25.2–53.3\,\mu g\,g^{-1}$ for Zn (mean $= 35.0\,\mu g g^{-1}$) in the wheat grown in Mexico. Sufficient genetic variation exits within the wheat germplasm to substantially increase Fe and Zn concentrations in wheat grain. There was a high correlation between grain-Fe and grain-Zn concentrations in the wheat lines studied. These findings indicated that it should be possible to improve Fe and Zn levels in wheat grain simultaneously through plant breeding. Many studies exploring required variability for grain iron and zinc content in landraces, wild and related germplasm, suggested that wild relatives had three to four times higher grain iron and zinc content than the modern hexaploid wheat cultivars, and the sitopsis section of *Aegilops* species possesses the highest variability for grain iron and zinc content (Chhuneja et al., 2006; Rawat et al., 2009).

Under the biofortification program, synthetic wheat was developed by crossing tetraploid emmer wheat with *T. taushii* accessions with high iron and zinc content (Monasterio and Graham, 2000; Calderini and Monasterio, 2003). In an attempt to improve nutritional qualities in wheat, Uauy et al. (2006) cloned a high grain protein (*Gpc-B1*) locus using wild wheat *T. turgidum* ssp. *dicoccoides* which induced early senescence and increased sequestration of iron and zinc from leaves to grains. Exploiting useful variability from wild relatives is an uphill, difficult task due to sterility and reduced chromosome pairing in the interspecific hybrids. In the case of wheat, extensive back crossing is required to get BC_1 plants, which can be avoided by synthetic amphiploids (Rawat et al., 2009; Tiwari et al., 2009). It has been reported that chromosomes of group 2 and 7 of wild wheats and *Aegilops* species carry genes for high grain iron and zinc content. In a diploid wheat recombinant inbred lines (RIL) population, three quantitative trait loci (QTLs) have already been mapped for grain Fe and Zn content on chromosomes 2A and 7A (Tiwari et al., 2009, 2010). Peleg et al. (2009) also mapped three major QTLs for grain iron and zinc concentrations in a tetraploid wheat *T. durum–T. dicoccoides* RIL population, and the marker interval for a QTL common for zinc and iron on chromosome 7A was the same in both the above-cited studies. Shi et al. (2008) reported seven QTLs on chromosomes 1A, 2D, 3A, 4A, 4D, 5A, and 7A for zinc content in a double haploid wheat population of Hanxuan10 and Lumai14. Four zinc QTLs were also identified on chromosomes 3D, 4B, 6B, and 7A in a double haploid wheat population by Genc et al. (2009). Table 9.2 includes successful examples of mapping of some QTLs for high grain micronutrients in cereals.

TABLE 9.2 Tagging and Mapping of QTLs for High Grain Micronutrients Across Cereals

Cereal Crop	Micronutrient	Number of QTLs	Chromosome	Reference
Wheat	Zinc	11	4A, 4D, 2D and 3A	Shi et al (2008)
Wheat	Iron	2	2A and 7A	Tiwari et al (2009)
Wheat	Zinc	1	7A	Tiwari et al (2009)
Rice	Iron	3	2S, 8L, and 12L	Stangoulis et al (2007)
Rice	Zinc	2	1L and 2L	Stangoulis et al (2007)
Wheat	Iron	1	4	Peleg et al (2008)
Wheat	Iron and zinc	2	7	Peleg et al (2008)
Wheat	Zinc	4	3D, 4B, 6B, and 7A	Genc et al (2009)
Rice	Iron and zinc	14	1, 3, 5, 7 and 12	Anuradha et al (2012)
Wheat	Iron zinc and protein content	1	6B	Distelfeld et al (2007)
Wheat	Iron, zinc, copper, and manganese	1	5	Ozkan et al (2006)
Wheat	Iron and zinc	3	7A	Peleg et al (2009)

TABLE 9.3 The Mean and Range in Concentrations (Dry Weight Basis) of Fe and Zn in Six Sets of Brown Rice Germplasm (939 Genotypes) Grown under Similar Conditions at IRRI, Los Bãnos, Philippines

Genetic sets	Fe (μg g^{-1})		Zn (μg g^{-1})	
	Mean ± SE	Range	Mean ± SE	Range
Traditional and improved lines	13 ± 2.6	9.1–22.6	24.0 ± 4.7	13.5–41.6
IRRI breeding lines	10.7 ± 1.6	7.5–16.8	25.0 ± 7.6	15.9–58.4
Traditional and improved lines	12.9 ± 3.1	7.8–24.4	24.4 ± 4.7	16.5–37.7
Tropical japonicas	12.9 ± 1.5	8.7–16.5	26.3 ± 3.8	17.1–40.1
Popular lines and donors	13.0 ± 2.5	7.7–19.2	25.7 ± 4.6	15.3–37.3
Traditional and improved lines	13.8 ± 2.3	10.8–18.0	24.2 ± 4.1	19.9–33.3

SE, standard error of the mean.
Adapted from Graham et al. (1999).

Similarly for rice, researchers at IRRI have been evaluating the genetic variability of Fe concentration in rice grain since 1992. The range of Fe and Zn concentrations within the six sets of rice genotypes ($n = 939$) tested in a study was 7.5–24.4 μg g^{-1} for Fe, and 13.5–58.4 μg g^{-1} for Zn (Graham et al., 1999) (Table 9.3). Thus, within those genotypes tested, there was about a four-fold difference in Fe and Zn concentrations, suggesting some genetic potential to increase the concentrations of these micronutrients in rice grain.

Rice varieties Jalmagna, Zuchem, Xua Bue Nuo, Madhukar, IR64, and IR36 were found to have the highest iron and zinc contents. Moreover Jalmagna, which is a traditional variety of eastern India, shows twice the Fe and 40% more Zn content than IR36. Grain Fe and Zn content was found to be four times higher in some aromatic rice lines than in popular cultivars (Graham et al., 1999; Gregorio et al., 2000). These results indicate that there is significant genetic diversity in the rice genome also to allow substantial increase in Fe and Zn concentrations in rice grain. QTL mapping for high grain iron and zinc on various rice chromosomes is shown in Table 9.2.

Maize, which is a staple crop of southern and eastern Africa, is also low in Fe and Zn content (CIMMYT, 2000). However, genetic variability for Fe and Zn has been reported by Banziger and Long (2000) in white grained tropical maize germplasm: in the range 16.4–22.9 μg g^{-1} for Fe and 14.7–24.0 μg g^{-1} for Zn, respectively. They also evaluated 1814 accessions in 13 trials over 6 years and reported a range in grain of 9.6–63.2 mg-Fe/kg and 12.9–57.6 mg-Zn/kg. In many developing countries of Latin America, Africa, and Asia, maize is the major staple food and often the only source of protein. Screening for maize lines with better amino acid profile led to discovery of *opaque 2* mutant maize in Connecticut, USA (Mertz et al., 1964) with high lysine and tryptophan content. Through interdisciplinary research efforts, scientists at International Maize and Wheat Improvement Center (CIMMYT), Mexico developed quality protein maize (Vasal, 1999). Several QPM varieties and hybrids have been released in the areas where maize is a staple food, and studies have shown a beneficial impact on growth (height and weight) in children.

2.2 Genetic Engineering Approaches

Genetic engineering approaches have been successfully applied for biofortification of a few traits in cereals (Naqvi et al., 2009; Wirth et al., 2009; Masuda et al., 2012). This has been tried by introduction of genes that code for micronutrient-binding proteins, overexpression of storage proteins already present, and/or increased expression of proteins that are responsible for micronutrient uptake into plants (Lonnerdal, 2003). The bioavailability of the existing or biofortified micronutrients can be also enhanced through reduction of anti-nutritional factors or increasing the amount of enhancers in the plant products through genetic engineering of the requisite pathways or using the same during processing.

Two novel iron-binding proteins have been incorporated into rice (Suzuki et al., 2003; Nandi et al., 2002; Goto et al., 1999). Suzuki et al. (2003) reported very high expression level (5 g of lactoferrin/kg of grain; 6% of total protein) of human lactoferrin, the major iron-binding protein in breast milk, in rice. Each molecule of lactoferrin binds two atoms of ferric iron, and the recombinant lactoferrin was shown to be fully iron saturated. Similarly, Goto et al. (1999) reported the insertion of soybean ferritin gene in rice, and the iron content in a few transformants was found to be two- to three-fold higher than that of wild-type. Another research group, by using *Agrobacterium*-mediated transformation and the rice glutelin promoter, transformed the ferritin gene from *Phaseolus vulgaris* into rice (Lucca et al., 2001). They were able to double the iron content of dehusked rice. Recently, Wirth et al. (2009) reported two rice genes playing key roles in mobilization and storage of iron. One gene encodes nicotianamine synthase, the enzyme that produces nicotianamine. Nicotianamine chelates iron temporarily and facilitates its transport in the plant. The second gene encodes the protein ferritin, which comprises a sink for iron in the center of the endosperm, since the ferritin gene was expressed under the control of an endosperm-specific promoter. Masuda et al. (2012), by using three transgenic approaches involving iron storage protein ferritin with endosperm specific promoter, overproduction of metal chelator nicotianamine and an iron (II) nicotianamine transporter *OsYSL2* under sucrose transporter promoter, enhanced iron concentration to 4.4-fold higher in transgenic polished rice than the non-transgenic seeds.

Genetic engineering also offers an effective way to increase the vitamin content of staple crops. Naqvi et al. (2009) developed inbred South African transgenic maize plants in which the levels of three vitamins were increased specifically in the endosperm through the simultaneous modification of three separate metabolic pathways. The transgenic kernels contained 169-fold the normal amount of β-carotene, 6-fold the normal amount of ascorbate, and double the normal amount of folate. In the case of rice, the removal of aleurone layers during processing removes most of the oil-rich compounds, including provitamin A. The endosperm is devoid of vitamin A, thus people depending solely on rice as staple crop suffer from vitamin A deficiency. Burkhardt et al. (1997) first identified the C20 precursor molecule geranylgeranyl diphosphate in the wild type rice endosperm and transferred the phytoene synthase (*PSY*) gene from daffodils (*Narcissus pseudonarcissus*) into japonica rice variety Taipei 309. The transgenic phytoene synthase could condense geranylgeranyl diphosphate to phytoene, but carotenoid compounds could not be obtained. Ye et al. (2000) cloned the carotene desaturase gene (*CRT1*) from *Erwinia uredonora* along with the *PSY* gene. The *CRT1* gene catalyzed the conversion of phytoene to lycopene. Further cloning of lycopene β-cyclase (*β LCY*) DNA proved to be dispensable as rice transgenic plants with a two-gene (*PSY* and *CRT1*) and with a three-gene construct (*PSY*, *CRT1*, and *β LCY*) produced the same amount of carotenes. These lines became the prototype for lines for golden rice 1 production and accumulated up to $1.6 \mu g\,g^{-1}$ of carotenoids in the endosperm (Al-Babili and Beyer, 2005). Soon after Paine et al. (2005) could identify the rate-limiting nature of the phytoene synthase catalysis step by cloning *PSY* genes from maize. The carotenoid synthase could be increased to $37 \mu g\,g^{-1}$ when maize *PSY* genes were used. These experiments led to development of golden rice 2.

2.3   Physiological and Molecular Basis for Micronutrient Accumulation in Grains

For increased micronutrient metal uptake by roots, the levels of the available micronutrients in the root–soil interface must be increased to allow for more absorption by root cells. It could be achieved by stimulating certain root cell processes that alter micronutrient solubility and movement to root surfaces, such as the rate of root cell efflux of H and metal-chelating compounds and reductants. The Strategy-I dicotyledonous plants influx protons into the rhizosphere, followed by reduction of Fe^{3+} to usable Fe^{2+} form through the activity of ferric chelate reductases, such as FRO2. FRO2 is expressed specifically in roots and, along with other membrane-bound proteins including FRO6, FRO7, and FRO8b, takes part in iron transport in shoots (Mukherjee et al., 2006; Palmer and Geurinot, 2009). In the next step, specific iron-regulated transporter (IRT) proteins transport the readily soluble Fe^{2+} ions and Zn, Mn, and Cd into the plant roots (Rogers et al., 2000; Takahashi et al., 2011).

Members of the Poaceae family come under Strategy-II plants. Cultivation of rice, wheat, maize, and other cereals in conditions of limiting mineral availability (calcareous and saline sodic soil which contains iron in highly non-available form, i.e. Fe^{3+}) leads to lower iron uptake. Under these conditions, roots of various graminaceous plants secrete mugineic acids (MAs) called phytosiderophores (PS) into the rhizosphere, which chelate metals and are then absorbed by the plant as metal-chelator complexes (Mori, 1999; Takagi et al., 1998). Significant qualitative and quantitative differences for the production of mugineic acids have been observed among cereals. Among cereals, barley has been extensively explored for studying mineral uptake using phytosiderophores. Barley, as a Strategy-II plant, secretes mainly four types of mugineic acid family derivatives: mugineic acid (MA), 2'-deoxymugineic acid (DMA), 3-epihydroxymugineic acid (epi-HMA), and 3-epihydroxy-2-hydroxy mugineic acid (epi-HDMA) (Mori, 1999). The different forms of mugineic acid act as enhancers for the uptake of iron and zinc, and this provides tolerance to barley plants against low iron availability. Rice, wheat, and maize secrete only DMA in relatively low amounts, which makes them sensitive to low iron availability (Mori et al., 1990). Biosynthesis of MAs begins with S-adenosylmethionine, which is the precursor for the pathway. The biosynthetic pathway for mugineic acids has been studied extensively, and all the genes involved in this system have been cloned and characterized (Okumura et al., 1994; Takahashi et al., 1999; Bashir et al., 2006). Nozoye et al. (2011) reported that in barley and rice the efflux of deoxymugineic acid under stress conditions involves the *TOM1* and *HvTOM1* genes. In the next step the chelated metal-phytosiderophore complexes are taken up by an integral membrane protein of roots and shoots called Yellow Stripe 1 (YS1) transporter (Curie et al., 2001). Araki et al. (2011) reported HvYSL2, a metal–PS transporter that preferentially transports Fe (III)–PS. It plays a unique role in delivering essential metals in barley (Araki et al., 2011). Lee et al. (2012) reported a two-fold increase in iron content in seeds by over-expressing *OsNAS2* whereas Ishimaru et al., (2010) over-expressed *OsYSL2* using phloem specific *OsSUT1* promoter (phloem loading) and reported a four-fold increase in iron content of polished rice. These studies suggested that metal transport and sequestration from soil can be increased by coupling increased synthesis and release of phytosiderophores along with increased expression of genes encoding YSL proteins (Takahashi et al., 2001; Ishimaru et al., 2010; Lee et al., 2012). Screening of wild germplasm under iron sufficient and deficient conditions suggested 3–4 fold higher release of MA in some of the sitopsis sections of *Aegilops* species, including *Aegilops kotschyi* and *Ae. longissima* (Neelam et al., 2011). It would be interesting to see whether these wild sitopsis species are releasing only DMA or they are releasing all forms of MAs. Neelam et al. (2012) reported that group 2 wheat-*Aegilops* addition lines released higher MA in the rhizosphere under sufficient and deficient growth conditions which was positively correlated with high grain iron and zinc concentrations.

2.4   Sequestration of Mineral in Endosperm

Most of the mineral content present in the grains is confined to the aleurone layer and embryo, which is removed during milling and processing. A useful approach towards increasing the iron and zinc status of cereal-based diets would be to increase the sequestration of minerals to the endosperm, but little is known about transporters facilitating distribution of minerals across the grains (Mori, 1999; Curie et al., 2009; Palmer and Geurinot. 2009; Conn and Gilliham, 2010). Recently, with the development of techniques such as laser capture microdissection, it has been possible to study the expression of different genes across different tissues of grains (Borg et al., 2009; Tauris et al., 2009; Schiebold et al., 2011).

Transfer cells are the interface between the phloem of the maternal tissue and the endosperm tissue and have numerous wall ingrowths increasing the surface area by more than 20-fold. Highly expressed in transfer cells in barley were: heavy metal ATPases (HMA), zinc-regulated iron-regulated protein (ZIP), cation diffusion facilitator (CDF), natural resistance associated macrophage proteins (Nramp), vacuolar iron transporter 1 (VIT1), cation exchanger (CAX), yellow stripe like (YSL), metallothionins (MT), nicotinamine synthase (NAS), and nicotinamine amino transferase (NAAT) (Tauris et al., 2009). It has been proposed that Zn combines with NA or MA and then is taken up from the phloem and stored in vacuoles by CDF, VIT1, ZIP1, and CAX family transporters. YSL and ZIP transporters have been proposed to capture zinc–NA complexes from flowing back to the apoplast. The aleurone and embryo expression profiles were similar. HMA8, ZIP, and CDF were expressed in both the tissues at about the same level, although Nramp3, ZIP1, CAX1a, VIT1, NAS9, and NAATB were expressed at a higher rate in aleurone than in embryo. Nramp3 has been proposed to mediate efflux of Zn from the aleurone cells, while others control movement to the aleurone cells where the Zn is stored chiefly in vacuoles. The expression of transporter genes in the endosperm tissue was limited. Ramesh et al. (2004) over-expressed Zn transporter *AtZIP1* of *Arabidopsis* in barley under an ubiquitin promoter. The transgenic lines produced smaller seeds with high Zn concentration. Manipulating the expression of genes regulating CAX transporters has been proposed as an approach to increase Zn concentrations in the edible tissues of transgenic plants (Shigaki et al., 2005).

2.5   Bioavailability of Micronutrients

Bioavailability is the absorption and utilization of nutrients by human beings. Some common inhibitors in staple crops such as phytic acid, fibers, lignins, tannins, oxalic acid, and lectins (Graham et al., 2001) inhibit bioavailability. There are also some promoters, such as ascorbic acid, citric acid, fumaric acid, sulfur-containing amino acids, long-chain fatty acids, and selenium, which facilitate rapid uptake of micronutrients by the intestinal cells. Phytic acid is myo-inositol hexakisphosphate, the negative charges of which chelates divalent minerals such as Ca^{2+}, Mg^{2+}, Zn^{2+} and Fe^{2+} strongly. Phytate constitutes 1–3% of seed weight and accounts for 60–90% of total phosphorus in seeds (Graf, 1983). During germination, the enzyme phytase is activated to release phosphates from phytate. Bioavailability can be enhanced by lowering phytic acid in low-phytic-acid mutants of food crops and by transgenic expression of phytic acid-degrading enzyme, phytase, in the seeds.

2.5.1 Lowering Phytic Acid in Grains and Products

Phytic acid synthesis in plants starts from conversion of glucose 6-phosphate to inositol-3-phosphate by myo-inositol-3-Pi synthase (MIPS) followed by inositol phosphate kinases. Therefore mutations in genes encoding MIPS and inositol polyphosphate kinases are the targets for production of low-phytic-acid (*lpa*) mutants. The seeds of *lpa* mutants have been found to be viable and normal. Low-phytic-acid mutants are being developed either by chemical- or radiation-induced mutagenesis. These *lpa* mutants include *lpa1* mutant of maize (Raboy et al., 2000), barley (Larson et al., 1998; Rasmussen and Hatzack, 1998), and rice (Larson et al., 2000). Breeding programs for low phytic acid and high inorganic phosphate are associated with some undesirable characteristics of nutritional quality, disease susceptibility, and yield, due to inhibition of inositol metabolism in the vegetative tissue. Thus the selection of seed-specific *lpa* mutants which have normal vegetative phytic acid content can be more useful in the breeding programs. The maize *lpa3* mutant and the barley *lpa1* mutant are embryo and aleurone specific, respectively, without having any deleterious effect on the other characteristics, and have a larger amount of free phosphate. Antisense technology has been employed for deriving seed-specific *lpa* mutants. Shi et al. (2007) showed that maize *lpa1* mutants are defective in a multidrug-resistance-associated protein (MRP) ATP-binding cassette (ABC) transporter that is highly expressed in embryos, and within immature endosperm, germinating seed, and vegetative tissues. Silencing the gene in an embryo-specific manner in normal maize resulted in low phytic acid and high Pi (inorganic phosphate) content without any significant alteration in seed dry weight. In rice, attempts have been made to generate low-phytic-acid transgenics by silencing the rice Ins (3) P (1) synthase gene, *RINO1*. Kuwano et al. (2006), working in the University of Tokyo, for the first time could produce a transgenic rice line with low phytic acid and high inorganic phosphate level by silencing the *RINO1* gene under the control of the rice major storage protein glutelin GluB-1 promoter. In another study, Kuwano et al. (2009) reported production of transgenic rice by silencing the *RINO1* with 68% less seed phytic acid, having no negative effect on seed weight, germination, or plant growth.

Phytases, both microbial and plant sources, are the enzymes for sequential removal of phosphates from the phytic acid molecule. Action of phytase thus makes most of the micronutrients available for absorption by the intestinal cells.

Two categories of phytases have been identified by the International Union of Biochemistry: 3′ phytases that initiate the removal of phosphate from the 3′ position, and 6′ phytases that act on the 6′ position. The former type is predominant in microbes and the latter in plants. Many approaches have been employed for lowering levels of phytic acid in food. These methods include soaking, cooking, germination, and fermentation. Fermentation covers both microbial and enzymatic processing. These methods of food processing are not fully helpful, as phytate is not fully hydrolyzed by the phytases occurring naturally in plants and microorganisms. Addition of phytase preparations during food production is an alternative to optimize phytate dephosphorylation. The added phytases must be thermotolerant and show wide pH range variation in activity to be applied in food processing. Microbial phytases are highly thermophilic with a wide range of pH. *Aspergillus niger, A. fumigatus*, and *Rhizopus oligosporus* are the major source of industrial phytases (Greiner and Konietzny, 2006). A transgenic approach has been employed to transfer thermophilic microbial phytases to cereals to enhance bioavailability of micronutrients. Many model and crop plants, including tobacco, canola, soybean, wheat, rice alfalfa, *Arabidopsis*, maize, and others have been successfully transformed for microbial phytase production (Gontia et al., 2012). Lucca et al. (2001) achieved high expression of a heat stable fungal (*Aspergillus fumigatus*) phytase into rice, which was expected to remain active after food processing such as cooking; however, to the contrary, the *in-planta*-produced phytase showed decreased enzymatic activity during cooking. The *phy A* gene from *A. niger* has also been introduced into wheat by Brinch-Pederson et al. (2000, 2007). Holme et al. (2012) have reported cisgenic barley with improved phytase activity. Various approaches for higher expression of microbial phytases in transgenic plants, such as codon usage modifications, use of secretory signals, and glucosylation, and their localization to target regions, are being used (Gontia et al., 2012).

2.5.2   Animal Feed Trials and Cell Line Studies

The phytic acid in seeds chelates substantial amounts of phosphate, calcium, iron, and zinc, and hence the feeds for monogastric animals have to be fortified for production of healthy animals incurring higher costs and eutrophication. Ertl et al. (1998) first used maize *lpa1* mutant as a chick feed and found 46% and 49% greater blood phosphate and calcium level, respectively, than for chicks fed with wild maize. Transgenic plants over-expressing phytases have also been tried in various feed trials, showing enhanced absorption of various minerals by lowering the phytic acid content (Gontia et al., 2012). Han et al. (1994), using the Caco-2 cell assay, reported the inhibitory effect of phytic acid on iron and zinc absorption. A 50% increase in iron absorption by Caco-2 cells from *lpa1-1* maize was reported as compared with wild type maize (Raboy, 2007). Recently, Salunke et al. (2011, 2012), using Caco-2 cell lines, found negative correlation of iron and zinc uptake from high iron and zinc wheat derivatives with phytic acid content. Iron absorption was found to be 49% greater in men when fed with tortillas from *lpa 1-1* maize over the wild type maize (Mendoza et al., 1998). Adams et al. (2002) found 30% zinc absorption when healthy adults were fed with *lpa1-1* mutant as compared with 17% for wild type maize. These studies indicate that the lowering of phytic acid in the diet enhances micronutrient availability. Thus low-phytic-acid crops and phytase over-producing transgenic plants can be very helpful in combating micronutrient malnutrition in both traditional and biofortified crops, provided that the plant growth associated disadvantages can be addressed.

3. MICRONUTRIENT CONCENTRATION AND GRAIN YIELD

A number of studies have been carried out to find the effect of increased grain minerals content on grain yield (Graham et al., 1999). The wild relatives of wheat and land races have smaller seeds than modern wheat cultivars and higher concentration of minerals in grain. Modern elite wheat cultivars with larger seeds and higher starch content in grains have lower mineral concentration, which might be due to the dilution effect of a certain amount of micronutrients taken up for grain sequestration. The Zn concentration was found to be negatively correlated with grain yield in many field trials involving bread and durum wheat cultivars, whereas no significant correlation was observed between increased Fe concentration and grain yield (Oury et al., 2006; Zhao et al., 2009; Ficco et al., 2009). Welch and Graham (2004) reported no significant linkage between increased grain Fe and Zn concentration and grain yield. Many other reports have shown a consistently positive association between high grain protein content and Fe and Zn concentration (Distelfeld et al., 2007). Similarly, using two diverse populations of pearl millet, Gupta et al. (2009) reported no significant penalty of enhanced grain Fe and Zn concentration on grain yield and 1000 grain weight. In several studies among cereals the higher protein content has been reported to be negatively correlated with yield, whereas Regvar et al. (2011) found the protein globoids in the aleurone layer of wheat to be the preferential storage site for various micronutrients, suggesting that the higher micronutrient content may lead to yield penalty.

4. CONCLUSION

The global food system is failing to deliver adequate quantities of healthy, nutritionally balanced food, especially to resource-poor underprivileged people, leading to micronutrient malnutrition. The malnutrition of minerals (Fe, Zn) and vitamin A is a major food-related primary health problem among populations of the developing world where there is a heavy dependence on cereal-based diets and limited access to meat, fruits, and vegetables. Hence, in addition to the traditional objectives of disease resistance, yield, drought tolerance, etc., plant breeders have to exploit biofortification as a key objective for the enrichment of staple crops, as this approach offers the best possible and economical solution for eradication of micronutrient malnutrition or hidden hunger. Existing variability in staple crops' germplasm offers enormous possibilities of biofortification in these crops through conventional and molecular breeding approaches. Genetic engineering is also another obvious alternative to enhance the micronutrient levels in staple crop plants. Enhanced bioavailability of the micronutrients in traditional as well as biofortified crops is equally important. Even once biofortification has been shown to be efficacious, successful introduction of biofortified foods will still require the addressing of various socioeconomic and sociopolitical challenges, to popularize their cultivation by farmers and ultimately to gain consumer acceptance for biofortified crops, thereby increasing the intake of the target nutrients. A multi-tier coordinated strategy will play a pivotal role in overcoming hidden hunger.

REFERENCES

ACC/SCN (UN Administrative Committee on Coordination, Subcommittee on Nutrition), 2000. Fourth Report on the World Nutrition Situation. ACC/SCN in Collaboration with the International Food Policy Research Institute, Geneva.

Adams, C.L., Hambidge, M., Raboy, V., Dorsch, J.A., Sian, L., Westcott, J.I., et al., 2002. Zinc absorption from a low-phytic acid maize. Amer. J. Clin. Nutr. 76, 556–559.

Al-Babili, S., Beyer, P., 2005. Golden rice—five years on the road—five years to go. Trends Plant Sci. 10, 565–573.

Anuradha, K., Agarwal, S., Rao, V., Viraktamath, B.C., Sarla, N., 2012. Mapping QTLs and candidate genes for iron and zinc concentrations in unpolished rice of Madhukar × Swarna RILs. Gene Available at: dx.doi.org/10.1016/j.gene.2012.07.054.

Araki, R., Murata, J., Murata, Y., 2011. A novel barley yellow stripe 1-like transporter (HvYSL2) localized to the root endodermis transports metal–phytosiderophore complexes. Plant Cell Physiol. 52, 1931–1940.

Banziger, M., Long, J., 2000. The potential for increasing the iron and zinc density of maize through plant-breeding. Food Nutr. Bull. 21, 397–400.

Bashir, K., Inoue, H., Nagasaka, S., Takahashi, M., Nakanishi, M., Mori, S., et al., 2006. Cloning and characterization of deoxymugineic acid synthase genes from graminaceous plants. J. Biol. Chem. 281, 32395–32402.

Borg, S., Brinch-Pedersen, H., Tauris, B., Holm, P.B., 2009. Iron transport, deposition and bioavailability in the wheat and barley grain. Plant Soil 325, 15–24.

Bouis, H., 1996. Enrichment of food staples through plant breeding: a new strategy for fighting micronutrient malnutrition. Nutr. Rev. 54, 131–137.

Bouis, H.E., 2000. Special issue on improving human nutrition through agriculture. Food Nutr. Bull. 21, 351–576.

Bouis, H.E., 2003. Micronutrient fortification of plants through plant breeding: can it improve nutrition in man at low cost? Proc. Nutr. Soc. 62, 403–411.

Brinch-Pedersen, H., Olesen, A., Rasmussen, S.K., Holm, P.B., 2000. Generation of transgenic wheat (Triticum aestivum L.) for constitutive accumulation of an Aspergillus phytase. Mol. Breed. 6, 195–206.

Brinch-Pederson, H., Borg, S., Tauris, B., Holm, P.B., 2007. Molecular genetic approaches to increasing mineral availability and vitamin content of cereals. J. Cereal Sci. 46, 308–326.

Burkhardt, P.K., Beyer, P., Wunn, J., Kloti, A., Armstrong, G.A., Schledz, M., et al., 1997. Transgenic rice (*Oryza sativa*) endosperm expressing daffodil (*Narcissus psuedonarcissus*) phytoene synthase accumulates phytoene, a key intermediate of provitamin A biosynthesis. Plant J. 11, 1071–1078.

Cakmak, I., 2008. Enrichment of cereal grains with zinc: agronomic or genetic biofortification? Plant Soil 302, 1–17.

Cakmak, I., Ozkan, H., Braun, H.J., Welch, R.M., Romheld, V., 2000. Zinc and iron concentrations in seeds of wild, primitive, and modern wheats. Food Nutr. Bull. 21, 401–403.

Calderini, D.F., Monasterio, I., 2003. Are synthetic hexaploides a means of increasing grain element concentrations in wheat? Euphytica 134, 169–178.

Chhuneja, P., Dhaliwal, H.S., Bains, N.S., Singh, K., 2006. Aegilops kotschyi and Aegilops tauschii as sources for higher levels of grain iron and zinc. Plant Breed. 125, 529–531.

CIMMYT (2000). World Maize Facts and Trends: Meeting World Maize Needs: Technological Opportunities and Priorities. Pingali, P.L. (ed.). Mexico, D.F.: CIMMYT.

Conn, S., Gilliham, M., 2010. Comparative physiology of elemental distributions in plants. Annal. Bot. 105, 1081–1102.

Curie, C., Panaviene, Z., Loulergue, C., Dellaporta, S.L., Briat, J.F., Walker, E.L., 2001. Maize yellow stripe1 encodes a membrane protein directly involved in Fe (III) uptake. Nature 409, 346–349.

Curie, C., Cassin, G., Couch, D., Divol, F., Higuchi, K., Jean, M.L., et al., 2009. Metal movement within the plant: contribution of nicotianamine and yellow stripe 1-like transporters. Ann. Bot. 103, 1–11.

DellaPenna, D., 1999. Nutritional genomics: manipulating plant micronutrients to improve human health. Science 285, 375–379.

Distelfeld, A., Cakmak, I., Peleg, Z., Ozturk, L., Yazici, A.M., Budak, H., et al., 2007. Multiple QTL-effects of wheat Gpc-B1 locus on grain protein and micronutrient concentrations. Physiol. Plantarum 129, 635–643.

Ertl, D.S., Young, K.A., Raboy, V., 1998. Plant genetic approaches to phosphorus management in agricultural production. J. Environ. Qual. 27, 299–304.

FAO, (2004). Cereals and other starch-based staples: are consumption patterns changing? Joint meeting of the Intergovernmental Group on grains (30th session) and the Intergovernmental Group on rice (41st session) Rome, Italy, 10-11 February CCP:GR-RI/04/4-Sup.1.

Ficco, D.B.M., Riefolo, C., Nicastro, G., De Simone, V., Di Gesù, A.M., Beleggia, R., et al., 2009. Phytate and mineral elements concentration in a collection of Italian durum wheat cultivars. Field Crops Res. 111, 235–242.

Genc, Y., Verbyla, A.P., Torun, A.A., Cakmak, I., Willsmore, K., Wallwork, H., et al., 2009. Quantitative trait loci analysis of zinc efficiency and grain zinc concentration in wheat using whole genome average interval mapping. Plant Soil 314, 49–66.

Gibson, R.S., 2006. Zinc: the missing link in combating micronutrient malnutrition in developing countries. Proc. Nutr. Soc. 65, 51–60.

Gomez-Galera, S., Rojas, E., Sudhakar, D., Zhu, C., Pelacho, A.M., Capell, T., et al., 2010. Critical evaluation of strategies for mineral fortification of staple food crops. Transgenic Res. 19, 165–180.

Gontia, I., Tantwai, K., Rajput, L.P.S., Tiwari, S., 2012. Transgenic plants expressing phytase gene of microbial origin and their prospective application as feed. Food Technol. Biotechnol. 50, 3–10.

Goto, F., Yoshihara, T., Shigemoto, N., Toki, S., Takaiwa, F., 1999. Iron fortification of rice seed by the soybean ferritin gene. Nat. Biotechnol. 17, 282–286.

Graf, E., 1983. Applications of phytic acid. J. Am. Oil Chem. Soc. 60, 1861–1867.

Graham, R.D., Welch, R.M., 1996. Breeding for Staple-Food Crops with High Micronutrient Density. International Food Policy Research Institute, Washington, D.C, 72 pp.

Graham, R.D., Senadhira, D., Beebe, S.E., Iglesias, C., 1998. A strategy for breeding staple-food crops with high micronutrient density. Soil Sci. Plant Nutr. 43, 1153–1157.

Graham, R., Senadhira, D., Beebe, S., Iglesias, C., Monasterio, I., 1999. Breeding for micronutrient density in edible portions of staple food crops: conventional approaches. Field Crop. Res. 60, 57–80.

Graham, R.D., Welch, R.M., Bouis, H.E., 2001. Addressing micronutrient malnutrition through enhancing the nutritional quality of staple foods: Principles, perspectives and knowledge gaps. Adv. Agron. 70, 77–142.

Gregorio, G.B., Senadhira, D., Htut, T., Graham, R.D., 2000. Breeding for trace mineral density in rice. Food Nutr. Bull. 21, 382–386.

Greiner, R., Konietzny, U., 2006. Phytase for food application. Food Technol. Biotechnol. 44, 125–140.

Grotz, N., Guerinot, M.L., 2006. Molecular aspects of Cu, Fe and Zn homeostasis in plants. Biochim. Biophys. Acta. 1763, 595–608.

Grusak, M., DellaPenna, D., 1999. Improving the nutrient composition of plants to enhance human nutrition and health. Ann. Rev. Plant Physiol. Plant Molec. Biol. 50, 133–161.

Grusak, M.A., DellaPenna, D., Welch, R.M., 1999. Physiologic processes affecting the content and distribution of phytonutrients in plants. Nut. Rev. 57, 27–33.

Gupta, S.K., Velu, G., Rai, K.N., Sumalini, K., 2009. Association of grain iron and zinc content with grain yield and other traits in pearl millet (Pennisetum glaucum (L.) R.BR). Crop Improvement 36, 4–7.

Han, O., Failla, M.L., Hill, A.D., Morris, E.R., Smith, J.C., 1994. Inositol phosphates inhibit uptake and transport of iron and zinc by a human intestinal-cell line. J. Nutr. 124, 580–587.

Holme, I.B., Dionisio, G., Brinch-Pedersen, H., Wendt, T., Madsen, C.K., Vincze, E., et al., 2012. Cisgenic barley with improved phytase activity. Plant Biotechnol. J. 10, 237–247.

Hotz, C., Brown, K.H., 2004. Assessment of the risk of zinc deficiency in populations and options for its control. Food Nutr. Bull. 25, S91–S204.

Ishimaru, Y., Masuda, H., Bashir, K., Inoue, H., Tsukamoto, T., Takahashi, M., et al., 2010. Rice metal-nicotianamine transporter, OsYSL2, is required for the long-distance transport of iron and manganese. Plant J. 62, 379–390.

Johns, T., Eyzaguirre, P.B., 2007. Biofortification, biodiversity and diet: a search for complementary applications against poverty and malnutrition. Food Policy 32, 1–24.

Kim, S.A., Geurinot, M.L., 2007. Mining iron: iron uptake and transport in plants. FEBS Lett. 581, 2273–2280.

King, J.C., Keen, C.L., 1999. Zinc. In: Shills, M.E., Olsem, J.A.S., Shike, M.A., Ross, C. (Eds.), Modern Nutrition in Health and Disease, ninth ed. Williams and Wilkins, Baltimore, pp. 223–239.

Kuwano, M., Ohyama, A., Tanaka, Y., Fumio, M., Yoshida, T.T., 2006. Molecular breeding for transgenic rice with low-phytic-acidphenotype through manipulating myo-inositol 3-phosphate Synthase gene. Mol. Breed. 18, 263–272.

Kuwano, M., Mimura, T., Takaiwa, F., Yoshida, K.T., 2009. Generation of stable 'low phytic acid' transgenic rice through antisense repression of the 1D-myo-inositol 3-phosphate synthase gene (RINO1) using the 18-kDa oleosin promoter. Plant Biotechnol. J. 7, 96–105.

Larson, S.R., Young, K.A., Cook, A., Blake, T.K., Raboy, V., 1998. Linkage mapping: two mutations that reduce phytic acid content of barley grain. Theor. Appl. Genet. 97, 141–146.

Larson, S.R., Rutger, J.N., Young, K.A., Raboy, V., 2000. Isolation and genetic mapping of a nonlethal rice low phytic acid 1 mutation. Crop Sci. 40, 1397–1405.

Lee, S., Kim, Y.S., Jeon, U.S., Kim, Y.K., Schjoerring, J.K., An, G., 2012. Activation of rice nico-tinamine synthase 2 (OsNAS2) enhances iron availability for biofortification. Proc. Natl. Acad. Sci. U.S.A. 106, 22014–22019.

Lombi, E., Smith, E., Hanson, T.H., Paterson, D., de Jonge, M.D., Howard, D.L., et al., 2011. Megapixel imaging of (micro)nutrients in mature barley grains. J. Exp. Bot. 62, 273–282.

Lonnerdal, B., 2003. Genetically modified plants for improved trace element nutrition. J. Nutr. 133, 1490S–1493S.

Lucca, P., Hurrell, R., Potrykus, I., 2001. Genetic engineering approaches to improve the bioavail-ability and the level of iron in rice grains. Theor. Appl. Genet. 102, 392–397.

Mason, J.B., Lotfi, M., Dalmiya, N., Sethuraman, K., Deitchler, M., 2001. The micronutri-ent Report: Current progress and Trends in the Control of Vitamin A, Iodine, and Iron Deficiencies. The Micronutrient Initiative, Ottawa, Canada.

Masuda, H., Ishimaru, Y., Aung, M.S., Kobayashi, T., Kakei, Y., Takahashi, M., et al., 2012. Iron biofortification in rice by the introduction of multiple genes involved in iron nutrition. Sci. Rep. 2, 543. doi: 10.1038/srep00543.

Mendoza, C., Viteri, F.E., Lonnerdal, B., Young, K.A., Raboy, V., Brown, K.H., 1998. Effect of genetically modified, low-phytic acid maize on absorption of iron from tortillas. Am. J. Clin. Nutr. 68, 1123–1127.

Mertz, E.T., Bates, L.S., Nelson, O.E., 1964. Mutant gene that changes protein composition and increases lysine content of maize endosperm. Science 145, 279.

Monasterio, I., Graham, R.D., 2000. Breeding for trace minerals in wheat. Food Nutr. Bull. 21, 392–396.

Monasterio, I., Palacios-Rojas, N., Meng, E., Pixley, K., Trethowan, R., Pena, R.J., 2007. Enhancing the mineral and vitamin content of wheat and maize through plant breeding. J. Cereal Sci. 46, 293–307.

Mori, S., 1999. Iron acquisition by plants. Curr. Opin. Plant Biol. 2, 250–253.

Mori, S., Nishizawa, N., Fujizaki, J., 1990. Identification of Rye chromosome 5R as a carrier of genes for mugineic acid synthetase and 3-hydroxy mugineic acid synthetase using wheat-rye addition lines. Jpn. J. Genet. 65, 343–352.

Mukherjee, I., Campbell, N.H., Ash, J.S., Connolly, E.L., 2006. Expression profiling of the Arabidopsis ferric chelate reductase (FRO) gene family reveals differential regulation by iron and copper. Planta 223, 1178–1190.

Nandi, S., Suzuki, Y., Huang, J., Yalda, D., Pham, P., Wu, L., et al., 2002. Expression of human lac-toferrin in transgenic rice grains for the application in infant formula. Plant Sci. 163, 713–722.

Naqvi, S., Zhu, C., Farre, G., Ramessar, K., Bassie, L., Breitenbach, J., et al., 2009. Transgenic multivitamin corn through biofortification of endosperm with three vitamins representing three distinct metabolic pathways. Proc. Nat. Acad. Sci. U.S.A. 106, 7762–7767.

Neelam, K., Rawat, N., Tiwari, V.K., Malik, S., Tripathi, S.K., Randhawa, G.S., et al., 2011. Molecular and cytological characterization of high grain iron and zinc Wheat- Aegilops pereg-rina derivatives. Mol. Breed. 28, 623–634.

Neelam, K., Rawat, N., Tiwari, V.K., Tripathi, S.K., Randhawa, G.S., Dhaliwal, H.S., 2012. Evaluation and identification of wheat- Aegilops addition lines controlling high grain iron and zinc content and mugineic acids production. Cer. Res. Communic. 40, 53–61.

Nozoye, T., Nagasaka, S., Kobayashi, T., Takahashi, S., Sato, Y., Uozumi, N., et al., 2011. Phytosiderophore efflux transporters are crucial for iron acquisition in graminaceous plants. J. Biochem. 286, 5446–5454.

Okumura, N., Nishizawa, N.K., Umehara, Y., Ohata, T., Nakanishi, H., Yamaguchi, T., et al., 1994. A dioxygenase gene (Ids2) expressed under iron deficiency conditions in the roots of Hordeum vulgare. Plant Mol Biol. 25, 705–719.

Oury, F.X., Leenhardt, F., Rémésy, C., Chanliaud, E., Dupperrier, B., Balfourier, F., et al., 2006. Genetic variability and stability of grain magnesium, zinc and iron concentrations in bread wheat. Eur. J. Agron. 25, 177–185.

Ozkan, H., Brandolini, A., Torun, A., Altintas, S., Eker, S., Kilian, B., et al. (2006). Natural variation and identification of microelements content in seeds of Einkorn Wheat (*Triticum monococcum*). In Proceedings of the Seventh International Wheat Conference. 27 November–2 December 2005. Mar del Plata, Argentina. pp. 455–462.

Paine, J.A., Shipton, C.A., Chaggar, S., Howells, R.M., Kennedy, M.J., Vermon, G., et al., 2005. Improving the nutritional value of Golden Rice through increased pro-vitamin A content. Nat. Biotechnol. 23, 482–487.

Palmer, C.M., Guerinot, M.L., 2009. Facing the challenges of Cu, Fe and Zn homeostasis in plants. Nat. Chem. Biol. 5, 333–340.

Peleg, Z., Saranga, Y., Yazici, A., Fahima, T., Ozturk, L., Cakmak, I., 2008. Grain zinc, iron and protein concentrations and zinc-efficiency in wild emmer wheat under contrasting irrigation regimes. Plant Soil 306, 57–67.

Peleg, Z., Cakmak, I., Ozturk, L., Yazici, A., Jun, Y., Budak, H., et al., 2009. Quantitative trait loci conferring grain mineral nutrient concentrations in durum wheat x wild emmer wheat RIL population. *Theor. Appl. Genet.* 119, 353–369.

Pfeiffer, W.H., McClafferty, B., 2007. HarvestPlus: breeding crops for better nutrition. Crop Sci. 47, S88–S105.

Prasad, A.S., 2003. Zinc deficiency has been known of for 40 years but ignored by global health organisations. Brit. Medic. J. 326, 409–410.

Prasad, A.S., Halsted, J.A., Nadimi, M., 1961. Syndrome of iron deficiency anemia, hepatosplenomegaly, hypogonadism, dwarfism and geophagia. Am. J. Med. 31, 532–546.

Raboy, V., 2007. Seed phosphorus and the development of low-phytate crops. In: Turner, B.L. (Ed.), Inositol Phosphates Linking Agriculture and the Environment CABI, Oxfordshire, UK, pp. 111–132.

Raboy, V.P., Gerbasi, K.A., Stoneberg, Y.S., Pickett, S.G., Bauman, A.T., Murthy, P.P.N., et al., 2000. Origin and seed phenotype of maize low phytic acid 1-1 and low phytic acid 2-1. Plant Physiol. 124, 355–368.

Ramesh, S.A., Choimes, S., Schachtman, D.P., 2004. Over-expression of an *Arabidopsis* zinc transporter in *Hordeum vulgare* increases short-term zinc uptake after zinc deprivation and seed zinc content. Plant Mol. Biol. 54, 373–385.

Rasmussen, S.K., Hatzack, F., 1998. Identification of two low-phytate barley (*Hordeum vulgare* L.) grain mutants by TLC and genetic analysis. Hereditas 129, 107–112.

Rawat, N., Tiwari, V.K., Singh, N., Randhawa, G.S., Singh, K., Chhuneja, P., et al., 2009. Evaluation and utilization of *Aegilops* and wild *Triticum* species for enhancing iron and zinc content in wheat. Genet. Resour. Crop. Evol. 56, 53–64.

Regvar, M., Eichert, D., Kaulich, B., Gianoncelli, A., Pongrac, P., Vogel-Mikus, K., et al., 2011. New insights into globoids of protein storage vacuoles in wheat aleurone using synchrotron soft X-ray microscopy. J. Exp. Bot. 60, 3929–3939.

Rogers, E.E., Eide, D.J., Geurinot, M.L., 2000. Altered selectivity in an Arabidopsis metal transporter. *Proc. Nat. Acad. Sci.* U.S.A. 97, 12356–12360.

Salunke, R., Neelam, K., Rawat, N., Tiwari, V.K., Randhawa, G.S., Dhaliwal, H.S., et al., 2011. Bioavailability of iron from wheat aegilops derivatives selected for high grain iron and protein contents. J. Agric. Food Chem. 59, 7465–7473.

Salunke, R., Neelam, K., Rawat, N., Tiwari, V.K., Randhawa, G.S., Dhaliwal, H.S., et al., 2012. Determination of bioavailable-zinc from biofortified wheat using a coupled in vitro digestion/Caco-2 reporter-gene based assay. J. Food Comp. Anal. 25, 149–159.

Schiebold, S., Tschiersch, H., Borisjuk, L., Heinzel, N., Radchuk, R., Rolletschek, H., 2011. A novel procedure for the quantitative analysis of metabolites, storage products and transcripts of laser microdissected seed tissues of *Brassica napus*. Plant Methods 7, 19.

Shi, J., Wang, H., Schellin, K., Li, B., Faller, M., Stoop, J.M., et al., 2007. Embryo-specific silencing of a transporter reduces phytic acid content of maize and soybean seeds. Nat. Biotechnol. 25, 930–937.

Shi, R., Li, H., Tong, Y., Jing, R., Zhang, F., Zou, C., 2008. Identification of quantitative trait locus of zinc and phosphorus density in wheat (*Triticum aestivum* L.) grain. Plant Soil 306, 95–104.

Shigaki, T., Barkla, B.J., Miranda-Vergara, M.C., Zhao, J., Pantoja, O., Hirschi, K.D., 2005. Identification of a crucial histidine involved in metal transport activity in the *Arabidopsis* cation/H+ exchanger CAX1. J. Biol. Chem. 280, 30136–30142.

Stangoulis, J.C.R., Huynh, B., Welch, R.M., Choi, E., Graham, R.D., 2007. Quantitative trait loci for phytate in rice grain and their relationship with grain micronutrient content. Euphytica 154, 289–294.

Stein, A.J., 2010. Global impacts of human malnutrition. Plant Soil. 335, 133–154.

Suzuki, Y.A., Kelleher, S.L., Yalda, D., Wu, L., Huang, J., Huang, N., et al., 2003. Expression, characterization and biological activity of recombinant human lactoferrin in rice. J. Pediatr. Gastroenterol. Nutr. 36, 190–199.

Takagi, S., Kamei, S., Yu, M.H., 1998. Efficiency of iron extraction by mugeneic acid family phytosiderophores. J. Plant Nutr. 11, 643–650.

Takahashi, M., Yamaguchi, H., Nakanishi, H., Shioiri, T., Nishizawa, N.K., Mori, S., 1999. Cloning two genes for nicotinamine aminotransferase, a critical enzyme in iron acquisition (strategy II) in graminaceous plants. Plant Physiol. 121, 947–956.

Takahashi, M., Nakanishi, H., Kawasaki, S., Nishizawa, N.K., Mori, S., 2001. Enhanced tolerance of rice to low iron availability in alkaline soils using barley nicotianamine aminotransferase genes. Nat. Biotechnol. 19, 466–469.

Takahashi, R., Ishimaru, Y., Senoura, T., Shimo, H., Ishikawa, S., Arao, T., et al., 2011. The OsNRAMP1 iron transporter is involved in Cd accumulation in rice. J. Exp. Bot. 62, 4843–4850.

Tauris, B., Borg, S., Gregersen, P.L., Holme, P.B., 2009. A roadmap for zinc trafficking in the developing barley grain based on laser capture microdissection and gene expression profiling. J. Exp. Bot. 60, 1333–1347.

Thacher, T.D., Fischer, P.R., Strand, M.A., Pettifor, J.M., 2006. Nutritional rickets around the world: causes and future directions. Ann. Trop. Paediatr. 26, 1–16.

Tiwari, V.K., Rawat, N., Chhuneja, P., Neelam, K., Aggarwal, R., Randhawa, G.S., et al., 2009. Mapping of quantitative trait loci for grain iron and zinc concentration in diploid A genome wheat. J. Hered. 100, 771–776.

Tiwari, V.K., Rawat, N., Neelam, K., Kumar, S., Randhawa, G.S., Dhaliwal, H.S., 2010. Substitution of 2S and 7U chromosomes of *Aegilops kotschyi* in wheat enhances grain iron and zinc concentration. Theor. Appl. Genet. 121, 259–269.

Uauy, C., Distelfeld, A., Fahima, T., Blech, A., Dubcovsky, J., 2006. A NAC gene regulating senescence improves grain protein, zinc and iron content in wheat. Science 314, 1298–1301.

Vasal, S.K. (1999). Quality Protein Maize Story. In Improving Human Nutrition through Agriculture: The Role of International Agricultural Research. A Workshop hosted by the International Rice Research Institute, Los Baños, Philippines organized by the International Food Policy Research Institute. October 5-7.

Welch, R.M., Graham, R.D., 2002. Breeding crops for enhanced micronutrient content. Plant Soil 245, 205–214.

Welch, R.M., Graham, R.D., 2004. Breeding for micronutrients in staple food crops from a human nutrition perspective. J. Exp. Bot. 55, 353–364.

White, P.J., Broadley, M.R., 2005. Biofortifying crops with essential mineral elements. Trends Plant Sci. 10, 586–593.

White, P.J., Broadley, M.R., 2009. Biofortification of crops with seven mineral elements often lacking in human diets – iron, zinc, copper, calcium, magnesium, selenium and iodine. New Phytol. 182, 49–84.

Wintergerst, E.S., Maggini, S., Hornig, D.H., 2007. Contribution of selected vitamins and trace elements to immune function. Ann. Nutr. Metab. 51, 301–323.

Wirth, J., Poletti, S., Aeschlimann, B., Yakandawala, N., Drosse, B., Osorio, S., et al., 2009. Rice endosperm iron biofortification by targeted and synergistic action of nicotianamine synthase and ferritin. Plant Biotechnol. J. 7, 631–644.

World Health Organization (2002). World health report, 2002. Available at: <http://www.who.int/whr/2002/>.

Ye, X., Al-Babili, S., Klöti, A., Zhang, J., Lucca, P., Beyer, P., et al., 2000. Engineering the provitaminA (β-carotene) biosynthetic pathway into (carotenoid-free) rice endosperm. Science 287, 303–305.

Zhao, F.J., Su, Y.H., Dunham, S.J., Rakszegi, M., Bedo, Z., McGrath, S.P., et al., 2009. Variation in mineral micronutrient concentrations in grain of wheat lines of diverse origin. J. Cereal Sci. 49, 290–295.

Zimmerman, M.B., Hurrel, R.F., 2007. Nutritional iron deficiency. Lancet 370, 511–519.

Chapter 10

Nutrient-focused Processing of Rice

Nadina Müller-Fischer

Bühler, Uzwil, Switzerland

1. INTRODUCTION

Rice, the staple food of Asia, provides on average 30% of the daily calorie intake across the continent. It accounts for about a quarter of the world's cereal production volume per annum—672 Mtonne paddy rice (FAOSTAT 2010; Table 10.1). Rice production ranks second to maize (844 Mtonne) while being comparable to that of wheat (651 Mtonne). Importantly, and in contrast to these two crops, the vast majority of rice goes to human consumption (>80%), with little used as seed or for animal feed.

The Asian subcontinent accounts for ≥90% of the world's rice production and consumption (Figure 10.1), China and India alone for 50%. The gross domestic product (GDP) growth and population magnitude of these two countries alone challenge the capability to supply. Greater than 50% of the daily caloric intake in Indonesia, Vietnam, Bangladesh, Cambodia, Laos, and Myanmar comes from rice (International Rice Research Institute, 1999). The International Monetary Fund predicts GDP growth rates of >5% (2010–2015) for these countries and increasing population.

The importance of rice in the Asian region as a staple food, energy and nutrient provider of economic and political leverage will increase in the coming years. Agricultural practices will be modernized and paddy farm capacities increased. However, supply will only match demand with improvements in logistics management and processing efficiencies to eliminate waste and spoilage, increase yield, and decrease processing costs per kg. A sustainable value chain requires more than capacity and efficiency improvement, as it also needs to deliver the highest quality both in sensorial pleasure and in food safety.

Agricultural Sustainability. DOI: http://dx.doi.org/10.1016/B978-0-12-404560-6.00010-1

TABLE 10.1 Production of Different Cereals Worldwide in 2010

Cereal	Cereal Crop Production 2010	
	(Mtonne)	(% Total Production)
Maize	844	34
Paddy rice	672	28
Wheat	651	27
Barley	124	5
Sorghum	56	2
Millet	29	1
Oats	20	1
Other cereals	37	2

Source: FAOSTAT.

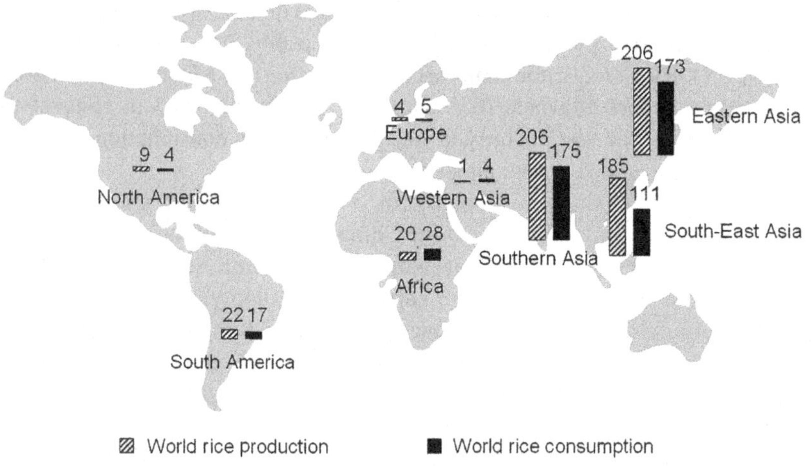

☒ World rice production   ■ World rice consumption

FIGURE 10.1 Rice production and consumption in different regions of the world in 2007, shown as Mtonne paddy rice equivalent. *Source: FAOSTAT.*

2. NUTRIENT COMPOSITION OF RICE FRACTIONS

Rice is predominantly consumed in dehulled and milled form, i.e., as white head rice. Figure 10.2 shows the typical microstructures of a paddy rice kernel (left) as well as a detail view of the endosperm, bran layers, and hull. Compared

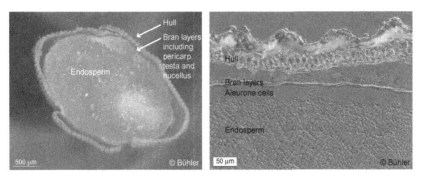

FIGURE 10.2   Macroscope image of rough rice kernel cut transversely and photographed with a Leica DC500 on a Photomakroskop Wild M400 in incident light (left). Differential Interference Contrast Microscopy of typical structures in rice hull, bran layers, and endosperm including the aleurone layer (right).

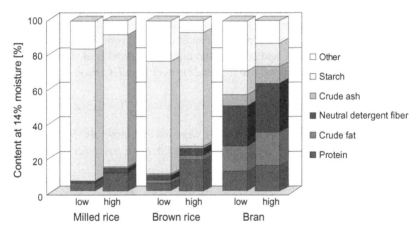

FIGURE 10.3   Low and high values of macronutrients found in different edible fractions of the rice kernel. *Data source: Champagne et al. (2004).*

with milled rice, brown rice contains 2–10 times more minerals, 2–3 times more fiber, and 5 times more lipids (Champagne et al., 2004).

2.1   Macronutrients

The rice hull represents about 20% of the rough rice grain and is not edible. Its major components are non-starch carbohydrates, and it contains about 20% of silica, 9–20% lignin as well as 2–6% cutin. The main constituent of white rice is starch (~78%, see Figure 10.3), followed by protein (5–11%). Fat, fiber, and ash content are low at 0.3–0.5%, 0.7–2.3%, and 0.3–0.8%, respectively. In contrast, bran as one of the co-products of milling is nutritionally valuable and contains significant amounts of fat (15–20%), neutral detergent fiber (24–29%), and ash (7–10%), as well as slightly elevated amounts of protein (11–15%). The starch content of rice bran is about 14% only, with some variation depending on the degree of milling (Champagne et al., 2004). Fat from rice bran is nutritionally valuable since it contains high fractions of mono- and polyunsaturated fatty acids (38.4% oleic acid C18:1, 34.4% linoleic acid C18:2, 2.2% α-linolenic acid C18:3) and only about 25% saturated fatty acids consisting of myristic acid C14:0, palmitic acid C16:0, and stearic acid C18:0 (Orthoefer and Eastman, 2004). γ-oryzanol, a unique mixture of triterpene alcohols and sterol ferulates present in rice bran oil, has been reported to have hypocholesterolemic activity in various animal and human studies (Patel and Naik, 2004).

Rice is an important source of protein in Asia, especially in tropical Asia, where it accounts for 35–40% of the dietary protein (Juliano, 1993). The protein content in rice is relatively low, but its quality and bioavailability are nevertheless good (Bhattacharya, 2011). However, similar to other cereals, rice protein is deficient in the essential amino acid lysine but contains the essential sulfur-containing amino acids cysteine and methionine in abundance. A diet combining rice and legumes balances out amino acid composition quite well (Rand et al. 1984, Eggum et al., 1987) because pulses are rich in lysine but poor in sulfur amino acids.

2.2   Micronutrients

Milled rice contains little in the way of vitamins and minerals (Figure 10.4). Rice bran on the other hand contains considerable amounts of thiamin

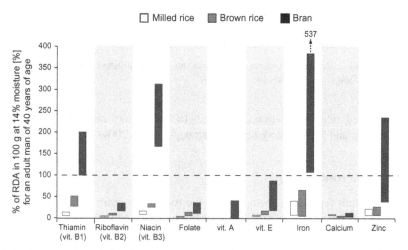

FIGURE 10.4   Micronutrient content of different edible rice fractions. Bars represent range in contents. *Data sources: for content in vitamins and minerals, Champagne et al. (2004); for recommended dietary allowances (RDA), USDA (2010).*

(vit. B1), niacin (vit. B3), iron and zinc, with values in 100 g rice bran clearly exceeding the recommended daily allowances. In contrast, rice contains moderate amounts of vitamin E, very little vitamin A, folate, riboflavin and calcium and nearly no vitamin C and iodine (values not shown). Phosphorus content is substantial (1.7–4.3 mg/g in brown rice at 14% moisture), with half of it in the form of phytic acid which is only partially bioavailable and acts as a potent inhibitor of iron, zinc, and calcium absorption (Hurrell et al., 1992).

3. HEALTH PROBLEMS IN RICE CORE REGIONS

According to Juliano (1993), the most important nutritional problem in rice-consuming countries is an inadequate and unbalanced dietary intake of nutrients. Protein-energy malnutrition and unsatisfactory levels of fat consumption are the nutritional challenges on the macronutrient side. With respect to micronutrient status, nutritional anemia is common, particularly from iron deficiency. The prevalence of iron deficiency is estimated to be about 30% of the world's population (WHO, 1992). This makes iron the most widespread nutrient deficiency worldwide by far. Poor pregnancy outcomes, including increased mortality of mother and children, reduced psychomotor and mental development in infants, decreased immune function, tiredness, and poor work performance can be the consequences (Cook et al., 1994). Absorption of iron depends on its source and is worse for non-heme iron from plant sources than for heme iron contained in meat. In developing countries, iron is mostly consumed in the form of non-heme iron from grain and legume staples in which phytic acid acts as a potent inhibitor of absorption (Lucca et al., 2001). Additional nutritional concern stems from vitamin A and iodine deficiency disorders, the latter due to naturally low iodine levels in soil in some regions or leaching of iodine by rainwater and floods (Dexter, 1998). Deficiencies in thiamine, riboflavin, calcium, vitamin C, and zinc are also prevalent.

Not all of the nutritional concerns are directly caused by the consumption of rice *per se* but reflect an overall impact of multiple causative factors similar to those in other developing countries where rice is not a major staple. Beriberi, however, caused by thiamine deficiency, is a characteristic disease of communities that consume polished rice as a staple. As shown in Figure 10.4, milling leads to a drastic loss in thiamine. Further losses occur during washing and cooking. Akroyd et al. (1940) first showed that brown rice contains sufficient thiamine (3–4 µg/g) whereas machine-milled raw rice contains little (0.5–1.3 µg/g). The same authors showed that the thiamine content of milled rice is significantly higher after parboiling (2.0–2.5 µg/g). These observations were corroborated by the relatively lower prevalence of beriberi in populations in India consuming parboiled rice (Kik and Williams, 1945).

4. RICE PROCESSING

Rice milling differs from wheat and maize milling in so far as the intactness of the rice kernel after milling is of utmost importance. The intact white rice kernel, the so-called head rice, has a much higher monetary value than broken kernels. Depending on taste, storage, cooking, way of eating, and nutritional preferences, rice is milled fresh, goes through an aging phase, is steamed, or parboiled before milling.

The milling result is strongly affected by pre-existing natural factors like tightness of husk interlocking or imperfect grains. Indudhara Swamy and Bhattacharya (1984) showed that whatever the grain type and quality and whatever the milling conditions, the total grain breakage almost never exceeded the total count of imperfect grains. The entire processing chain, including pre-milling steps, is of utmost importance with respect to end product quality as well. This can be exemplarily shown by the fact that head rice yield can be significantly improved by proper grading of grains into groups of uniform length and thickness (Sun and Siebenmorgen, 1993).

The harshness of rice milling itself determines whether fissured kernels stay intact or break apart. Milling can follow various flow paths for different reasons, of which one is rice variety. The flow path described in this section is usually applied to non-sticky, long grain *Indica* rice varieties cultivated throughout tropical Asia. Figure 10.5 shows an overview of the typical processing steps for *Indica* rice from harvest to white rice. Rice is dried, stored, optionally parboiled or steamed (not included in the figure), cleaned using sieves and classifiers, hulled, whitened (4 passes), polished (1–3 passes), sorted and graded. If the dried paddy rice is taken as the reference input material (100%), an average white head rice yield of 49% can be achieved. Edible products are rice bran (~10%) and brokens of different kernel length (~15%). Head-rice yield is in general significantly higher for parboiled rice and can be close to 64% if pre-processing and drying after parboiling are properly done (Bhattacharya, 1969).

In contrast to the flow path shown in Figure 10.5, *Japonica* rice usually passes through huller, abrasive mill (1 pass), degerminator (2–3 passes), and polisher (1–3 passes). *Japonica* varieties are usually short grain with a sticky cooked texture. It is the type of rice cultivated in temperate East Asia, in upland areas of Southeast Asia, and at high elevations in South Asia. Degerminators are necessary during milling of *Japonica* rice since the germ is more deeply embedded in the kernel than in *Indica* varieties. Vertical friction machines, in which the rice moves upwards, are usually applied as degerminators. Sometimes horizontal friction machines are used instead.

The processing steps drying, parboiling, hulling, whitening, polishing, sorting, and grading are described in more detail in the following sections.

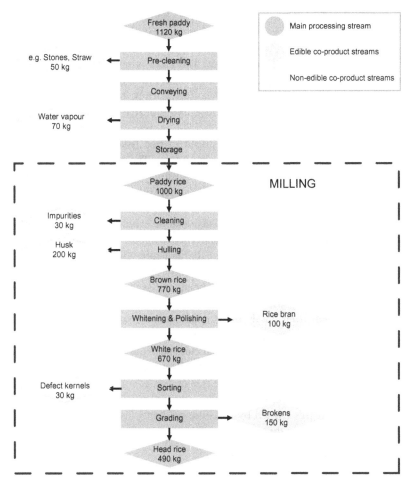

FIGURE 10.5 Mass stream of rice during processing from field to freshly milled *Indica* rice, including main product stream, edible co-products, and inedible co-products. Numbers are valid for raw rice.

4.1 Drying

Paddy rice is typically harvested with 16–24% (wb) moisture, depending on the season and on weather conditions before and during harvest. This moisture content is well above critical limits with respect to spoilage, i.e., enzymes, molds, yeasts, and even some bacteria are active (Labuza et al., 1970). Rice is, thus, dried to a moisture content of about 12% (wb) which is safe for long-term storage. Swedish scientists (Gustavsson et al., 2011) found that, in countries such as those in south and southeast Asia, food loss mostly occurs in the early and middle steps of the food supply chain, namely during agricultural production and post-harvest handling, where in each step up to 6% losses occur. Since rice kernels are predominantly eaten as intact white kernels, any breakage during milling is undesirable. The grain is not easily susceptible to structural damage by thermal stress, but is susceptible to moisture stress upon hydration or dehydration (Bhattacharya, 2011). Thus, besides having a major impact on spoilage-related losses, correct drying of rice is an essential step for ensuring milling quality.

Improper drying and cooling may lead to rice fissuring which promotes breakage during milling (Bhattacharya, 2011). The critical moisture content for fissuring is related to the glass transition temperature, the threshold between a rubbery and glassy state of the kernel (Figure 10.6, right). Rice that is cooled directly after drying develops fissures some time after the drying process is complete. However, grain breakage can be avoided by hot tempering at the drying temperature for several hours. The reason for this fissuring phenomenon is that during the drying process the moisture content of the surface of the grain decreases faster than that in the centre of the kernel (see Figure 10.6, left). As a consequence, the moisture dependent glass transition temperature T_g of the surface layer is higher than that in the center. Therefore, if the product is cooled immediately after the drying is concluded, the surface enters the glassy state while the center remains rubbery. The resulting change in state evokes differential stresses within the grain that can cause the grain to fissure (Perdon et al., 2000; Cnossen and Siebenmorgen, 2000). If, on the other hand, drying is followed by a tempering step in the rubbery state, continued moisture diffusion is facilitated. As a consequence, the moisture within the kernel gradually equalizes, and stresses caused by moisture gradients disappear. Cooling down becomes uncritical once the moisture within the kernel is homogeneous.

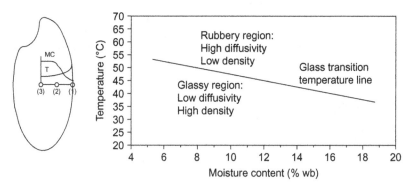

FIGURE 10.6 Left: Hypothetical temperature (T) and moisture content (MC) distribution within a rice kernel during drying, plotted onto a rice kernel. Points 1, 2, and 3 on the x-axis of the plot correspond respectively to the surface, mid-point between surface and center, and center of the kernel. Right: Glass transition relationship for Bengal brown rice (Sun et al., 2002). *Reprints with permission from (left) Cnossen and Siebenmorgen (2002), (right) Cnossen et al. (2002).*

The type of rice drying depends on the extent of mechanization. Traditionally, rice is dried in batches in the open air. Industrial drying methods include batch drying in bins as well as continuous drying in a flow column, rotary bed, or fluid bed. Methods where temperature and moisture can be controlled are clearly advantageous with respect to kernel fissuring. Two-stage or multistage drying that includes tempering between stages, where the rice is held for hours at the drying temperature, is usually applied in industrial processes to avoid structural damage to the kernels.

4.2   Parboiling

The term parboiling refers to a partial boiling of rice, mostly done while still in the husk and sometimes in the form of brown rice. About a fifth of all rice is parboiled before milling, and 90% of all parboiled rice is produced in South Asia. Diverse ways of parboiling are known, which can be grouped into three main categories (Bhattacharya, 2004):

- "conventional parboiling" including the steps soaking, draining, cooking, and drying,
- "low-moisture parboiling" where partial soaking is followed by high-pressure steaming, and
- "dry-heat parboiling" with soaking followed by a combined conduction heating/drying step.

Parboiled rice differs from raw rice in many ways (Bhattacharya and Ali, 1985). In contrast to raw rice, parboiled rice appears glassy, translucent and has an amber color before cooking. After cooking, it is firmer, fluffier, and less sticky. Even though parboiling leads to a decrease in the thiamine content of brown rice, milled parboiled rice contains more thiamine than milled raw rice at the same degree of milling. This observation is commonly explained by an inward diffusion of the vitamin during parboiling (Padua and Juliano, 1974). Similar trends were observed for nicotinic acid and riboflavin (Akroyd et al., 1940), whereas fat-soluble vitamins are not introduced into the kernel during parboiling but migrate outwards. As an additional positive consequence of parboiling, head rice yield is increased to near-maximum if parboiling and subsequent drying are properly done (Bhattacharya, 1969). The probable explanation is that the swelling of the starchy endosperm during cooking heals pre-existing defects like chalky parts or fissures.

4.3   Rice Milling

The term milling usually subsumes the core processing steps—i.e., hulling, whitening, and polishing—in the case of *Indica* rice processing. The principal aim is the separation of hull and bran from the endosperm while keeping the white rice kernel intact.

4.3.1 Hulling of Paddy Rice

The purpose of hulling is to remove the husk from the paddy grain without damaging the bran layer and limiting kernel breakage to a minimum. The efficiency of the huller is determined by the percentage of kernels dehulled at one pass. This is called the degree of hulling. Under-runner disk hullers were the most popular machines before the introduction of the rubber roll huller. Under-runner disk hullers consist of two plates with adjustable gap size, one fixed and the other rotating. Paddy is fed in at the center of the fixed plate. Due to an abrasive coating and the pressure between the plates, the husks are sheared off. Since the abrasive material is coated on the moving plate and husk is also highly abrasive, an uneven surface develops on the plate over time. As a consequence, the percentage of broken kernels rises and the degree of hulling declines. The rubber roll huller has now replaced this machine almost completely.

The rubber roll huller consists of two rolls that rotate in opposite directions with a difference in speed of 1:1.25 to 1:1.35. The paddy is fed in between the rolls and the shear force exerted tears the husk off. The rubber wears off due to the abrasive nature of the husk and, hence, the gap has to be adjusted to maintain the degree of hulling. In modern-day machines this is done automatically. Hulling efficiency with rubber roll hullers is much better than with under-runner disk hullers. In addition, rubber rollers do not damage the bran layer, which explains why the number of broken kernels is much lower. Following hulling, loose husks are removed in a husk separator using aspiration. Figure 10.7 shows the principles of the rubber roll huller and husk separator.

Hulling degree is maintained at 90% on modern hullers for most rice varieties except in long grains, where the efficiency is reduced to 80% to minimize the brokens generated. As a consequence of the incomplete hulling, separation of unhulled paddy from brown rice is necessary. This is done in paddy separators. There are different types of these machines—tray type, compartment type, and screen separators—of which only tray type separators are described here. Tray type separators are nowadays preferred for their high efficiency of separation. They consist of inclined, oscillatory moving trays with an indented surface. The mechanism of separation is based on differences in specific gravity, grain length, and friction coefficients. Brown rice is forced to move towards the upper end, while paddy rice floats on top and moves towards the lower end of the table.

4.3.2 Abrasive Milling or Whitening

Abrasive milling, also called whitening, is used to separate the major part of the bran layers as well as the germ from the starch-rich core of the kernel. The principle of abrasive milling is that brown rice travels through a gap between an inner rotating abrasive cylinder and an outer perforated metal cage

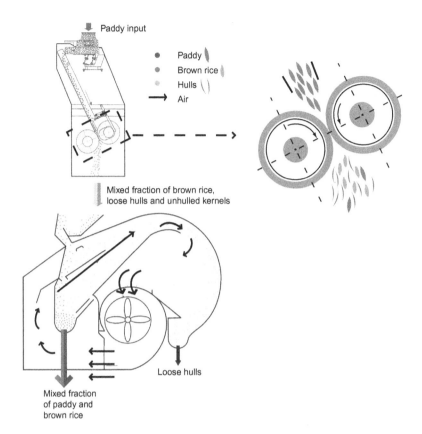

Paddy input

- Paddy
- Brown rice
- Hulls
- → Air

Mixed fraction of brown rice, loose hulls and unhulled kernels

Loose hulls

Mixed fraction of paddy and brown rice

FIGURE 10.7   Principle of Bühler rubber roll huller and husk separator (left). Schematic drawing of rubber rollers (right).

(Figure 10.8, right). Bran is abraded from the kernel while passing through the gap. Different types of abrasive rice mills are available. Their main differences are in the design of the milling stone and the orientation of the milling zone: i.e., horizontal versus vertical setting and conical versus cylindrical stone geometry.

Vertical abrasive cones were the first machines to be invented, in the late nineteenth century by Douglas and Grant in Scotland. Rice is fed from the top into the gap and flows downwards under the influence of gravity. Rubber brakes slow down the flow of rice and build up pressure. Any loose bran is removed through a wire mesh by suction. The cone diameter is largest on top. Peripheral speed is, hence, highest at the rice inlet, and the resulting quick acceleration of rice kernels leads to physical stresses and high numbers of broken kernels. Another disadvantage of this set-up is that underloading results in non-uniform whitening and uneven wear and tear on the machine.

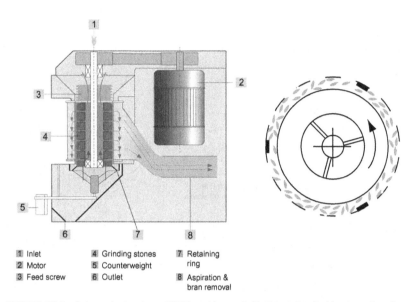

1	Inlet	4	Grinding stones	7	Retaining
2	Motor	5	Counterweight		ring
3	Feed screw	6	Outlet	8	Aspiration &
					bran removal

FIGURE 10.8   Schematic drawing of Bühler whitener (left). Principle of whitening where bran abrasion is achieved between an abrasive rotating inner cylinder and a perforated cage (right).

In horizontal whiteners with cylindrical stone geometry, the circumferential speed is the same throughout the length of the machine. A higher and more uniform degree of whiteness can be achieved. However, as a result of the horizontal alignment, rice is only discharged by the back pressure of incoming rice. Another drawback is that only the bottom half of the machine is completely filled. Friction is high in this area, which results in an undesired heating of the rice. To counter this heating, air is blown into the machine, with the result that head rice yield decreases. Modern vertical whiteners (Figure 10.8, left) prefer cylindrical stones in a vertical setting and include a feed screw. This design guarantees uniform rice distribution and limited acceleration of the kernels at the inlet. The rotor is assembled from individual stones. Angled slots in the sieve enclosure mirror the spiral path taken by rice in the milling gap and ensure maximum bran removal with a minimum of broken kernels. The whitening degree is controlled by a counter weight arrangement at the outlet which ensures that the filling degree of the milling chamber is optimized and optimal back pressure is generated. In modern rice mills, rice passes through one (for short grains, e.g. China), two (for some specific varieties in Europe), three (short and medium grain), to four (long grain) consecutive abrasive mills to minimize the number of brokens while maximizing the throughput. After whitening, the kernel surface is still rough. This is why in most industrial rice mills, abrasive milling is followed by friction milling or polishing steps.

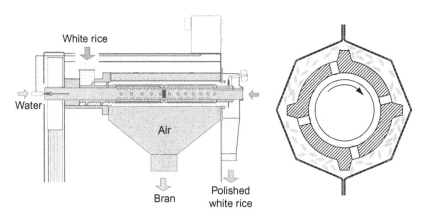

FIGURE 10.9   Schematic drawing of Bühler water mist polisher (left). Principle of polisher, where friction between kernels is created by ribbed cams (right).

4.3.3   Friction Milling or Polishing

In contrast to whitening, where grains are abraded between a rotating inner cylinder and an outer cage, polishing uses friction between the kernels created by ribbed cams (Figure 10.9, right). The remaining traces of bran are removed and any scratches created during whitening filled. As a consequence, the shelf life of rice is increased and a glossy finish achieved.

Polishing machines are almost always horizontal and include the option of water addition. Older machines did not add water and relied solely on friction to achieve a lustrous look and feel. Due to the very high heat generated, they also generated a large number of brokens. To counteract this, a hollow shaft was added, through which air was blown in to cool the rice, but this proved ineffective. Modern water mist polishers use atomized water to humidify the rice grain and thereby increase friction. Temperatures remain lower in this type of machine and the rice surface does not dry out. The addition of water also helps to create a slip layer between the bran fragments and the rice kernel, thus ensuring better bran removal. The result is a much better look, longer shelf life of rice, and a higher head rice yield. The application of one to three consecutive polishing machines of this type is now the industry standard worldwide (Figure 10.9, left).

Figure 10.10 shows the surface and near surface layers of a Basmati rice kernel before and after different milling steps, achieved using a Bühler Basmati line. The images clearly show how first the hull is removed. The resulting brown rice kernel is still covered by intact bran layers. Following this, four passes of whitening were undertaken. With each whitening pass, more of the bran layers are removed ending with a near-white kernel with a jagged surface. Polishing then evens out the surface by removing residual bran

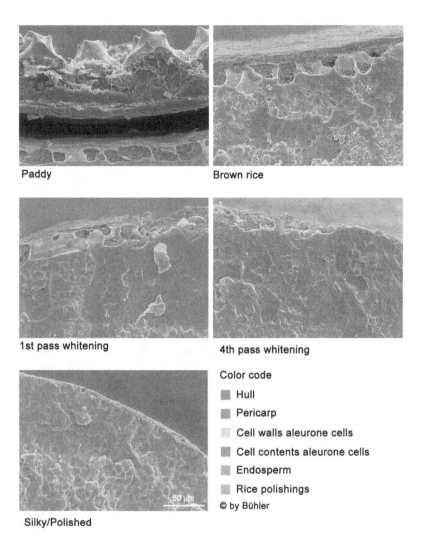

Paddy

Brown rice

1st pass whitening

4th pass whitening

Silky/Polished

Color code

▨ Hull

▨ Pericarp

░ Cell walls aleurone cells

▨ Cell contents aleurone cells

▨ Endosperm

▨ Rice polishings

© by Bühler

FIGURE 10.10   Scanning electron microscope images of the surface layers of a Basmati rice kernel before and after the different milling steps: hulling, whitening, and polishing. Rice kernels were defatted with petrol ether and dried at room temperature. Cross sections were produced by cutting the kernels transversely with a razor blade. The samples were then mounted on SEM object holders with conductive carbon cement and, prior to the examination with SEM, sputtered with gold. *Images were subsequently colored with Photoshop CS4.*

parts and filling the indentations with loose material which mainly consists of starch. The end product is a white kernel with a silky finish.

4.3.4   Sorting and Grading

After milling, rice is optically sorted and graded, although in some cases rice is sorted after grading. Optical sorting improves the product quality by

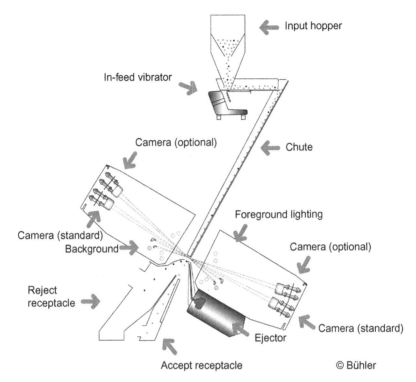

FIGURE 10.11 Schematic drawing of a Bühler Sortex Z optical sorter.

identifying and removing defective grains such as "Yellows", "Peck Defects", "Brown Defects", "Purples" and "Chalkies". The optical sorter will also remove foreign material such as weed seeds and mud balls.

A typical optical sorter is made up of four elements:

- the feed system consisting of a vibrator and inclined chute arrangement which presents the product to the vision system as a cascading plane of rice of uniform speed and distribution;
- a vision system where high-speed cameras and lighting create a continuously rolling image of the product which is digitized and processed;
- the sorting electronics in which digital images are processed relative to user defined criteria and grains are classified as either "accept" or "reject"; and
- the pneumatic ejector array where grains or objects identified as reject material are removed from the product stream with a short blast of air.

Figure 10.11 shows a schematic drawing of a Bühler Sortex Z optical sorter.

Grading for removal of broken grains after rice milling is achieved with indent cylinders. The indent cylinder is a constantly rotating drum. Broken grains collect in the indents as the drum is rotated and are lifted from the product stream. Acceptable grains are too large to get caught in the indents and simply pass through the cylinder.

5. POTENTIAL USAGES OF EDIBLE CO-PRODUCTS

Whitening and polishing of brown to white rice results in about 30% edible co-products in the form of broken kernels and rice bran. These are valuable sources of macro- and micronutrients and should be considered as raw materials for further food processing.

5.1   Rice Brokens—Case Study: Reconstituted Rice

Broken kernels are comparable to head rice with respect to storage properties and composition, and re-enter the food chain to a large extent. Brokens are either sold as low-value rice, added to compound products like breakfast cereals, or are ground to flour. Rice flour can then be used as an ingredient for sauces and soups or to create unique products like rice pasta or reconstituted rice kernels. The following information on processes and principles of the formation of reconstituted rice is also true for gluten-free pasta, even if the same variability in end product properties is not desired.

Reconstituted kernels are commonly produced using cold or hot extrusion. Pasta equipment with or without the addition of steam as well as twin-screw extruders are commonly applied. Rice flour of varying granulation plus an optional vitamin/mineral premix is extruded to form rice kernels closely resembling natural rice. Rice does not contain any gluten, and thus no networks can be formed which support and hold the structure together during cooking. Instead, starch acts as the main structuring agent. The starch matrix can be tailored by varying the degree of starch gelatinization. Clear differences in the state of starch during processing, between cold and hot extrusion, can be seen in Figure 10.12. During cold extrusion, temperature and moisture conditions are such that amylopectin melts to a very limited extent only, whereas it is mostly in its molten state during hot extrusion. Color and transparency adjustment of the kernels can be influenced by selecting the raw material characteristics amylose/amylopectin ratio and granulation as well as adapting the mechanical and thermal energy input during processing.

Figure 10.13 shows a selection of reconstituted rice kernels produced from raw material of different granulation processed on a Bühler twin-screw extruder with different screw configurations resulting in different specific mechanical energy inputs. In addition to the differences in color, surface properties, and transparency shown here, different shapes can be achieved by exchanging dies. The microstructure of natural and reconstituted kernels differs significantly. Natural rice kernels exhibit a pronounced cellular microstructure with native starch kernels arranged in endosperm cells. In addition, there is a clear gradient in starch–protein distribution between kernel core and surface. Reconstituted kernels, on the other hand, consist of a near-completely gelatinized starch

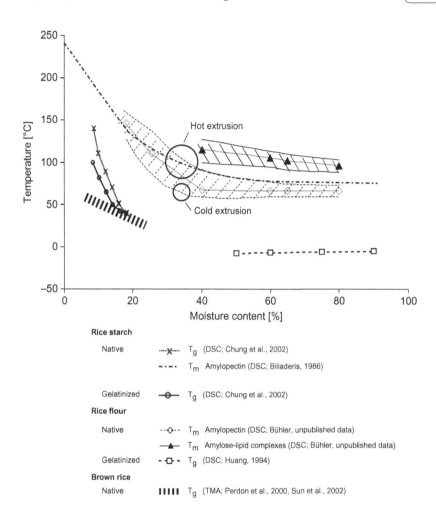

FIGURE 10.12   State of starch during cold and hot extrusion, based on literature data and on measurements undertaken by Bühler. (Biliaderis et al. (1986); Chung et al. (2002); Huang et al. (1994))

matrix throughout the kernel that merely contains some endosperm fragments and protein coagulates (Mueller-Fischer and Conde-Petit, 2009).

5.2   Rice Bran

Nowadays, rice bran is still mostly burnt, a small part added to feed, and an even smaller part used for food. Rice bran is applied in food in its entirety or after partial or full fractionation. Typical fractions of rice bran that are commercially sold are rice bran oil, rice bran solubles, and rice bran protein. In addition, high-value bioactive components like γ-oryzanol can be separated (see section 2.1).

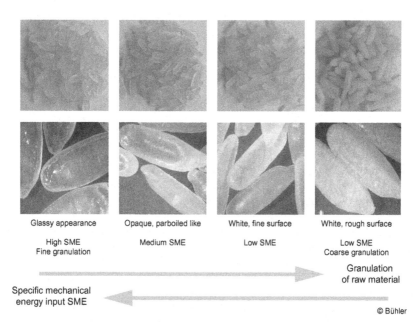

Glassy appearance	Opaque, parboiled like	White, fine surface	White, rough surface
High SME Fine granulation	Medium SME	Low SME	Low SME Coarse granulation

Granulation
of raw material

Specific mechanical
energy input SME

© Bühler

FIGURE 10.13   Visual appearances of reconstituted rice kernels, varying from translucent and smooth to opaque and rough. All rice kernels were produced on a Bühler twin-screw extruder. Raw material granulation was varied and different screw configurations applied, the latter resulting in different specific mechanical energy (SME) inputs.

Rice bran is a food material which is difficult to handle. It contains about 15–20% of oil (Luh et al., 1991) which is rich in unsaturated fatty acid prone to lipid oxidation. If not properly stabilized, rice bran deteriorates within hours due its very active lipolytic enzyme system from both endogenous and microbial origin. Oxidative rancidity and bran build-up in pipes due to the material's stickiness are additional concerns. Stabilization needs to be done within hours after milling and, to achieve the highest quality rice bran, within 1 hour.

Most industrially available systems for enzyme inactivation apply combined heat and moisture in the form of steam to inactivate both lipase and lipoxygenase (Orthoefer and Eastman, 2004). The most common processes are single- or twin-screw extrusion, where mechanical impact on the protein denaturation plays an additional role. Heat stabilization of this kind is a trade-off between stabilization against enzymatic oxidation on the positive side, losses in vitamins, darkening of bran and oil color as well as initiation of oxidative rancidity on the possible negative side. This can be seen in Figure 10.14 where data for free fatty acid, hexanal, and vitamin E development over time is shown. In addition, oil color of unstabilized rice bran and of hydrothermally stabilized rice bran either treated at atmospheric pressure with no shear or at elevated pressures with shear is also given.

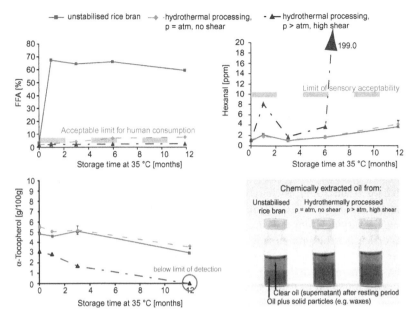

FIGURE 10.14 Development of enzymatic rancidity (free fatty acids, FFA, as indicator), oxidative rancidity (hexanal as indicator), vitamin E content (here represented by α-tocopherol) over the course of one year storage at 35°C. In addition, the color of chemically extracted oil is shown. The performance of unstabilized rice bran is compared with two hydrothermally stabilized rice bran samples: one treated at atmospheric pressure without shear, one at pressures above atm and high shear. All results are based on measurements undertaken by Bühler.

Newer systems under development for rice bran stabilization apply alternative physical as well as biological and chemical principles or a combination thereof (e.g., Ramezanzadeh et al. 2000, Gangadhara and Prakash, 2009). Promising technologies for the future should keep the nutritional quality of the product intact while protecting the lipids against oxidation and should also guarantee food safety with respect to pathogen microorganisms.

Contaminants are an additional issue in rice bran. Similar to other cereals, vegetative microorganism and spore counts are higher in bran than in the endosperm. Mycotoxins, as major biological contaminants with fungal origin, can be present, while the heavy metal arsenic is the most critical contaminant from the chemical side.

Mycotoxins are secondary metabolites of fungal origin. Mycotoxin-producing molds grow if agricultural practices are subpar, climatic conditions adverse, or if grain is not properly dried before storage. Mycotoxins can induce a variety of toxic and carcinogenic effects when contaminated food or feed is ingested (Matić et al., 2008). In the case of rice, aflatoxins are the most frequently reported mycotoxins. Overall, the extent of grain contaminated with mycotoxin is less frequent for rice than for other cereals (Reddy et al.,

2008). Various techniques are available for the detection of mycotoxin in commodities at typical levels in the ppb to ppm range, including rapid methods (Goryacheva and Saeger, 2011).

In contrast to mycotoxins, arsenic contamination is not directly dependent on climatic or storage conditions but on geographic origin and agricultural practices. Rice readily takes up arsenic from the soil and irrigation water and deposits the element in the rice grain. Transfer of arsenic from soil to grain is ten times higher in rice than in wheat and barley (Williams et al., 2007a). Speciation and concentration depend on the origin. The major component species of total arsenic in the rice grain is inorganic arsenic in the form of arsenate and arsenite. Inorganic arsenic is a class 1, non-threshold carcinogenic (Meharg et al., 2009). Meharg et al. analyzed 901 polished (white) rice samples originating from 10 countries from 4 continents and found total arsenic contents varying from 0.04 mg/kg to 0.28 mg/kg. Inorganic arsenic levels in white rice can be problematic for population groups that consume most of their daily calories in the form of white rice. This is the case in large parts of Asia where the daily rice consumption lies between 200 and 500 g with extremes going up to 650 g. Assuming daily ingestions of white rice of 200 g, arsenic levels of <0.05 mg/kg are necessary to stay below the WHO limit for water of 0.01 mg/L (Stone, 2008). At 500 g daily consumption, the values should consequently stay below 0.02 mg/kg polished rice (Meharg et al. 2009). Further risk groups are infants and sufferers of Celiac disease (Williams et al., 2007b). Inorganic arsenic levels in commercially purchased as well as freshly milled rice bran are 10–20 fold higher than concentrations found in polished rice (Sun et al., 2008). Future solutions envisage a reduction of arsenic in rice grains by a combination of breeding and adapted agricultural techniques. Until then, the risk of elevated arsenic levels has to be taken into account when developing products from rice bran and its fractions.

6.   FUTURE SCENARIOS OF NUTRIENT-FOCUSED RICE PROCESSING

Macro- and micronutrient contents available in brown rice and rice bran have the potential to lessen some of the nutritional deficiencies in "the rice countries". Therefore, optimized, "nutrient-focused" rice processing could bring about a clear benefit. It should combine the maximization of macro- and micronutrient retention, increase nutrient bioavailability, and eliminate contaminants whilst keeping or enhancing the sensory experience.

Some established processes like parboiling and germination have a long tradition in parts of Asia. These processes take us in the right direction in the sense that they lead to an improved nutritional value of rice. Germination (also called "sprouting") is an interesting natural approach to improve the intrinsic nutritional value of brown rice. The germination process consists of soaking the grains in water, draining the water and letting the grain germinate until the sprout (also called "germ") emerges and grows. The term partial or pre-germination indicates that germination is stopped as soon as the nutritive value is enhanced but before the sprout is fully developed. Subsequent gentle drying makes the product stable for storage but does not affect the nutritional quality. Germinated grains are nutritionally superior to their respective "non-germinated" grains, i.e., they contain higher levels of nutrients (such as vitamins), lower levels of anti-nutrients (such as phytic acid), contain more easily digestible protein and starch, and have better bioavailable minerals. Studies focusing on the effects of germination on the nutrient content of rice are available (Kayahara and Tsukahara, 2000; Kayahara, 2001; Kayahara et al., 2001; Moongngarm and Saetung, 2010). Quantitative results vary depending on differences in soaking, germination, and drying procedures as well as on rice varieties used. The amount of lysine, an essential amino acid that rice is deficient in, is significantly increased. Enhanced levels of the vitamins E, B1, B3, and B6 and of the minerals magnesium, potassium, and zinc were found as well, while antinutritive factors like phytic acid are lowered. Of special interest are the elevated levels of the bioactive ingredients γ-oryzanol and γ-aminobutyric acid (GABA). In comparison with untreated brown rice, the sensory properties are positively affected. Untreated brown rice is often not desired by consumers due to its hard and chewy texture. Germinated rice requires less cooking time and is sweeter, more mellow, and softer than cooked regular brown rice (Jiamyangyuen and Ooraikul, 2008; Patil and Kahn, 2011). The study of the effect of partial germination seems to be mostly limited to brown rice. Only one source reports increases in α-tocopherol and tocopherol of 13% in germinated white rice (Chattopadhyay and Banerjee, 1952).

In response to the growing market interest in intrinsically fortified products both in the form of whole grains and as ingredients for innovative formulations, industrial solutions based on the principle of germination have been developed. Bühler's "pargem®" is an example of an innovative process and technology for controlled partial or full germination of cereals, pulses, and other grains.

While partial germination can not yet fulfill the ambitious goal of combining all the goodness of brown rice with the pleasure of eating white rice, it is nevertheless a step in the right direction. A step from which we can learn and upon which we can build on our path towards nutrient-focused rice processing.

ACKNOWLEDGEMENTS

I thank all my colleagues from Bangalore, London, Beilngries, and Uzwil who kindly supplied me with detailed information and support, or critically read and improved the manuscript. I want to especially mention Srinivas Duvvuri, Stefania Bellaio, Matthew Kelly, Matthias Gräber, and Ian Roberts, who each wrote a section of this manuscript. I furthermore thank Gudrun Hugelshofer and Sujit Pande, who provided me with brilliant microscope images of rice kernels and with illustrative schematic drawings of Bühler rice processing, respectively. Many thanks as well to my review team Béatrice Conde-Petit, Ian Roberts, Nick Wilkins, Dipak Mane, Eliana Zamprogna, Satish Satyarthi, Niels Blomeyer, and Peter Böhni, who improved the chapter with their valuable comments.

REFERENCES

Akroyd, W.R., Krishnan, B.G., Passmore, R., Sundararajan, A.R., 1940. The rice problem in India. Indian Med. Res. Memoir. 32, 84.

Bhattacharya, K.R., 1969. Breakage of rice during milling and effect of parboiling. Cereal Chem. 46, 478–485.

Bhattacharya, K.R., Ali, S.Z., 1985. Changes in rice during parboiling and properties of parboiled rice. In: Pomeranz, Y. (Ed.), Advances in Cereal Science and Technology, vol. VII AACC, St. Paul, pp. 105–167.

Bhattacharya, K.R., 2004. Parboiling of rice. In: Champagne, E.T. (Ed.), Rice Chemistry and Technology, third ed. AACC, St. Paul, pp. 329–404.

Bhattacharya, K.R., 2011. Rice Quality: A guide to Rice Properties and Analysis. Woodhead Publishing Limited, Cambridge.

Biliaderis, C.G., Page, C.M., Maurice, T.J., Juliano, B.O., 1986. Thermal characterisation of rice starches: a polymeric approach to phase transitions of granular starch. J. Agric. Food Chem. 34, 6–14.

Champagne, E.T., Wood, D.F., Juliano, B.O., Bechtel, D.B., 2004. The rice grain and its gross composition. In: Champagne, E.T. (Ed.), Rice Chemistry and Technology, third ed. AACC, St. Paul, pp. 77–107.

Chattopadhyay, H., Banerjee, S., 1952. Effect of germination of the total tocopherol content of pulses and cereals. J. Food Sci. 17, 402–403.

Chung, H.-J., Lee, E.J., Lim, S.-T., 2002. Comparison in Glass transition and enthalpy relaxation between native and gelatinised rice starch. Carbohydr. Polym. 48, 287–298.

Cnossen, A.G., Siebenmorgen, T.J., 2000. The glass transition temperature concept in rice drying and tempering: effect on milling quality. Trans. Am. Soc. Agric. Eng. ASAE 43, 1661–1667.

Cnossen, A.G., Siebenmorgen, T.J., Yang, W., 2002. The glass transition temperature concept in rice drying and tempering: effect on drying rate. Trans. Am. Soc. Agric. Eng. ASAE 45, 759–766.

Cook, J.D., Skikne, B.S., Baynes, R.D., 1994. Iron deficiency: the global perspective. Adv. Exp. Med. Biol. 356, 219–228.

Dexter, P.B.1998. Rice fortification for developing countries, OMNI / USAID.

Eggum, B.O., Juliano, B.O., Ibabao, M.G.B., Perez, C.M., Carangal, V.R., 1987. Plant foods. Hum. Nutr. 37, 237.

FAOSTAT. Available at: <http://faostat.fao.org>.

Gangadhara, P.R.K., Prakash, V., 2009. Inhibition of rice bran lipase by azadirachtin from azadirachta indica. J. Sci. Food Agric. 89, 1642–1647.

Goryacheva, I.Y., de Saeger, S., 2011. Immunochemical methods for rapid mycotoxin detection in food and feed. In: de Saeger, S. (Ed.), Determining Mycotoxins and Mycotoxigenic Fungi in Food and Feed. Woodhead Publishing Limited, Cambridge, pp. 135–167.

Gustavsson, J., Cederberg, C., Sonesson, U., 2011. Global food losses and food waste: extent Causes and Prevention. FAO, Rome.

Huang, R.M., Chang, W.H., Chang, Y.H., Lii, C.Y., 1994. Phase transitions of rice starch and flour gels. Cereal Chem. 71, 202–207.

Hurrell, R.F., Juillareat, M.A., Reddy, M.B., Lynche, S.R., Dassenko, S.A., Cook, J.D., 1992. Soy protein, phytate and iron absorption in humans. Am. J. Clin. Nutr. 56, 573–578.

Indudhara Swamy, Y.M., Bhattacharya, K.R., 1984. Breakage of rice during milling. Effect of sheller, pearler and grain types. J. Food Sci. Technol. 21, 8–12.

International Rice Research Institute 1999. World Rice Statistics. Retrieved from: <http://www.irri.org>.

Jiamyangyuen, S., Ooraikul, B., 2008. The physico-chemical, eating and sensorial properties of germinated brown rice. J. Food Agric. Environ. 6, 119–124.

Juliano, B.O., 1993. Rice in Human Nutrition. FAO / IRRI. Food and Agriculture Organization of the United Nations, Rome, Food and Nutrition Series No. 26.

Kayahara, H., 2001. Functional components of pre-germinated brown rice, and their health promotion and disease prevention and improvement. Weekly Agric. Forest 1791, 4–6.

Kayahara, H., Tsukahara, K., 2000. Flavor, health and nutritional quality of pre-germinated brown rice. Presented at 2000 International Chemical Congress of Pacific Basin Societies, Hawaii.

Kayahara, H., Tsukahara, K., Tatai, T., 2001. In: Spanier, A.H. (Ed.), Flavor, Health and Nutritional Quality of Pre-Germinated Brown Rice. Royal Society of Chemistry, Cambridge.

Kik, M.C., Williams, R.R., 1945. The nutritional improvement of white rice. National Research Council, Bull. 112, National Academy of Sciences, Washington DC.

Labuza, T.P., Tannenbaum, S.R., Karen, M., 1970. Water content and stability of low moisture and intermediate moisture foods. J. Food Technol. 24, 543–550.

Lucca, P., Hurrell, R., Potrykus, I., 2001. Genetic engineering approaches to improve the bioavailability and the level of iron in rice grains. Theor. Appl. Genet. 102, 392–397.

Luh, B.S., Barber, S., de Barber, C.B., 1991. In: Luh, B.S. (Ed.), Rice Bran Chemistry and Technology, vol. II Van Nostrand Reinhold, New York.

Matić, J., Mandić, A., Mastilović, J., Mišan, A., Beljkaš, B., Milovanović, I., 2008. Contaminations of raw materials and food products with mycotoxins in Serbia. Food Process. Qual. Saf. 35, 65–70.

Meharg, A.A., Williams, P.N., Adamako, E., Lawgali, Y.Y., Deacon, C., Villada, A., et al., 2009. Geographical variation in total and inorganic arsenic content of polished (white) rice. Environ. Sci. Technol. 43, 1612–1617.

Moongngarm, A., Saetung, N., 2010. Comparison of chemical compositions and bioactive compounds of germinated rough rice and brown rice. Food Chem 122, 782–788.

Mueller-Fischer, N., Conde-Petit, B., 2009. Tailoring the structure of starch based food matrices through starch processing. In: Fischer, P., Windhab, E. (Eds.), Proceedings of ISFRS 2009. ETH Zurich, Zurich.

Orthoefer, F.T., Eastman, J., 2004. Rice bran and oil. In: Champagne, E.T. (Ed.), Rice Chemistry and Technology, third ed. AACC, St. Paul, pp. 569–593.

Padua, A.B., Juliano, B.O., 1974. Effect of parboiling on thiamin, protein and fat of rice. J. Sci. Food Agric. 25, 597–701.

Patil, S.B., Kahn, K., 2011. Germinated brown rice as a value added rice product: a review. J Food Sci Technol 48, 661–667.

Patel, M., Naik, S.N., 2004. γ-Oryzanol from rice bran oil – a review. J. Sci. Ind. Res. 63, 569–578.

Perdon, A., Siebenmorgen, J., Mauromoustakos, A., 2000. Glassy state transition and rice drying: development of a brown rice state diagram. Cereal Chem. 77, 708–713.

Ramezanzadeh, F.M., Rao, R.M., Prinyawiwatkul, W., Marshall, W.E., Windhauser, M., 2000. Effects of microwave heat, packaging, and storage temperature on fatty acid and proximate compositions in rice bran. J. Agric. Food Chem. 48, 464–467.

Rand, W.M., Uauy, R., Scrimshaw, N.S. 1984. Protein-energy-requirement studies in developing countries: results of international research. Food Nutr. Bull. Suppl. UN University, Tokyo.

Reddy, K.R.N., Reddy, C.S., Abbas, H.K., Abel, C.A., Muralidharan, K., 2008. Mycotoxigenic fungi, mycotoxins and management of rice grains. Toxin Rev. 27, 287–317.

Stone, R., 2008. Arsenic and paddy rice: a neglected cancer risk. Science 321, 184–185.

Sun, G.-X., Williams, P.N., Carey, A.-M., Zhu, Y.-G., Deacon, C., Raab, A., et al., 2008. Inorganic arsenic in rice bran and its products are an order of magnitude higher than in bulk grain. Environ. Sci. Technol. 42, 7542–7546.

Sun, H., Siebenmorgen, T.J., 1993. Milling characteristics of various rough rice kernel thickness fractions. Cereal Chem. 70, 727–733.

Sun, Z., Yang, W., Stelwagen, A.M., Siebenmorgen, T.J., Cnossen, A.G., 2002. Thermomechanical transitions of rice kernels. Cereal Chem. 79, 349–353.

USDA, 2010. Recommended dietary allowances for vitamins and element. Food and Nutrition Board, USA. <http://fnic.nal.usda.gov/dietary-guidance/dietary-reference-intakes/dri-tables>.

WHO. 1992. National strategies for overcoming micronutrient malnutrition, Geneva, Switzerland.

Williams, P.N., Villada, A., Deacon, C., Raab, A., Figuerola, J., Green, A.J., et al., 2007a. Greatly enhanced arsenic shoot assimilation in rice leads to elevated grain levels compared to wheat and barley. Environ. Sci. Technol. 41, 6854–6859.

Williams, P.N., Raab, A., Feldmann, J., Meharg, A.A., 2007b. Market basket survey shows elevated levels of As in South Central U.S. processed rice compared to California: consequences for human dietary exposure. Environ. Sci. Technol. 41, 2178–2283.

Chapter 11

Virus Resistance Breeding in Cool Season Food Legumes: Integrating Traditional and Molecular Approaches

Shalu Jain[*], Kevin McPhee[*], Ajay Kumar[*], Reyazul Rouf Mir[†], and Ravinder Singh[**]

[*]*Department of Plant Sciences, North Dakota State University, Fargo, ND, USA,* [†]*Department of Plant Breeding and Genetics, SKAUST, Jammu, India,* [**]*Department of Biotechnology, SKAUST, Jammu, India*

1. COOL SEASON FOOD LEGUMES

Faba bean (*Vicia faba* L.), chickpea (*Cicer arietinum* L.), lentil (*Lens culinaris* Medik.) and field pea (*Pisum sativum* L.) are referred to as cool season food legumes (CSFLs) because of their adaptation to temperate climates. CSFLs are members of the Fabaceae family and cultivated around the world in more than 100 countries as important sustainable crops even after domestication over 9,000 years ago (Zohary and Hopf, 1973). Center of origin, chromosome number, genome size and number of available expressed sequence tags (ESTs), worldwide cultivation area, and production of the cool season food legumes are provided in Table 11.1. Worldwide gross value of the CSFLs is ca. US$9,000 million (FAOSTAT, 2010) annually, and their export provides means of income at the household level and foreign currency earnings in many countries. CSFLs are a very important part of the human diet worldwide, particularly in terms of nutrition, because of their high protein content which can substitute for animal products in vegetarian or poor populations. They are an excellent source of dietary fiber and micronutrients such as zinc, iron, and calcium as well as folate. CSFLs also play an important role as rotation crops for breaking disease cycles and fixing atmospheric nitrogen in an endo-symbiotic association with root nodule rhizobium.

Cool season food legumes are vulnerable to a number of viral diseases which represent at least seven plant viral groups—potyviridae, enamoviridae,

Agricultural Sustainability. DOI: http://dx.doi.org/10.1016/B978-0-12-404560-6.00011-3

TABLE 11.1 General Information, Total Harvested Area, and Production Worldwide of Cool Season Food Legumes

Plant	Common Name	Origin	Chromosome No.	Genome Size (Mb)	Genomic Resource (EST[a])	Harvested Area[b] (ha)	Production[b] (Mt)
Cicer arientum	Chick pea	Middle east	2n = 16	740	45,784	11,982,140	10,918,081
Lens culinaris	Lentil	Near East	2n = 14	4,063	9,516	4,189,502	4,585,439
Pisum sativum	Field pea	Near East, Mediterranean region	2n = 14	4,400	21,837	6,313,839	10,208,812
Vicia faba	Faba bean	Middle East, Mediterranean region	2n = 12	~13,000	60,683	2,559,773	4,312,871

[a] Source: http://www.ncbi.nlm.nih.gov/dbEST.
[b] Source: FAOSTAT (2010).

tobraviridae, carlaviridae, luteoviridae, cucumoviridae, and alfalfa mosaic virus—resulting in severe damage to yield and quality every year worldwide. Viruses infecting CSFLs are transmitted by vectors, generally aphids, in either a persistent or non-persistent manner. Most common aphids are the pea aphid (*Acyrthosiphon pisum* Harris), cowpea aphid (*Aphis craccivora* Koch), green peach aphid (*Myzus persicae* Sulzer), and the bean aphid (*Aphis fabae* Scopoli). Virus infection can cause up to 100% yield loss, especially if infection occurs at early stages of plant development. Pulse crops damaged by viruses develop fewer and smaller seeds with reduced seed quality. If a virus infection occurs at low levels or late in the season, visible symptoms and yield reduction are unlikely (Bosque-Perez and Buddenhagen, 1990); however, seeds can be infected by some viruses such as cucumber mosaic virus (CMV), bean yellow mosaic virus (BYMV), alfalfa mosaic virus (AMV), and pea seed-borne mosaic virus (PSbMV), which can result in virus infection if that seed is planted.

2. METHODS OF DETECTING PLANT VIRUS DISEASES

Plant viruses are acellular and ultramicroscopic organisms consisting of nucleoproteins which can multiply only in the host cell, resulting in disease. Disease symptoms vary widely depending on the virus and crop combination, ranging from mild mosaics and yellowing to severe leaf deformation, wilting, and stunting. Description of symptoms is crucial in distinguishing diseases caused by viruses from those caused by other agents. Various viruses affecting CSFLs are listed, with visual symptoms and geographical distribution, in Table 11.2. Detailed description of these viruses can be found at http://www. agls.uidaho.edu/ebi/vdie/refs.htm. However, phenotypic evaluation of disease can lead to misdiagnosis because of similar symptoms for different viruses, and combinations of viruses may distort the expression of symptoms. It has been shown that visual field inspection tends to underestimate viral disease incidence in both legume and cereal fields and leads to severe economic crop loss caused by viral diseases (Kumari et al., 2008). A sensitive, rapid, and early detection method for viral diseases is essential for effective management of legume crops. In recent years, there has been a significant improvement in the sensitivity of the methods used to detect plant viruses by means other than physical symptoms.

2.1   Protein-Based Detection Methods

Rapid detection and identification of plant viruses has been accomplished by enzyme-linked immunosorbent assay (ELISA) (Clark and Adams, 1977) and its variants. ELISA is the most common method for certification of plants and planting materials to assess the extent of infection, especially in asymptomatic plants. This technology was further improved with the development

TABLE 11.2 Visible Symptoms of Some Economically Important Virus Diseases of Cool Season Food Legumes

Virus	Abbreviation	Group	Genome	Symptoms	Geographical Distribution
Pea seed-borne mosaic virus	PSbMV	Potyvirus	ssRNA	Transitory vein clearing, rosetting of stem and branches, leaflets folded adaxially, malformed flowers and seeds	Africa, Australia, Asia
Bean yellow mosaic virus	BYMV	Potyvirus	ssRNA	Mosaic, chlorosis, tip necrosis, leaf mottling	Africa, Australia, Asia
Bean leaf roll virus	BLRV	Luteovirus	ssRNA	Chlorotic or yellowed leaves rolled downward	Worldwide
Pea enation mosaic virus	PEMV	Luteovirus	ssRNA	Hyaline local lesions with enations, mosaic, puckering and stunting	America, Asia, Europe
Pea streak virus	PeSV	Carlavirus	ssRNA	Necrotic lesions or streaks	Europe, Asia, America, Canada
Bean common mosaic virus	BCMV	Potyvirus	ssRNA	Rugosity of lower leaves, mosaic, malformation of leaves and pods, vein necrosis, black root, and death	Worldwide
Pea mosaic virus	PeMV	Potyvirus	ssRNA	Marble or yellow mosaic, stunting, delayed unfolding of leaflets	Worldwide

TABLE 11.2 (Continued)

Virus	Abbreviation	Group	Genome	Symptoms	Geographical Distribution
Beet western yellow virus	BWYV	Luteovirus	ssRNA	Mild chlorotic spotting, yellowing, thickening and brittleness of older leaves	Worldwide
Clover yellow mosaic virus	ClYMV	Potexvirus	ssRNA	Mosaic with necrotic streaks	Canada, USA
Alfalfa mosaic virus	AMV	Alfamovirus	ssRNA	Wilting, necrosis, mosaic, mottling	Worldwide
Clover yellow vein virus	ClYVV	Potyvirus	ssRNA	Chlorotic or necrotic local lesions, systemic mosaic and necrosis	Worldwide
Cucumber mosaic virus	CMV	Cucumovirus	ssRNA	Necrosis, mosaic	Worldwide
Chickpea chlorotic dwarf virus	CpCDV	Mastrevirus	ssDNA	Systemic leaf necrosis, yellowing, mottling, and stunting	Asia

of monoclonal antibody technology and is applicable to many legume and cereal viruses (Makkouk and Kumari, 2009). The broad-spectrum monoclonal antibody 5G4 gives a positive reaction with all virus species in the family *Luteoviridae* which infects legume crops (Katul, 1992). In contrast, there are monoclonal antibodies available which react specifically with related viruses. However, it would be a better option to test the samples from field surveys with broad-spectrum antibodies. Only those samples that react positively are tested by the specific luteovirus monoclonal antibodies. To improve the sensitivity of ELISA, the TBIA (Tissue-blot immunoassay) has been used worldwide for virus surveys in legumes and for evaluating virus–host interactions

(Kumari and Makkouk, 2003). TBIA is a very sensitive, simple, and efficient method and is as reliable as ELISA, but requires no leaf grinding, and blotted membranes can be stored for bulk processing. The presence of bean leaf roll virus (BLRV) was tested in faba bean plants by tissue blot immunoassay using a BLRV-specific antibody (Makkouk et al., 2002). Serological means were used in virus surveys of legume crops in several countries of West Asia and North Africa (WANA) for the occurrence of a range of viruses and to identify the causal agent(s) of yellowing and stunting symptoms in faba bean and chickpea plantings near Ambo, Ethiopia (Abraham et al., 2006) using double-antibody sandwich ELISA (DAS-ELISA) and triple-antibody sandwich ELISA (TAS-ELISA) formats.

2.2   Molecular Methods in Plant Virus Detection

Polymerase chain reaction (PCR)-based methods have been successfully used to detect virus infection in plants (Henson and French, 1993). During PCR, a fragment of the viral genome is amplified using specific primers. For RNA viruses, reverse transcription of the viral RNA to a cDNA strand is necessary prior to the PCR, using reverse transcriptase (RT). With the availability of pathogen sequence data in public databases like GenBank (http://www. ncbi.nlm.nih.gov/Genbank/) and COGEME (http://www.cogeme.man.ac.uk/), the trend of using molecular detection methods has significantly increased. Primers designed to detect persistently transmitted aphid-borne legume and cereal viruses by PCR have been described in an earlier review by Makkouk and Kumari (2009). Several variants of nucleic acid based methods such as multiplex RT-PCR, quantitative real-time RT-PCR (qRT-PCR), and microarrays have been successfully adapted for plant pathogen detection (Agindotan and Perry, 2007). In qRT-PCR, the amplified DNA is quantified, using fluorescent dyes, as it accumulates in the reaction mixture after each cycle of PCR. In comparison with normal PCR, qRT-PCR has several advantages, including reduced risk of sample contamination, and it provides quantitative data and simultaneous testing for multiple pathogens.

DNA microarrays are also of great use for simultaneous pathogen detection. This is important, as plants are often infected with several pathogens, some of which may act together to cause a disease complex. Microarrays consist of pathogen-specific DNA sequences immobilized onto a solid surface. Sample DNA is amplified by PCR, labeled with fluorescent dyes, and then hybridized to the array. Another variant of PCR called immunocapture (IC)-PCR combines the advantages of serology and PCR and is a very sensitive method of detection (Mulholland, 2009). Immunocapture PCR is similar to the coating step in DAS-ELISA in which virus particles adhere to the tube wall coated with virus-specific antibodies. This way a suitable template is obtained and is subjected to PCR/RT-PCR. To check the effectiveness of various virus detection methods, different parts of plants showing mosaic symptoms were tested for the presence of bean yellow mosaic virus (BYMV) in gladiolus plants using either DAS-ELISA, one step RT-PCR, real time (rt)-RT-PCR or IC-rt-RT-PCR (Duraisamy et al., 2011). DAS-ELISA and one-step RT-PCR were able to detect BYMV in leaf samples only, while rt-RT-PCR and IC-rt-RT-PCR detected the virus in both leaves and corms or cormlets. Therefore, rt-RT-PCR and IC-rt-RT-PCR are valuable tools for reliable detection of viruses in different tissues of plants, particularly when virus-free stocks are required. Some advanced molecular techniques such as molecular beacon, liquid chromatography (LC), matrix-assisted desorption-ionization (MALDI)-mass spectrometry, and nucleic acid sequence based amplification (NASBA) are very promising for plant virus detection but still not widely used (Lim et al., 2005).

3. SOURCE OF RESISTANCE TO VIRUSES IN CSFLS

Virus-induced diseases are responsible for major crop losses worldwide. A better understanding of plant defense mechanisms would lead to the development of novel strategies for effective plant protection. Breeding for resistance using conventional approaches remains a long process, and suitable resistance is not always available in germplasm collections. Moreover, commercial cultivars that possess virus resistance need to be as good if not better than susceptible cultivars for all the other agricultural traits and quality characteristics in order to be adopted by farmers, because a farmer would prefer to grow a high-yielding susceptible variety rather than a low-yielding resistant variety. An effective approach to achieve durable resistance could be the combination of classical breeding and transgenic mediated resistance (Fuchs et al., 1997). The development of transgenic plants with non-host- or host-derived resistance may provide a means to control viruses in the future, if allowed by official regulations and accepted by farmers and consumers. More efficient and durable virus control can also be achieved by deploying resistant cultivars aimed at preventing, delaying, or reducing virus spread in crops (Lecoq et al., 2004).

3.1 Utilization of Non-Host Resistance

Genetic engineering has opened up new opportunities for transferring virus resistance into existing desirable plant cultivars through incorporation of foreign DNA into plant cells that would regenerate into transgenic plants with improved virus resistance. Initial attempts to create transgenes conferring virus resistance were based on the pathogen-derived resistance (PDR) concept (Sanford and Johnston, 1985) using a portion of a pathogen's genetic material for host defense against the pathogen itself. The underlying rationale was that certain pathogen-derived molecules may be critical for viral pathogenesis, and non-functional forms of such molecules could act in a dominant negative manner to interfere with virus replication, assembly, or movement. A host may confer virus resistance through transgenic expression of pathogen-derived molecules including coat protein (CP), replicase, or movement protein, etc. (Anderson et al., 1992; Prins et al., 2008).

Plants conferring virus resistance can be engineered using transgenes from plant-derived natural R genes, pathogen-derived transgenes, or non-plant- and non-pathogen-derived transgenes. Hill et al. (1991) generated transgenic alfalfa, expressing a large amount of the coat protein of alfalfa mosaic virus and were shown to be resistant to the virus. Macfarlane and Davies (1992) have shown that *Nicotiana benthamiana* plants transformed with a 54 kDa region of the 201 kDa replicase gene of pea early browning virus (PEBV) are resistant to PEBV as well as broad bean yellow band virus infection. Pea plants were transformed *in planta* by injection/electroporation of axillary meristems with a chimeric pea enation mosaic virus (PEMV) coat protein gene construct (Chowrira et al., 1998). Transgenic R2, R3, and R4 plants displayed delayed or transient PEMV multiplication and attenuated symptoms as compared with control inoculated individuals. Transgenic pea lines carrying the replicase (Nib) gene of Pea seed-borne mosaic potyvirus showed a highly resistant state upon infection by the homologous isolate (Jones et al., 1998). Induced resistance in pea lines was associated with a loss of both viral and transgenic RNA, which is indicative of a mechanism based on post-transcriptional gene silencing. Partial resistance to alfalfa mosaic virus (AMV) in transgenic peas was achieved by transformation with two chimeric gene constructs encoding the coat protein (CP) of AMV under greenhouse and field conditions (Timmerman-Vaughan et al., 2001).

Other strategies exploit the plant's innate defense mechanisms to combat invading viral pathogens: RNA-based resistance makes use of the plant post-transcriptional gene silencing (PTGS) mechanism to degrade viral RNAs. PTGS is a conserved mechanism for mRNA regulation in plants, animals, and fungi (Borgio, 2009). PTGS is characterized by the degradation of endogenous/transgenic mRNAs in the cytoplasm, resulting in reduced gene expression. In plants, PTGS controls numerous developmental processes and is required for innate immunity. It regulates virus accumulation through silencing of target mRNAs in the cytoplasm and generates small RNA molecules (~21–25 nt) from the silenced target mRNAs. Based on differences in their biogenesis, two types of small RNAs, siRNA and miRNA, have been identified. Recently, the expression of artificial microRNAs (amiRNAs) in transgenic plants has been shown to confer resistance against plant viruses (Niu et al., 2006; Qu et al., 2007). Taken together, future advancements in protein-mediated and RNA-mediated resistance for crop protection will bring significant economic benefits to agriculture. However, several concerns have been raised regarding the large-scale and long-term use of such transgenic resistance in the field (Tepfer, 2002). It is necessary to focus on research efforts to make this technology safe and publicly acceptable.

3.2 Screening Germplasm for Host Resistance to Develop Virus Resistant Cultivars

Although viruses are relatively simple genetic entities, mechanisms of virus resistance in plants are still largely unknown. Host resistance is the most effective and sustainable approach to prevent and control viral diseases in plants. It is also called genotypic resistance, occurring in some genotypes whereas other genotypes in the same gene pool are susceptible due to genetic polymorphism. A large collection of genetic resources of legumes (both cultivated and wild relatives) are conserved globally in gene banks. Germplasm collections comprise "primary" and "secondary" gene pools. The cultivars being grown, old local types or land races, breeding lines, and also wild forms of the cultivated species are included in the primary gene pool, while the secondary gene pool consists of related species that can also be evaluated for resistance, particularly when a thorough search among crop species has failed.

Screening for resistance is performed by inoculating a large number of accessions with virus, and by carefully studying the symptoms (type, localization, intensity, time of appearance) and reactions of each accession tested. Initial evaluation of a germplasm collection can be done in the field under conditions of natural infection when virus epidemics occur regularly and are severe. However, in these circumstances, the infecting virus populations may be uneven among the plants or may vary with location and year. Therefore, it is advisable to rescreen for resistance under controlled and reproducible conditions. Mechanical inoculation is possible for some viruses, but, for those viruses that are not mechanically transmissible, inoculation requires viruliferous vectors or grafting. The virus may or may not multiply in resistant individuals, but, in comparison with the susceptible host, spread of the pathogen through the plant is demonstrably restricted and disease symptoms generally are highly localized or are not evident. Standardized conditions including uniform plant age and condition, inoculum concentration, and growing conditions are preferable for resistance screening. Lines identified with resistance should be tested against a representative range of isolates from different hosts and geographical origins that cover the known virus variability as much as possible. If some virus isolates (referred to as virulent) can infect resistant plants, then the resistance is considered specific. If the resistance is effective against all the virus populations, then it can be defined as non-specific or comprehensive. Non-specific resistances may have different stability levels which depend on whether a new virus mutant emerges that can overcome the resistance following deployment in fields. Viral pathogens are prone to rapid evolution because of their high mutation rate in the case of RNA viruses (estimated to be ca. 10^{-4} mutations per replication cycle and per base), and due to their error-prone polymerase and the lack of a proof-reading mechanism during replication (Drake and Holland, 1999). Table 11.3 represents a summary of plant introduction (PI) lines and cultivars found to be resistant to various viruses in

TABLE 11.3 List of CSFL Germplasm Accessions, Breeding Lines and Cultivars with Some Level of Resistance/Tolerance to Various Viral Pathogens

Virus	Plant	Resistant Germplasm
PEMV	Pea	PI 140295, Lifter, Franklin, OSU 559-6, OSU 564-3, OSU 584-16, OSU 589-12
	Lentil	PI 472547, PI 472609, PI 606691, PI 606609, PI 633926, PI 533691, PI 472525, PI 299198 and PI 577172
	Chickpea	PI 315826, PI 450843, PI 450975, PI 450693, PI 439829, PI 450906, PI 451594, PI 450867, PI 450977, PI 450870, PI 450763
BLRV	Faba bean	BPL 5272, BPL 5274, BPL 5276, BPL 5277, BPL 5278, BPL 5279 and BPL 5280
	Pea	Abador, Alderman, Almota, Centurion, Champ, Climax, Cobri, Coquette, Elf, Frisky, Jubilee, Juwel, OSU 559-6, OSU 564-3a, OSU 584-16a, OSU 589-12a, Perfected 400, Rika, Sparkle, Splendor, Superlaska, Surpass, Telephone, Wando
	Lentil	ILL 74, ILL 75, ILL 85, ILL 213, ILL 5480, ILL 324, ILL 6816, ILL 7201, PI 212610, PI251786, PI 297745, PI 368648
AMV	Chickpea	Gully
	Pea	PI 121977, PI 164148, PI 166129, PI 180701, PI 184131, PI 193838, PI 197044, PI 197449, PI 197988, PI 197989, PI 201391, PI 210684, PI 244116, OSU 33, OSU 176-2, OSU 709-4
BYMV	Lentil	ILL7163
	Pea	OSU 559-6, OSU 564-3, OSU 584-16, OSU 589-12
	Chickpea	ICC 607, ICC 1468, ICC 2162, ICC 2342, ICC 3440, ICC 3598, ICC 4045, ICC 6999, ICC 11550
CMV	Lentil	L-4, PAK 3598, ILL1983, Mizia, Medovina 1-G, ILL788, ILL7700, ILL7163, ILL5405
	Chickpea	ICC 1781, ICC 8203, BG1100, BG 1101, PUSA 1103, BGD 112, Gully
FBNYV	Lentil	PI 612870, PI 612871, PI 612872, PI 612873), PI 612874, PI 612875
PSbMV	Lentil	PI 212610, PI 251786, PI 207745, PI 368648, Crimson, Palouse, Red Chief, ILL 6198
	Pea	OSU 559-6, OSU 564-3, OSU 584-16, OSU 589-12, VR 74-410-2, VR 74-1492-1
PeSV	Pea	OSU B442-15, OSU B445-66, OSU 644-2, OSU 663, OSU 668, OSU 709-4, PI 140297, PI 195405, PI 203066, PI 212029

cool season food legumes, which can be utilized to make crosses to transfer resistance in elite germplasm.

3.3   Mode of Inheritance of Resistance

Another crucial step in the study of genetics of viral resistance is to determine whether the resistant response is inherited, and, if so, the number of genes involved and their mode of inheritance. In plants, most of the disease resistance genes (R genes) are isolated and characterized to date fall into a series of related categories that involve proteins containing nucleotide-binding site (NBS) and leucine-rich repeat (LRR) regions (NBS-LRR domains), similar to those that control a wide array of other plant pathogens (Meyers et al., 1999). A number of studies have used PCR with degenerate oligonucleotide primers designed from conserved motifs in these regions, to amplify multiple DNA sequences, termed resistance-gene analog (RGA) candidates, from legumes (Yaish et al., 2004; Kang et al., 2005; Palomino et al., 2006).

Disease resistance genes tend to be clustered in the genome. One type of R gene cluster contains a set of genes showing similar inheritance and resistance phenotypes that control very closely related viral genotypes. This type of pattern occurs in *Pisum sativum* where recessive resistance has been mapped to two R gene clusters on two different linkage groups. Six very tightly linked monogenically inherited recessive loci—*bcm*, *cyv1*, *mo*, *pmv*, *sbm2*, and *wmv*— for resistance to bean common mosaic virus (BCMV), clover yellow vein virus (ClYVV), bean yellow mosaic virus (BYMV), pea mosaic virus (PMV), pea seed-borne mosaic virus (PSbMV-L1), and watermelon mosaic virus, respectively, occur in one cluster, but crossing over may occur between them. Five distinct loci—*cyv2*, *sbm1*, *sbm3*, *sbm4*, and *wlv*—conferring resistance to ClYVV, PSbMV-P1, PSbMV-L1 or -P2, PSbMV-P4, and white lupin mosaic virus (WLMV), respectively, were tightly linked in another cluster (Provvidenti and Alconero 1988; Provvidenti, 1990; Provvidenti and Hampton, 1992).

Resistance to pea enation mosaic virus (PEMV) is controlled by a single dominant gene, *En* (Schroeder and Barton, 1958). However, in comparison with single-gene resistance, polygenic resistances are generally durable and without clear strain-specific effects. Polygenic resistance can combat different virulence factors in the pathogen, thus limiting the breakdown of resistance through virus mutation (Parlevliet, 2002). These different genes can also control resistance at different stages of the infection process, or resistance through different mechanisms, which could also increase durability. Identifying markers for direct detection of genes conferring resistance to viral diseases in CSFLs would aid breeding programs, particularly when several resistance genes are selected simultaneously. Table 11.4 describes a number of virus resistance genes and their mode of inheritance.

Some recessively inherited genes are eukaryotic translation factors, including eIF4E (Kang et al., 2005). The effects of pea genes *sbm-1*, *sbm-2*, *sbm-3*,

TABLE 11.4 Mode of Inheritance of Some Known Resistance Genes and their Linked Markers

Plant	Gene	Allele	Virus	Closest Marker (type, distance)	Reference
Pea	sbm 1 sbm-2 sbm-3 sbm-4	Recessive	PSbMV-P1 PSbMV-L1, PSbMV-L1 or -P2 PSbMV-P4	G05_2537 (RAPD, 4 cM from sbm-1) ZG10 (AFLP, 0.7 cM from sbm-1)	Gao et al. (2004a,b) Frew et al. (2002) Bruun-Rasmussen et al. (2007) Keller et al. (1998) Johansen et al. (2001) Smykal et al. (2010)
	En	Dominant	PEMV	Adh1 (Isozyme, 10 cM) Cngc (CAPS, >3 cM)	Weeden and Provvidenti (1988) Randhawa and Weeden (2011)
	mo	Recessive	BYMV	(RFLP, 15 cM)	Dirlewanger et al. (1994)
	lr	Recessive	BLRV	–	Baggett and Hampton (1991)
Lentil	sbv	Recessive	PSbMV	–	Haddad et al. (1978)
Faba bean	bym-1 bym-2	Recessive	BYMV	–	Provvidenti and Hampton (1992)

and *sbm-4* are among the best studied for pea seed-borne mosaic virus resistance conferred by recessive alleles; *sbm1* and *sbm4* confer resistance to pathotypes P-1 and P-4, respectively, while *sbm2* and *sbm3* both confer resistance to pathotype P-2 (the L and L1 isolates). *sbm-1* and *sbm-2* have similarities to eIF4E and eIF(iso)4E, respectively (Gao et al., 2004a). Because all the known resistances to viruses are specific in nature, commercial cultivars with these genes may become susceptible to new pathogenic strains. Knowledge of the genetics of resistance in the host, together with information on genetic variability or potential genetic variability in the pathogen, should help the plant breeder and plant pathologist develop breeding strategies that will provide effective and stable disease control through genetic resistance.

4. MOLECULAR TOOLS FOR ACCELERATING VIRUS RESISTANCE BREEDING

Molecular breeding makes use of genomic tools in breeding programs through marker-assisted selection for desirable genes to improve the efficiency and effectiveness of conventional breeding programs. Unadapted cultivars, lines, or wild relatives need to be screened for desirable traits and these traits are then introduced into available high-yielding commercial lines/varieties. It is becoming a common practice to use molecular markers in conventional plant breeding programs, due to recent developments in genetic and genomic resources. Only a small amount of DNA is required to run the molecular markers, which can be extracted from any part of the plant without destroying it. A known set of primers are used to amplify the DNA through polymerase chain reaction (PCR). Genotype of the individual plant for resistance or susceptibility could then be determined through the polymorphic banding pattern. A genome database for CSFLs (www.gabcsfl.org) has been developed to facilitate breeders' access to genetic and genomic information about these crops. Successful application of biotechnology to select for disease resistance in legume crops will require framework and dense molecular maps comprising different types of molecular markers to localize the gene of interest. The progress made in molecular breeding of legume crops in general has been highlighted in recent reviews emphasizing the mapping of genes that control agronomically important traits (Varshney et al., 2010; Torres et al., 2010; Kumar et al., 2011).

4.1   Molecular Markers

Marker systems employed in modern plant breeding programs include restriction fragment length polymorphism (RFLP), randomly amplified polymorphic DNA (RAPD) markers, amplified fragment length polymorphisms (AFLPs), simple sequence repeats (SSRs or microsatellites), sequence characterized amplified regions (SCARs), cleaved amplified polymorphic sequences (CAPS), derived cleaved amplified polymorphic sequences (dCAPS), sequence tagged sites (STS), single strand conformational polymorphism (SSCP), and single nucleotide polymorphisms (SNPs). ESTs also serve as a source for identifying candidate genes or QTLs involved in the genetic control of specific traits. Various ESTs available for CSFLs (www.ncbi.nlm.nih.gov/dbEST) are mentioned in Table 11.1. These ESTs can also be exploited in the development of molecular markers like EST-SSRs, EST-SNPs, etc. Sequences of resistance gene analogues (RGAs) have been transformed into molecular markers such as dCAPS and CAPS, and are being used to detect the presence of SNPs and their subsequent mapping using the available mapping populations of faba bean, pea, and chickpea (Palomino et al., 2009; Torres et al., 2010). Diversity array technology (DArT) markers have also been developed for CSFLs and have been found useful for enhancing the density of genetic maps and the pool of markers for MAS because of their low cost and ultra-high throughput nature (Upadhyaya et al., 2011). In parallel gene cloning, BAC-based physical maps are being produced around singleton and clustered disease resistance gene homologs, while BAC end sequencing has provided the basis for the development of SSR and SNP genetic markers. Next generation sequencing technologies such as 454/FLX (Roche Inc.), ABI

SOLiD (Applied Biosystems), Solexa (Illumina Inc.), etc. are high through-put and cost effective for generating large-scale sequence data (Varshney et al., 2009), but CSFL crops are still lagging behind in the exploration of these techniques. Research efforts need to be focused on development of molecular genetic resources for candidate disease resistance genes in each of the target species, providing tools for molecular breeding as well as more fundamental studies of resistance gene evolution.

4.2   Genetic Linkage Maps

Development of saturated genetic linkage maps provides a valuable tool in plant genetics and breeding, to estimate the number of loci controlling genetic variation in a segregating population and to estimate the effects of these loci. Substantial research efforts are needed to identify and utilize polymorphic DNA markers as a framework around which the gene/quantitative trait locus (QTL) for a desirable trait could be located. Genetic variation between the parents of a cross is prerequisite for the generation of a mapping population. To map a desirable trait the best way is to select parents that are genetically divergent from each other, including for the traits to be mapped, so that a large set of polymorphic markers is well distributed across the genome. Sometimes because of limited polymorphism within a species, mapping in inbreeding species requires the selection of parents that are distantly related or belong to different subspecies or even species. For example, because of low genetic diversity in lentil, crosses were made between wild species and cultivars to develop mapping populations (Muehlbauer et al., 1989; Weeden et al., 1992). Similarly, genetic maps of chickpea were developed employing populations from crosses between *C. arietinum* and *C. reticulatum* (Rakshit et al., 2003; Pfaff and Kahl 2003; Abbo et al., 2005), *C. arietinum* × *C. echinospermum* (Collard et al., 2003), and intraspecific populations (Flandez-Galvez et al., 2003; Cobos et al., 2005). However, lower recombination rates and smaller map sizes may be due to use of divergent parents (Tadmor et al., 1987). It is also noted that the size of a mapping population can greatly impact the ultimate resolution of a map. Both framework and dense molecular maps comprising more than one type of molecular marker have been constructed in CSFLs (Nayak et al., 2010; Hamwieh et al., 2005; Loridon et al., 2005).

A composite genetic map can be generated by using the polymorphism data from different mapping populations having some anchor markers: e.g., data obtained from three different mapping populations were used to build a map comprising 239 microsatellite markers and spanning a map distance of 1430 cM in pea (Loridon et al., 2005). Bordat et al. (2011) prepared a new consensus functional map of pea spanning a map distance of 1389 cM with 97% of marker intervals below 10 cM. This map includes 3 morphological markers, 180 SSR markers, 133 RAPD, 6 RFLP, and 214 gene-based markers. This map has the advantage of incorporating markers used in different published maps (Weeden et al., 1998; Laucou et al.,1998; Choi et al., 2004; Loridon et al., 2005; Aubert et al., 2006; Jing et al., 2007; Prioul-Gervais et al., 2007). A composite map of the *V. faba* genome was constructed incorporating data from 11 F_2 families sharing the common female parent Vf6 (Román et al., 2004). The joint segregation analysis revealed 14 major linkage groups, and total map length obtained was 1559 cM, making it a comprehensive genetic map of faba bean. Ellwood et al. (2008) constructed a genetic map of faba bean with 151 markers revealing seven major and five small linkage groups and spanning a total length of 1685.8 cM.

Pfaff and Kahl (2003) tapped the resources of the databanks to create gene-specific markers based on genes whose products are involved in defense responses in chickpea, and generated a map covering 2500 cM. A consensus genetic map consisting of 555 loci involving different markers has also been developed in chickpea from 10 mapping populations (Millan et al., 2010). Bacterial artificial chromosome (BAC)-end sequences (BES)-derived SSR and diversity arrays technology (DArT) markers were used to construct a high-density genetic map based on a recombinant inbred line (RIL) population derived from the cross ICC 4958 (*C. arietinum*) × PI 489777 (*C. reticulatum*) comprising 1,291 markers on eight linkage groups (LGs) spanning a total of 845.56 cM distance (Thudi et al., 2011). The number of markers per linkage group ranged from 68 (LG 8) to 218 (LG 3) with an average inter-marker distance of 0.65 cM.

A gene-based genetic linkage map of lentil (*Lens culinaris* ssp. *culinaris*) was constructed using a F_5 population developed from a cross between the cultivars Digger (ILL5722) and Northfield (ILL5588), using 79 intron-targeted amplified polymorphic (ITAP) and 18 genomic simple sequence repeat (SSR) markers (Phan et al., 2007). Linkage analysis revealed seven linkage groups (LGs) comprising 5–25 markers that varied in length from 80.2 to 274.6 cM. Another linkage map of lentil was developed by Tullu et al. (2008) for mapping of earliness and plant height traits. Their map consisted of 207 markers (AFLP, RAPD, and SSRs) and covered the genome (1868 cM) with an average marker density of 8.9 cM. Tanyolac et al. (2010) constructed a molecular linkage map of lentil with 166 markers consisting of 11 linkage groups covering 1396.3 cM with an average map distance between framework markers of 8.4 cM. In summary, since molecular markers allow the indirect selection of interesting genotypes, segregating mapping populations and genetic maps are the means for identifying closely linked markers.

4.3 Towards Marker-Assisted Selection (MAS) for Virus Resistance Breeding

After initial hybridization, it takes several years in field selection and yield trials to develop a variety, but this time can be reduced using molecular markers to select for the trait of interest in early generations. MAS utilizes establishment of tight linkage between a molecular marker and the chromosomal location of the gene(s) to select desirable traits in the laboratory prior to confirmation of the phenotype. Closely linked markers may serve as efficient tools for virus resistance breeding in CSFLs, since they facilitate the selection of resistant plants prior to disease evaluation in the field. MAS has its greatest impact when phenotypic evaluations are inconsistent or unpredictable: e.g., in the case of PEMV and BLRV, virus infection relies on optimum climatic conditions or on viruliferous aphids. In practice, the availability of appropriate molecular markers allows breeding material to be screened *in vitro*, and only those plants that carry the resistance allele will be transferred to the field or greenhouse.

Knowledge of the inheritance of agronomic characters is a basic requirement to identify and localize interesting genes in linkage maps and to utilize these maps for MAS of these characters to accelerate the development of new cultivars. Due to the recessive mode of inheritance of BLRV/PSbMV resistance, a selfing generation is needed after each backcross to allow the identification of homozygous recessive genotypes on the phenotypic level. However, using co-dominant markers, such as SSRs, heterozygous carriers of a recessive resistance allele can be readily identified. Once a tightly linked marker has been developed, selection can be conducted in the laboratory and may not require its selection under field conditions for disease resistance in early generations and early stages of plant development. Table 11.4 gives a brief summary of mode of resistance and linked markers to various virus resistance genes. However, the scarce genomic resources developed for CSFLs and the limited saturation of the genomic regions bearing a candidate gene or putative QTL make it difficult to identify the most tightly linked markers for selecting a QTL. The single dominant gene for PEMV resistance (*En*) has been placed on linkage group III based on association with anchor loci *st*, *uni*, and *Adh 1* in pea (Gritton and Hagedorn 1980; Marx et al., 1985). The isozyme locus *Adh-1* is about 5 cM from the *En* gene (Weeden and Provvidenti, 1988) but it is not an ideal marker for MAS. Later, two RAPD markers, P256_900 and B500_400 were linked with *En* for resistance to PEMV (Yu et al., 1995a). Once molecular linkage maps with high marker density are constructed, the chances of finding a marker tightly linked to a desirable trait is increased, even when the chromosomal location of a breeding character/gene is unknown: e.g., three anchor loci—*wb*, *k*, and *Pgm-p*—were linked to a group of virus resistance genes *mo*, *sbm2*, *pmv*, *wmv2*, and *cyv1* on linkage group II in pea (Weeden et al., 1984; Provvidenti and Alconero, 1988; Provvidenti, 1990; Provvidenti and Hampton, 1993). A group of potyvirus resistance genes are found in a cluster on linkage group VI in pea and include *sbm1*, *sbm3*, *sbm4*, *cyv2*, and *wlv* (Provvidenti and Muehlbauer, 1990; Provvidenti and Hampton, 1993) based on the demonstration of linkage with *wlo* (Gritton and Hagedorn, 1975), *p* and *art1* (Skarzynska, 1988), and *Prx3* (Weeden et al., 1991). Yu et al. (1995b) described the development of an allele-specific associated primer

(ASAP) assay based on RAPD BC3021200, which is closely linked to *mo* (3 cM) and the linkage group II resistance gene cluster. The clustering of potyvirus resistance loci in two genomic regions has significant implications for virus resistance breeding. Two loci on linkage groups II and VI have shown association to resistance against PSbMV in pea (Gao et al., 2004b). A combination of parallel approaches was used to collate linked markers, particularly for *sbm-1* resistance on linkage group VI. Two random amplified polymorphic DNA markers (G05_2537 and L01_910) and one restriction fragment length polymorphism (P446) linked to *sbm1* have been identified and converted into three simple PCR-based STS markers (Frew et al., 2002). Linkage analysis in two F_2 populations showed that the G05-2537 is the most tightly linked of these three STS loci and is approximately 4 cM from *sbm1*. A parallel cDNA AFLP comparison of pairs of resistant and susceptible lines also identified an expressed tag marker just 0.7 cM from sbm-1 (Bruun-Rasmussen et al., 2007). Sequences derived from the genes for the eukaryotic translation initiation factors eIF4E and eIF(iso)4E were tightly linked to the resistance gene clusters on linkage groups VI and II, respectively. In a different mapping population, the gene *eIF(iso)4E* was also shown to be linked to *sbm-2* on linkage group II. This correlation strengthens the use of markers as valuable tools to assist in breeding for virus resistances into peas. Tolerance or partial resistance has been described for BLRV, pea streak virus, and red clover vein mosaic virus (Hampton, 1984), and alfalfa mosaic virus (Latham and Jones, 2001); however, the genetic loci underlying these resistances have not been characterized by linkage or quantitative trait loci (QTL) mapping.

The identification of molecular markers associated with favorable alleles of economically important genes provides more possibilities for the application of MAS. However, cross-overs may occur between the marker and the gene when the linked marker used for selection is at a distance away from the gene of interest, leading to a high percentage of false-positives/negatives in the screening process. Sometimes, a marker developed for a gene in one cross may not be useful in other crosses even though the same gene may be segregating in the second cross, unless the marker is from the gene itself. Likewise, markers developed for one pathotype or biotype may not have application to other locations in which different pathotypes or biotypes occur, unless resistance is controlled by the same gene. In the final analysis, success will depend on identifying marker(s) as close to the gene as possible for its utility across all populations. Substantial efforts in finding linked markers for MAS for virus resistance in CSFLs are still lacking and need to be addressed boisterously to avoid any setbacks.

4.4   Potential of Comparative Genomics for CSFLs

The large genome size and scarcity of genomic resources have hampered genomics-assisted virus resistance breeding in CSFLs. With the availability of whole genome sequences for some legume crops, there is some hope to accelerate molecular breeding in CSFLs through comparative mapping (Young and Bharti, 2012). Two legume species, *Medicago truncatula* and *Lotus japonicus*, have emerged as model plants to investigate the genetics of resistance or tolerance to stresses due to their small and diploid genomes, autogamous nature, short generation times, and prolific seed production (Handberg and Stougaard 1992; Cook 1999). *M. truncatula* SSRs were significantly transferable to CSFLs such as chickpea, pea, and faba bean (Gutierrez et al., 2005). To study gene function and genome evolution in legumes, an integrative database LegumeIP (http://plantgrn.noble.org/LegumeIP/) has been developed for comparative genomics. Another portal to get the genetic and genomic information about legumes is http://www.thelegumeportal.net/www/genetics.htm. Aubert et al. (2006) developed and mapped new gene-anchored markers in pea and *M. truncatula* that constitute 41 new links between the maps of two species. Sixty-six out of the 71 gene-based markers, which were previously assigned to *M. truncatula* genetic and physical maps, were found in regions syntenic between the *Lens* and *M. truncatula* genomes (Phan et al., 2007). This extensive conservation of macro- or microsynteny implies that genetic and genomic tools developed in Medicago can be readily applied to other legume species. A large set of markers from single- or low-copy-coding regions in *M. truncatula* have been created and used to identify potential orthologous genes in related legume crops such as pea, chickpea, faba bean, lentil, lupin, and clover (Zhu et al., 2005).

Certain comparisons between pea and lentil (Weeden et al., 1992), pea and chickpea (Simon and Muehlbauer, 1997), and between pea and Medicago (Choi et al., 2004; Kalo et al., 2004) provided some evidence for conserved gene order. Using the comparative genomics approach, a relatively high level of macrosynteny was observed within CSFLs in comparison with synteny between cool vs. warm season legumes or species of diverse legume clades (Kalo et al., 2004; Choi et al., 2004; Zhu et al., 2005). A direct and simple relationship was shown to exist between the *M. truncatula* and *Lens culinaris* ssp. *culinaris* chromosomes while difference in chromosome numbers was explained by the moderate level of chromosomal rearrangements (Phan et al., 2007). Results from a comparative mapping study among faba bean, lentil, and Medicago are in agreement with phylogenetic studies that place the genera *Vicia*, *Lens*, and *Pisum* within the tribe Viceae, while *Medicago* and *Melilotus* form a parallel tribe Trifolieae within the Galegoid or cool season legumes. Different levels of macrosynteny was observed between *M. truncatula*, *P. sativum*, *V. radiata*, *G. max*, and *Phaceolus vulgaris* dependent on phylogenetic distance with some chromosomal rearrangements (Choi et al., 2004; Ellwood et al., 2008). The identification of conserved features among CSFLs and model species is teaching us more about synteny and will allow identification of candidate genes in the model species corresponding to loci mapped in the less-studied crops. Bacterial artificial chromosome (BAC) libraries offer a powerful tool for comparison of physical and genetic information between model species and legume crops to search disease resistance genes: e.g., two *Hind III* BAC libraries of pea were constructed using germplasm accession PI 269818 to successfully amplify some of the previously known RGAs in pea (Coyne et al., 2007). A BAC library from chickpea line FLIP 84–92C was constructed to screen with a STMS marker Ta96, to tag Fusarium wilt resistance gene (Rajesh et al., 2004). Further screening of BAC library with markers positioned near the R genes will help to elucidate the organization of the R gene complex and generate more tightly linked markers. Different BAC libraries developed in pulse crops have been described in a recent review by Yu (2011). The ultimate objective is to develop dense molecular maps to facilitate faster and more detailed studies of gene/QTL synteny among related species and for validating the position of expression QTL (eQTL) or candidate gene(s) across variable genetic backgrounds.

ACKNOWLEDGEMENTS

Authors are grateful to all the collaborators of Legume Virus Project (http://www.cals.uidaho.edu/aphidtracker/) and funding (grant no. 2008-511010-4522) by the RAMP (Risk Assessment and Mitigation Program) of NIFA (National Institute for Food and Agriculture).

REFERENCES

Abbo, S., Molina, C., Jungmann, R., Grusak, M.A., Berkovitch, Z., Reifen, R., et al., 2005. Quantitative trait loci governing carotenoid concentration and weight in seeds of chickpea (*Cicer arietinum* L.). Theor. Appl. Genet. 111, 185–195.

Abraham, A.D., Menzel, W., Lesemann, D.E., Varrelmann, M., Vetten, H.J., 2006. Chickpea chlorotic stunt virus: a new polerovirus infecting cool-season food legumes in Ethiopia. Phytopathology 96, 437–446.

Agindotan, B., Perry, K.L., 2007. Macroarray detection of plant RNA viruses using randomly primed and amplified complementary DNAs from infected plants. Phytopathology 97, 119–127.

Anderson, J.M., Palukaitis, P., Zaitlin, M., 1992. A defective replicase gene induces resistance to cucumber mosaic virus in transgenic tobacco plants. Proc. Natl. Acad. Sci. USA 89, 8759–8763.

Aubert, G., Morin, J., Jacquin, F., Loridon, K., Quillet, M., Petit, A., et al., 2006. Functional mapping in pea, as an aid to the candidate gene selection and for investigating synteny with the model legume *Medicago truncatula*. Theor. Appl. Genet. 112, 1024–1041.

Baggett, J.R., Hampton, R.O., 1991. Inheritance of viral bean leaf roll tolerance in peas. J. Am. Soc. Hortic. Sci. 116, 728–731.

Bordat, A., Savois, V., Nicolas, M., Salse, J., Chauveau, A., Bourgeois, M., et al., 2011. Translational genomics in legumes allowed placing *in silico* 5460 unigenes on the pea functional map and identified candidate genes in *Pisum sativum* L. G3 (Bethesda) 1, 93–103.

Borgio, J.F., 2009. RNA interference (RNAi) technology: a promising tool for medicinal plant research. J. Medici. Plants Res. 3, 1176–1183.

Bosque-Perez, N.A., Buddenhagen, I.W., 1990. Studies on epidemiology of virus diseases of chickpea in California. Plant Dis. 74, 372–378.

Bruun-Rasmussen, M., Møller, I.S., Tulinius, G., Hansen, J.K.R., Lund, O.S., Johansen, I.E., 2007. The same allele of translation initiation factor 4E mediates resistance against two Potyvirus spp. in *Pisum sativum*. Mol. Plant Microbe Interact. 20, 1075–1082.

Choi, H.K., Mun, J.H., Kim, D.J., Zhu, H., Baek, J.M., Mudge, J., et al., 2004. Estimating genome conservation between crop and model legume species. Proc. Natl. Acad. Sci. USA 101, 15289–15294.

Chowrira, G.M., Cavileer, T.D., Gupta, S.K., Lurquin, P.F., Berger, P.H., 1998. Coat protein-mediated resistance to pea enation mosaic virus in transgenic *Pisum sativum* L. Transgenic Res. 7, 265–271.

Clark, M.F., Adams, A.N., 1977. Characteristics of the microplate method of enzyme-linked immunosorbent assay for the detection of plant viruses. J. Gen. Virol. 34, 475–483.

Cobos, M.J., Fernandez, M.J., Rubio, J., Kharrat, M., Moreno, M.T., Gil, J., et al., 2005. A linkage map of chickpea (*Cicer arietinum* L.) based on populations from Kabuli x Desi crosses: location of genes for resistance to fusarium wilt race 0. Theor. Appl. Genet. 110, 1347–1353.

Collard, B.C.Y., Pang, E.C.K., Taylor, P.W.J., 2003. Selection of wild *Cicer* accessions for the generation of mapping population segregating for resistance to *Ascochyta blight*. Euphytica 130 (1-9).

Cook, D., 1999. *Medicago truncatula*: a model in the making!. Curr. Opin. Plant Biol. 2, 301–304.

Coyne, C.J., McClendon, M.T., Walling, J.G., Timmerman-Vaughan, G.M., Murray, S., Meksem, K., et al. (2007) Construction and characterization of two bacterial artificial chromosome libraries of pea (*Pisum sativum* L.) for the isolation of economically important genes. 50: 871-875.

Dirlewanger, E., Isaac, P.G., Ranade, S., Belajouza, M., Cousin, R., Vienne, D.D., 1994. Restriction fragment length polymorphism analysis of loci associated with disease resistance genes and developmental traits in *Pisum sativum* L.. Theor. Appl. Genet. 88, 17–27.

Drake, J.W., Holland, J.J., 1999. Mutation rates among RNA viruses. Proc. Natl. Acad. Sci. USA 96, 13910–13913.

Duraisamy, G.S., Pokorný, R., Holková, L., 2011. Possibility of Bean yellow mosaic virus detection in Gladiolus plants by different methods. J. Plant Dis. Protect. 118 (1), 2–6.

Ellwood, S.R., Phan, H.T.T., Jordan, M., Hane, J., Torres, A.M., Avila, C.M., et al., 2008. Construction of a comparative genetic map in faba bean (*Vicia faba* L.); conservation of genome structure with *Lens culinaris*. BMC Genomics 9, 380.

FAOSTAT (2010) <http://faostat.fao.org/site/567/default.aspx#ancor/>.

Flandez-Galvez, H., Ford, R., Pang, E.C.K., Taylor, P.W.J., 2003. An intraspecific linkage map of the chickpea (*Cicer arietinum* L.) genome based on sequence tagged microsatellite site and resistance gene analog markers. Theor. Appl. Genet. 106, 1447–1456.

Frew, T.J., Russell, A.C., Timmerman-Vaughan, G.M., 2002. Sequence tagged site markers linked to the *sbm1* gene for resistance to pea seed borne mosaic virus in pea. Plant Breed. 121, 512–516.

Fuchs, M., McFerson, J.R., Tricoli, D.M., McMaster, J.R., Deng, R.Z., Boeshore, M.L., et al., 1997. Cantaloupe line CZW-30 containing coat protein genes of cucumber mosaic virus, zucchini yellow mosaic virus, and watermelon mosaic virus-2 is resistant to these three viruses in the field. Mol. Breed. 3, 279–290.

Gao, Z., Johansen, E., Eyers, S., Thomas, C.L., Ellis, T.H.N., Maule, A.J., 2004a. The potyvirus recessive resistance gene, *sbm1*, identifies a novel role for translation initiation factor eIF4E in cell-to-cell trafficking. Plant J. 40, 376–385.

Gao, Z., Eyers, S., Thomas, C.L., Eills, T.H.N., Maule, A.J., 2004b. Identification of markers tightly linked to the sbm recessive genes for resistance to pea seed-borne mosaic virus. Theor. Appl. Genet. 109, 488–494.

Gritton, E.T., Hagedorn, D.J., 1975. Linkage of the genes *sbm* and *wlo* in peas. Crop Sci. 11, 945–946.

Gritton, E.T., Hagedorn, D.J., 1980. Linkage of the *En* and *st* genes in pea. Pisum Newsl. 12, 26–27.

Gutierrez, M.V., Vaz Patto, M.C., Huguet, T., Cubero, J.I., Moreno, M.T., Torres, A.M., 2005. Cross-species amplification of Medicago truncatula microsatellite across three major pulse crops. Theor. Appl. Genet. 110, 1210–1217.

Haddad, N.I., Muehlbauer, F.J., Hampton, R.O., 1978. Inheritance of resistance to pea seed-borne mosaic virus in lentils. Crop Sci. 18, 613–615.

Hampton, R.O., 1984. Pea seed-borne mosaic. Compend. of Pea Dis., Am. Phytopathol. Soc., 34–35.

Hamwieh, A., Udupa, A.S.M., Choumane, W., Sarker, A., Dreyer, F., Jung, C., et al., 2005. A genetic linkage map of *Lens* sp. based on microsatellite and AFLP markers and the localization of fusarium vascular wilt resistance. Theor. Appl. Genet. 110, 669–677.

Handberg, K., Stougaard, J., 1992. *Lotus japonicus*, an autogamous, diploid legume species for classical and molecular genetics. Plant J. 2, 487–496.

Henson, J.M., French, R., 1993. The polymerase chain reaction and plant disease diagnosis. Annu. Rev. Phytopathol. 31, 81–109.

Hill, K.K., Jarvis-Eagan, N., Halk, E.L., Krahn, K.J., Liao, L.W., Mathewson, R.S., et al., 1991. The development of virus-resistant alfalfa, *Medicago sativa* L. Biotechnol. 9, 373–377.

Jing, R., Johnson, R., Seres, A., Kiss, G., Ambrose, M.J., Knox, M.R., et al., 2007. Gene based sequence diversity analysis of field pea (*Pisum*). Genetics 177, 2263–2275.

Johansen, I.E., Lund, O.S., Hjulsager, C.K., Laursen, J., 2001. Recessive resistance in *Pisum sativum* and potyvirus pathotype resolved in a gene-for-cistron correspondence between host and virus. J. Virol. 75, 6609–6614.

Jones, A.L., Johansen, I.E., Bean, S.J., Bach, I., Maule., A.J., 1998. Specificity of resistance to pea seed-borne mosaic potyvirus in transgenic peas expressing the viral replicase (NIb) gene. J. Gen. Virol. 79, 3129–3137.

Kalo, P., Seres, A., Taylor, S.A., Jakab, J., Kevei, Z., Kereszt, A., et al., 2004. Comparative mapping between *Medicago sativa* and *Pisum sativum*. Mol. Genet. Genomics 272, 235–246.

Kang, B., Yeam, I., Jahn, M.M., 2005. Genetics of plant virus resistance. Annu. Rev. Phytopathol. 43, 581–621.

Katul, L. (1992). Characterization by Serology and Molecular Biology of Bean Leaf Roll Virus and Faba Bean Necrotic Yellow Virus. Ph.D. Thesis, University of Göttingen, Germany, pp. 115.

Keller, K.E., Johansen, E., Martin, R.R., Hampton, R.O., 1998. Potyvirus genome linked protein (VPg) determines pea seed-borne mosaic virus pathotype-specific virulence in *Pisum sativum*. Mol. Plant Microbe. Interact. 11, 124–130.

Kumar, J., Choudhary, A.K., Solanki, R.K., Pratap, A., 2011. Towards marker-assisted selection in pulses: a review. Plant Breed. 130, 297–313.

Kumari, S.G., Makkouk, K.A., 2003. Differentiation among bean leafroll virus susceptible and resistant lentil and faba bean genotypes on the basis of virus movement and multiplication. J. Phytopathol. 151, 19–25.

Kumari, S.G., Makkouk, K.A., Loh., M.H., Negassi, K., Tsegay, S., Kidane, R., et al., 2008. Viral diseases affecting chickpea crops in Eritrea. Phytopathol. Mediterr. 47, 42–49.

Latham, L.J., Jones, R.A.C., 2001. Incidence of virus infection in experimental plots, commercial crops, and seed stocks of cool season crop legumes. Aus. J. Agri. Res. 52, 397–413.

Laucou, V., Haurogné, K., Ellis, N., Rameau, C., 1998. Genetic mapping in pea. 1. RAPD-based genetic linkage map of *Pisum sativum*. Theor. Appl. Genet. 97, 905–915.

Lecoq, H., Mourya, B., Desbiez, C., Palloix, A., Pitrat, M., 2004. Durable virus resistance in plants through conventional approaches: a challenge. Virus Res. 100, 31–39.

Lim, D.V., Simpson, J.M., Kearns, E.A., Kramer, M.F., 2005. Current and developing technologies for monitoring agents of bioterrorism and biowarfare. Clin. Microbiol. Rev. 18 (4), 583–607.

Loridon, K., McPhee, K., Morin, J., Dubreuil, P., Pilet-Nayel, M.L., Aubert, G., et al., 2005. Microsatellite marker polymorphism and mapping in pea (*Pisum sativum* L.). Theor. Appl. Genet. 111, 1022–1031.

Macfarlane, S.A., Davies, J.A., 1992. Plants transformed with a region of the 201-kilodalton replicase gene from pea early browning virus RNA1 are resistant to virus infection. Proc. Natl. Acad. Sci. USA 89, 5829–5833.

Makkouk, K.M., Kumari, S.G., 2009. Epidemiology and integrated management of persistently transmitted aphid-borne viruses of legume and cereal crops in West Asia and North Africa. Virus Res. 141, 209–218.

Makkouk, K.M., Kumari, S.G., van Leur, J.A.G., 2002. Screening and selection of faba bean (*Vicia faba* L.) germplasm resistant to bean leaf roll virus. Aust. J. Agric. Res. 53, 1077–1082.

Marx, G.A., Weeden, N.F., Provvidenti, R., 1985. Linkage relationships among markers in chromosome 3 and *En*, a gene conferring virus resistance. Pisum Newsl. 17, 57–60.

Meyers, B.C., Dickerman, A.W., Michelmore, R.W., Sivaramakrishnan, S., Sobral, B.W., Young, N.D., 1999. Plant disease resistance genes encode members of an ancient and diverse protein family within the nucleotide-binding superfamily. Plant J. 20, 317–332.

Millan, T., Winter, P., Jungling, R., Gil, J., Rubio, J., Cho, S., et al., 2010. A consensus genetic map of chickpea (*Cicer arietinum* L.) based on 10 mapping populations. Euphytica 175, 175–189.

Muehlbauer, F.J., Weeden, N.F., Hoffman, D.L., 1989. Inheritance and linkage relationships of morphological and isozyme loci in lentil (*Lens* Miller). J. Hered. 80, 298–303.

Mulholland, V., 2009. Immunocapture-PCR for plant virus detection. Methods Mol. Biol. 508, 183–192.

Nayak, S.N., Zhu, H., Varghese, N., Datta, S., Choi, H.K., Horres, R., et al., 2010. Integration of novel SSR and gene-based SNP marker loci in the chickpea genetic map and establishment of new anchor points with *Medicago truncatula* genome. Theor. Appl. Genet. 120, 1415–1441.

Niu, Q.W., Lin, S.S., Reyes, J.L., Chen, K.C., Wu, H.W., Yeh, S.D., et al., 2006. Expression of artificial microRNAs in transgenic *Arabidopsis thaliana* confers virus resistance. Nat. Biotech. 24, 1420–1428.

Palomino, C., Satovic, Z., Cubero, J.I., Torres, A.M., 2006. Identification and characterization of NBS-LRR class resistance gene analogs in faba bean (*Vicia faba* L.) and chickpea (*Cicer arietinum* L.). Genome 49, 1227–1237.

Palomino, C., Fernández-Romero, M.D., Rubio, J., Torres, A., Moreno, M.T., Millan, T., 2009. Integration of new CAPS and dCAPS-RGA markers into a composite chickpea genetic map and their association with disease resistance. Theor. Appl. Genet. 118, 671–682.

Parlevliet, J.E., 2002. Durability of resistance against fungal, bacterial and viral pathogens; present situation. Euphytica 124, 147–156.

Pfaff, T., Kahl, G., 2003. Mapping of gene-specific markers on the genetic map of chickpea (*Cicer arietinum* L.). Mol. Gen. Genomics 269, 243–251.

Phan, H.T.T., Ellwood, S.R., Hane, J.K., Ford, R., Materne, M., Oliver, R.P., 2007. Extensive macrosynteny between *Medicago truncatula* and *Lens culinaris* ssp. culinaris. Theor. Appl. Genet. 114, 549–558.

Prins, M., Laimer, M., Noris, E., Schubert, J., Wassenegger, M., Teofer, M., 2008. Strategies for antiviral resistance in transgenic plants. Mol. Plant Pathol. 9, 73–83.

Prioul-Gervais, S., Deniot, G., Receveur, E.M., Frankewitz, A.M., Fourmann, M., Rameaue, C., et al., 2007. Candidate genes for quantitative resistance to *Mycosphaerella pinodes* in pea (*Pisum sativum* L.). Theor. Appl. Genet. 114, 971–984.

Provvidenti, R., 1990. Inheritance of resistance to pea mosaic virus in *Pisum sativum*. J. Hered. 81, 143–145.

Provvidenti, R., Alconero, R., 1988. Inheritance of resistance to a lentil strain of pea seed-borne mosaic virus in *Pisum sativum*. J. Hered. 79, 45–47.

Provvidenti, R., Hampton, R.O., 1992. Sources of resistance to viruses in the Potyviridae. Arch. Virol. (Suppl. 5), 189–211.

Provvidenti, R., Hampton, R.O., 1993. Inheritance of resistance to white lupin mosaic virus in common pea. Hort. Sci. 28, 836–837.

Provvidenti, R., Muehlbauer, F.J., 1990. Evidence of a cluster of linked genes for resistance to pea seed-borne mosaic virus and clover yellow vein virus in *Pisum sativum*. Pisum Nwsl. 22, 43–45.

Qu, J., Ye, J., Fang, R., 2007. Artificial miRNA-mediated virus resistance in plants. J. Virol. 81, 6690–6699.

Rajesh, P.N., Coyne, C., Meksem, K., Sharma, K.D., Gupta, V., Muehlbauer, F.J., 2004. Construction of a *Hind III* Bacterial artificial chromosome library and its use in identification of clones associated with disease resistance in chickpea. Theor. Appl. Genet. 108, 663–669.

Rakshit, S., Winter, P., Tekeoglu, M., Juarez Muñoz, J., Pfaff, T., Benko-Iseppon, A.M., et al., 2003. DAF marker tightly linked to a major locus for *Ascochyta blight* resistance in chickpea (*Cicer arietinum* L.). Euphytica 132, 23–30.

Randhawa, H., Weeden, N.F., 2011. Refinement of the position of En on LG III and identification of closely linked DNA markers. Pisum Genet. in press.

Román, B., Satovic, Z., Pozarkova, D., Macas, J., Dolezel, J., Cubero, J.I., et al., 2004. Development of a composite map in *Vicia faba*, breeding applications and future prospects. Theor. Appl. Genet. 108, 1079–1088.

Sanford, J.C., Johnston, S.A., 1985. The concept of pathogen derived resistance. J. Theor. Biol. 113, 395–405.

Schroeder, W.T., Barton, D.W., 1958. The nature and inheritance of resistance to the pea enation mosaic virus in garden pea, *Pisum sativum* L. Phytopathology 48, 628–632.

Simon, C.J., Muehlbauer, F.J., 1997. Construction of chickpea linkage map and its comparison with the maps of pea and lentil. J. Herid. 88, 115–119.

Skarzynska, A., 1988. Supplemental mapping data for chromosome 6. Psium Newslett 20, 34–36.

Smykal, P., Safarova, D., Navratil, M., Dostalova, R., 2010. Marker assisted pea breeding: eIF4E allele specific markers to pea seed-borne mosaic virus (PSbMV) resistance. Mol. Breed. 26, 425–438.

Tadmor, Y., Zamir, D., Ladizinsky, G., 1987. Genetic mapping of an ancient translocation in the genus Lens. Theor. Appl. Genet. 73, 883–892.

Tanyolac, B., Ozatay, S., Kahraman, A., Muehlbauer, F.J., 2010. Linkage mapping of lentil (*Lens culinaris* L.) genome using recombinant inbred lines revealed by AFLP, ISSR, RAPD and some morphologic markers. J. Agri. Biotechnol. Sustainable Dev. 2, 01–06.

Tepfer, M., 2002. Risk assessment of virus-resistant transgenic plants. Annu. Rev. Phytopathol. 40, 467–491.

Thudi, M., Bohra, A., Nayak, S.N., Varghese, N., Shah, T.M., Penmetsa, R.V., et al., 2011. Novel SSR Markers from BAC-End sequences, DArT arrays and a comprehensive genetic map with 1,291 Marker Loci for chickpea (*Cicer arietinum* L.). PLoS ONE 6, e27275. doi: 10.1371/journal.pone.0027275.

Timmerman-Vaughan, G.M., Pither-Joyce, M.D., Cooper, P.A., Russell, A.C., Grant, J.E., 2001. Partial resistance of transgenic peas to alfalfa mosaic virus under greenhouse and field conditions. Crop Sci. 41, 846–853.

Torres, A.M., Avila, C.M., Gutierrez, N., Palomino, C., Moreno, M.T., Cubero, J.I., 2010. Marker-assisted selection in faba bean (*Vicia faba* L.). Field Crops Res. 115, 243–252.

Tullu, A., Tar'an, B., Warkentin, T., Vandenberg, A., 2008. Construction of an intraspecific linkage map and QTL analysis for earliness and plant height in lentil. Crop Sci. 48, 2254–2264.

Upadhyaya, H.D., Thudi, M., Dronavalli, N., Gujaria, N., Singh, S., Sharma, S., et al., 2011. Genomic tools and germplasm diversity for chickpea improvement. Plant Genetic Res. 9, 45–58.

Varshney, R.K., Close, T.J., Singh, N.K., Hoisington, D.A., Cook, D.R., 2009. Orphan legume crops enter the genomics era. Curr. Opin. Plant Biol. 12, 1–9.

Varshney, R.K., Glaszmann, J.C., Leung, H., Ribaut, J.M., 2010. More genomic resources for less-studied crops. Trends Biotechnol. 28, 452–460.

Weeden, N.F., Provvidenti, R., 1988. A marker locus, Adh-1, for resistance to pea enation mosaic virus in *Pisum sativum*. J. Hered. 79, 128–131.

Weeden, N.F., Provvidenti, R., Wolko, B., 1991. Prx-3 is linked to *sbm*, the gene conferring resistance to pea seed-borne mosaic virus. Pisum Genet. 23, 42–43.

Weeden, N.F., Muehlbauer, F.J., Ladizinsky, G., 1992. Extensive conservation of linkage relationships between pea and lentil genetic maps. J. Hered. 83, 123–129.

Weeden, N.F., Ellis, T.H.N., Timmerman-Vaughan, G.M., Swiecicki, W.K., Rozov, S.M., Berdnikov, V.A., 1998. A consensus linkage map for *Pisum sativum*. Pisum Genet. 30, 1–3.

Yaish, M.W.F., Sáenz de Miera, L.E., Pérez de la Vega, M., 2004. Isolation of a family of resistance gene analogue sequences of the nucleotide binding site (NBS) type from Lens species. Genome 47, 650–659.

Young, N.D., Bharti, A.K., 2012. Genome-enabled insights into legume biology. Annu. Rev. Plant Biol. 63, 283–305.

Yu, J., Gu, W.K., Provvidenti, R., Weeden, N.F., 1995a. Identifying and mapping two DNA markers linked to the gene conferring resistance to pea enation mosaic virus. J. Am. Soc. Hort. Sci. 120, 730–733.

Yu, J., Gu, W.K., Weeden, N.F., 1995b. Development of an ASAP marker for resistance to bean yellow mosaic virus in *Pisum sativum*. Pisum Genet. 28, 31–32.

Yu, K., 2011. Bacterial artificial chromosome libraries of pulse crops: characteristics and applications. J. Biomed. Biotech. doi: 10.1155/2012/493186.

Zhu, H., Choi, H.K., Cook, D.R., Shoemaker, R.C., 2005. Bridging model and crop legumes through comparative genomics. Plant Physiol. 137, 1189–1196.

Zohary, D., Hopf, M., 1973. Domestication of pulses in the old world. Science 182, 887–894.

Expert Advice on Policy and Developmental Aspects

Chapter 12

Talking Agricultural Sustainability Issues—an Interview with Dr. Gurdev Khush

Gurbir S. Bhullar

Swiss Federal Institute of Technology, Zurich, Switzerland

Dr. Gurdev Khush is an agricultural scientist well known to the world for his remarkable role in the development of high-yielding rice varieties, which greatly contributed towards ushering a number of countries (particularly in Asia) into the Green Revolution. While leading research at the International Rice Research Institute in the Philippines, as Principal Plant Breeder and Head of the Plant Breeding, Genetics and Biochemistry division, he contributed more than 300 rice varieties including mega-varieties such as IR36, which was planted over 11 million hectares in Asia in the 1980s and attained the status of the most widely planted variety of any food crop. With large-scale adoption of these high-yielding varieties, many rice-producing countries attained self-sufficiency in food grain production. Along with honorary doctorates and degrees from twelve universities, his marvelous contribution towards world food security has been honored with many international awards, including the Borlaug Award (1977), Japan Prize (1987), World Food Prize (1996), Wolf Prize for Agriculture (2000), Padma Shri Award – India (2001), and the Golden Sickle Award (2007), to mention a few. He has served as consultant to several national governments and as advisor to a number of institutions and foundations. Besides being an excellent scientist he is a great role model for young scientists. With gratitude to this great personality, I take pleasure in presenting an interview with Dr. Khush for the readers of this book.

Question: **Dr. Khush, which scientific issues related to agricultural sustainability concern you most?**

Dr. Khush: The most important scientific issues related to agricultural sustainability are depletion of water resources, overuse or misuse of agricultural chemicals, and the expected impacts of climate change on crop productivity,

Agricultural Sustainability. DOI: http://dx.doi.org/10.1016/B978-0-12-404560-6.00012-5

such as higher temperatures, new races of diseases and insects, drought and floods, etc.

Question: **A large proportion of the human population still suffers from hunger and malnutrition, particularly in underdeveloped parts of the world. What do you see as the major reason for this problem? Is it insufficient food production or something else?**

Dr. Khush: The main reason for hunger and malnutrition is economic disparity between the haves and have-nots. Poor countries and poor people do not have the resources to develop infrastructure for agricultural development, such as irrigation, roads, markets, electrification, and mechanization, resulting in insufficient food production. Lack of credit and unemployment are some of the other reasons for hunger and poverty.

Question: **A lot of different programs to eradicate hunger are in operation around the world. But the problem still persists. Do you think the efforts to fight hunger have been enough? In your opinion, are the current international policies to eradicate hunger sufficient? What major changes (if any) would you like to suggest?**

Dr. Khush: Donor countries and organizations have put considerable resources into eradicating hunger. Some have been successful and others less so. A successful example is the Green Revolution that saved a billion lives from hunger and starvation. However, the fight against hunger has not yet been won. Therefore, continued efforts are needed. There should be more investments in infrastructure development.

Question: **There is no doubt that the technologies that led to the Green Revolution in the second half of the past century are widely recognized and appreciated for their contribution towards increased yields and self-sufficiency in food grain production, particularly in agricultural economies such as that of India. At the same time, concerns over degradation of natural resources due to overexploitation and overuse of high-energy inputs and loss of biodiversity due to widespread monocultures, question modern farming practices. What is your opinion in this debate?**

Dr. Khush: There is no alternative to modern farming practices. That is the only way to feed 7 billion rising to 8 billion people. In some instances there has been over-exploitation of water resources due to incorrect government policies such as free electricity to farmers or over-use and misuse of chemicals, resulting in environmental pollution. Most of the critics of modern farming in Asia are from western countries, but all the western countries follow modern farming practices!

Question: **Genetically modified crop varieties are advocated as a breakthrough technology that has potential to play an important role in** agriculture, but there is also skepticism regarding the possible ecological impacts. Would you like to share your opinion on that discussion?

Dr. Khush: Genetically modified crop varieties can help us produce more food, are more nutritious, require less in the way of pesticides, are cost effective to grow, and are environmentally benign. They contribute to agricultural sustainability by reducing the need for harmful pesticides. Most of the concerns about possible ecological impacts are exaggerated and unjustified.

Question: **In a recent special report of the IPCC, it has been projected that the frequency and intensity of extreme weather events such as floods and droughts will increase in the coming decades, severely affecting agriculture and food security. Besides, there is an ever-increasing demand for food and energy caused by the rapidly growing population. It is predicted that underprivileged people, particularly in less developed nations, are going to suffer from a severe food shortage; the recent famine in the Horn of Africa is evidence. Considering this scenario, what kind of research priorities would you suggest to the agricultural research community?**

Dr. Khush: The most important research thrust should be to develop more-resilient crop varieties tolerant to higher temperatures, drought, floods, and resistant to new races of diseases and insects likely to evolve under changing environmental situations.

Question: **As you are providing consultancy to several national governments, would you like to suggest some changes to agricultural research policy, particularly in the case of countries falling into the "extreme risk" category?**

Dr. Khush: The risk factors in those countries are well known. Research and development policies should address those risk factors.

Question: **What are the key issues that need to be resolved for enhancing the institutional setup for agricultural research, particularly in developing countries? As you have made significant contributions both intellectually and financially towards betterment of agricultural institutions, particularly in India, perhaps you can share some examples from your experience.**

Dr. Khush: I am familiar only with agricultural universities in India. Any institution should have periodic reviews of its research programs and administrative set-up. I am not sure if any such reviews of agricultural universities have been undertaken and whether any redundant programs have been discontinued and new programs added. I understand there is lot of bureaucracy in the operations. Scientists spend a lot of time on mundane chores. When I was at IRRI, I was given a budget and I did not have to spend time to get approval from higher authorities to spend any amount.

Question: **Do you think that equipping young agricultural research-ers, especially in developing countries, with better scientific training may contribute towards agricultural sustainability? What kind of training programs would you suggest for that?**

Dr. Khush: Training of young agricultural scientists in developing countries is very important. Scientists should be equipped with knowledge to address problems of agricultural sustainability such as judicious use of agricultural chemicals, water-saving technologies, tillage practices to reduce soil erosion, proper rotations, use of IPM for pest control, and use of electronic media to reach the farmers.

Question: **Dr. Khush, now I would like to learn some more about you on a personal-professional note. Would you like to share any experiences that influenced you to pursue a career in agricultural research?**

Dr. Khush: I grew up on a wheat farm. My uncles were farmers, and I helped them in agricultural operations when I was growing up. Therefore, I developed a strong interest in agriculture. My father was the main person who encouraged me to go for agricultural education.

Question: **Did you face any major challenges at personal or professional level?**

Dr. Khush: I have been fortunate to have received coveted support for personal and professional advancement. My parents gave me guidance and support for my education up to college level and taught me a good value system. The University of California provided me with adequate resources for graduate studies, and my supervisor taught me the importance of scientific enquiry. I was lucky to be invited to join one of the premier international agricultural research institutes dedicated to eradication of poverty and hunger, which provided me the opportunities to grow professionally.

Question: **What would you be doing if you weren't involved in science?**

Dr. Khush: I would have joined politics and fought elections.

Question: **What do you recall as the best part of your work—the part that gave you the most satisfaction?**

Dr. Khush: The best part of my work was to be able to develop technologies for food production and alleviation of hunger, to be able to work with scientists in other countries and to train young scientists.

Question: **You have been honored with many prestigious awards for your contribution towards agricultural research. Would you like to share some memories and feelings from those moments?**

Dr. Khush: My first major award was the Japan Prize in 1987, which was presented by Prince Akihito (the present emperor of Japan). I sat next to

Princess Michiko (the present Empress of Japan) at the post-prize-awarding opera performance, and my wife sat next to the prince. Our entire family joined the private reception with the prince and the princess after the banquet. We also had an audience with the then Emperor Hirohito. We were overwhelmed by the hospitality. My second major prize was the World Food Prize in 1996, and it was presented by Nobel laureate Dr. Norman Borlaug, the father of the Green Revolution.

Question: **During your scientific career, you have worked on several projects. Is there anything that you can think of, on which you wanted to work but could not spare time for?**

Dr. Khush: I wanted to spend more time on improvement of *Basmati* rice but could not do so, due to time constraints and involvement with many other projects.

Question: **Dr. Khush, thanks for sparing precious time and for sharing your knowledge with us. In the end, what advice would you give to the emerging scientists and students who intend to shape their future with a career in agricultural research?**

Dr. Khush: The best advice I can give to young scientists is to work hard. There are no short cuts in research. Do not worry about rewards. Sincere work always pays.

Chapter 13

Economics and Politics of Farm Subsidies in India

S.S. Johl[1]

Chancellor, Central University of Punjab, Bathinda, Punjab, India

Economic policy prescriptions out of tune with political ground realities remain futile exercises and sterile in their impact on the political economy of the country. On the other hand, political decisions and policy dispensations not based on economic rationale wreak disaster on the financial health of the economy. There has to be, therefore, a functional synergy between rational economic analysis and political decision-making for healthy socio-economic growth and development as well as sturdy political values and governing systems of the country. Unfortunately, in almost all the governance systems of different countries, policy formation carries much deeper hues of political considerations, and economic realities are often blatantly ignored at the peril of the financial health of the economy and social equity.

There is no doubt that subsidies play an important role in a welfare state, both in respect of producer subsidies as well as consumer subsidies in terms of restoring equities and improving the income levels and living conditions of disadvantaged sections of society. Many years ago, in his book *Future Shock*, Irvin Toffler propounded the theory that, as the choice matrix expands with the development of technology, and multiple choices become available, fewer and fewer people make the right choices to succeed, and larger and larger

---

[1] Dr S.S. Johl is an agricultural economist of international repute. He has made significant contributions while serving in prominent positions such as consultant to the United Nations Food and Agriculture Organization (FAO), Economic Commission for Western Asia (ECWA), and the World Bank. In India, he has been Chairman of the Commission for Agricultural Costs and Prices, Director of the Reserve Bank of India, and Member of the Prime Minister's Economic Advisory Council. He is a fellow of a number of reputed academic, literary, and professional organizations. Recognizing his contributions, the Government of India granted him the prestigious Padma Bhushan Award in 2004.

Agricultural Sustainability. DOI: http://dx.doi.org/10.1016/B978-0-12-404560-6.00013-7

proportions of the society fail to make the right choices. Hence, the burden of welfare falls on a smaller and smaller proportion of the successful population with heavier and heavier diversions of tax money for welfare purposes in the form of consumer and producer subsidies and income support programs. It is an inevitable road map for a continuous development process both in the developed as well as developing economies. Whereas consumer subsidies are meant to improve the access of disadvantaged sections of society to the necessities of life, such as food, fiber, and fuel, producer subsidies have multiple objectives to be achieved.

1. CONSUMER SUBSIDIES

Consumer subsidies, as for other subsidies, must be focused, and commodities meant for the targeted populations must reach the beneficiaries identified with a discriminating intellect so that, in the name of the poor and needy, the subsidies do not gravitate to undeserving sections of society. In India, subsidies are distributed in the form of rationed goods such as wheat grains, flour, rice, pulses, cooking oils, sugar, kerosene, etc. on ration cards of various categories such as ordinary cards, below poverty line yellow cards, etc. It is a known fact that, in the first place, significant numbers of ration cards get issued to undeserving families; a large part of these consumer goods are also pilfered and sold in the open market by the depot holders in connivance with officials of the concerned government department. According to the Planning Commission of India Survey of 2008, not more than 42% of subsidized food grains reach the targeted populations. This means that 58% of the food grain meant for the Public Distribution System finds its way onto the open market and never reaches the targeted populations. The story of other rationed goods provided through the Public Distribution System (PDS) cannot be very different. In the state of Punjab, this diversion is estimated to be around 90%, which therefore makes Punjab the most corrupt in the country. Yet the food subsidy burden on the exchequer in India has reached Rupees (INR) 723.7 billion in 2011–12 and is expected to be around INR 750 billion in 2012–13 (budget estimates). Subsidy on food grains alone amounts to more than INR 150 billion. This means that, if the leakage is not plugged, food grains alone worth 85 billion rupees, meant for the poor, would not reach them in the year 2012–13. This financial liability is when the poverty line is defined at expenditure of INR 21 per person per day for rural populations and INR 27 per person per day for urban populations. From the angle of nutritional deficiency, three-quarters of the population in India needs to be provided with subsidized food compared with about one-third of the population being at present defined as below poverty line beneficiaries. Again, the government of India plans to provide 7 kg food grain at INR 1 per kg to the targeted below poverty line (BPL) families, and 3 kg at 50% price to the rest of the families in India, costing about INR 2 trillion, which would require about 65 million tons of food grain. This is a huge call on the physical and financial resources of the country. The government budget has provided for INR 3.5 trillion for the purpose, which also includes provision for enhancing production. It is a moot question as to what extent the governance system in place will be able to achieve the objectives of the scheme. All this requires unstinted commitment and a high level of efficiency at various levels of administration, of which there is a lack in the country.

Direct income support is advocated in many developing and developed countries, so that the benefit accrues to the targeted populations without any losses or pilfering at the hands of administrators and handlers of the products. We have examples of coupons and green stamps issued by food stores and other corporate depots as discounts, which can be used to buy the products of choice by the stamp/coupon holders. If direct income support is provided in hard cash, this can be easily used/misused for purposes other than food items. If, however, the direct income support is provided in the form of food stamps to be used only for specified food items, this will serve the purpose of providing the food to the needy. Even if these stamps are exchanged for money, the stamps cannot be used for any other purpose and ultimately will end up in the purchase of specified food items only. These food stamps can be encashed at specified banks after having been marked as used by the licensed depot holders and presented in the banks by these depot holders only. Banks in turn would claim the money from the government department(s) issuing these stamps. There will still be some diversion of the subsidy to other than food purposes, but its extent will be minimized through this system. Since a very large proportion of the BPL population, both farming and non-farming, lives in rural areas, this system will help the rural populations in improving their food security and will reduce considerably the extensive malnutrition particularly prevalent among vulnerable sections of rural society, such as children, women, and the elderly. Since the underprivileged sections of society, viz. schedule caste and backward class populations, are mainly in the rural areas, particularly so in Punjab, the system will help them considerably in respect of food security, access to food, and nutritional security.

No doubt food subsidies are an inevitable provision-claiming priority in the budget of any welfare state, yet a progressive increase in the subsidy bill and the number of persons covered in absolute and percentage terms should be a matter of grave concern, because it puts a heavy drain on the exchequer as is indicated in Table 13.1. This is a huge burden on the state exchequer and is fast escalating with the passing of time, which amounts to an average escalation of INR 23.14 billion per year.

TABLE 13.1 Expenditure on Food Subsidy, Government of India,1980–81 through 2011–12.

Year	Subsidy (billion INR)	Year	Subsidy (billion INR)
1980–81	6.50	2001–02	120.60
1985–86	16.50	2005–06	230.77
1990–91	24.50	2010–11	629.30
1995–96	53.77	2011–12	723.70

Source: Fertilizer Statistics, The Fertilizer Association of India, New Delhi.

2. AGRICULTURAL SUBSIDIES

Agricultural subsidies are given in several forms. Fertilizer subsidies come through the system of retention prices provided to the fertilizer producers (fertilizer factories). These prices are calculated on the basis of economic costs to the factories, based on their investments, running costs, and a margin of profit (say 12%), and the government then decides on the subsidized prices at which the fertilizers are sold to the farmers. Imported fertilizers are also provided to the farmers at subsidized prices. The factories have been resorting to the malpractices of excessive reporting of investments and working costs, known as *gold plating*, resulting in estimation of higher retention prices. Since every factory had a different level of investment and working costs, retention prices varied from factory to factory. Since the economic costs of every factory were paid, along with a profit margin, it sucked out the incentive to make improvements and introduce functional efficiency. Although on several occasions the international prices of fertilizers remained lower than the same fertilizers produced in India, the government was obliged to keep the domestic factories producing the fertilizers at higher costs due to the international prices being highly volatile in nature. Moreover, India being a major consumer of fertilizers, its entry into the market for purchase of fertilizers would instantly increase the international prices. Through time, however, the government has been trying to reduce these subsidies so that ultimately the market prices prevail and the fertilizer producers become more efficient, and farmers make rational use of the fertilizers. Yet, the effort has not succeeded and there remains a high element of subsidy in the supply of fertilizers to the farmers. Table 13.2 depicts the escalation in the subsidy bill on fertilizers in the country. It is estimated that about 8% of this subsidy goes to Punjab farmers, cultivating only 2.5% percent of the cultivated area in the country.

Another important kind of subsidy is institutional credit. Agricultural credit has remained the major emphasis of the agriculture policy in India, and the

TABLE 13.2 Central Fertilizer Subsidy in India, 1980–81 through 2010–11.

Year	Subsidy (billion INR)	Year	Subsidy (billion INR)
1980–81	5.05	2000–01	138.00
1985–86	19.24	2005–06	184.60
1990–91	43.89	2008–09	966.03
1995–96	67.35	2010–11	499.81

Source: Fertilizer Statistics, The Fertilizer Association of India, New Delhi.

agricultural sector has been prioritized for lending by the banks. A minimum of 18% of the banks' exposure is mandatory for farm sector credit. Some dispensations were given to banks that could not meet this mandatory target. The defaulting banks could place the balance with NBARD (National Bank for Agriculture and Rural Development) or deposit the money with the Reserve Bank of India at a lower rate. NBARD uses this money for agricultural and rural infrastructure and development projects. Also, the farmers under debt have been given loan waivers on several occasions as a relief up to a certain limit. Now the farm loans are available at a 7% interest rate, which is far below the prime lending rates of the banks. Recently, subvention of 3% has been provided for those borrowers who return the loans on time.

Punjab farmers are considered to be under the heaviest debt in the country. The reason is that the lands, being two-crop irrigated lands, provide a high level of eligibility for production credit (crop loans) to the farmers in the state. The marginal, small, and medium size farmers in many cases divert these loans to unproductive purposes, like marriage of their children, medical treatment, or to meet other social obligations. When not used for productive purposes, this credit money does not generate matching repaying capacity, and loans become overdue and turn into bad debt, which sometimes leads to suicides. Paradoxically, whereas farm debt is one of the highest in Punjab, the credit/deposit ratio remains around 40%, which means that for every INR 100 deposited in the banks, Punjab utilizes only about INR 40 as credit from the institutional sources (banks). This is because the depositors are mainly from secondary and tertiary sectors of the state economy, and a considerable amount of deposits are from domestic savings from the cities; the borrowers are mainly farmers and industrialists who are under huge debt in the state.

The most controversial item of subsidy in the state is for water and electric power for tube-wells in the state. This subsidy would form about 38% of the total direct debt of Punjab, amounting to more than INR 87 billion by the end of the financial year 2012–13. The details are given in Table 13.3.

TABLE 13.3 Revenue Expenditure on Subsidy to the Punjab State Electricity Board/Punjab State Power Corporation Limited (PSEB/PSPCL), 1996–97 through 2012–13.

Year	Power Subsidy (billion INR)	Year	Power Subsidy (billion INR)
1996–97	13.38	2004–05	21.70
1997–98	8.73	2005–06	15.51
1998–99	–	2006–07	14.24
1999–2000	4.04	2007–08	28.48
2000–01	6.05	2008–09	26.01
2001–02	4.48	2009–10	28.74
2002–03	7.50	2010–11	33.76
2003–04	13.51	2011–12	42.00

Source: Punjab budget documents, 2012–13

Budgetary estimates of power subsidy in 2012–13 are INR 60 billion. An important aspect of this power subsidy is that it is not reflected in the Minimum Support Prices (MSP) of the commodities that the national government procures from the farmers. These subsidies ultimately land in the pockets of the farmers only for those commodities that are sold in the open market, such as fruit and vegetables, pulses, oilseeds, maize, etc., which the government agencies do not procure. Even though MSP is prescribed for some 24 commodities in the country, effective procurement at the MSP is only for food grains. The Cotton Corporation of India does sometimes procure cotton, but it does so mainly for the government factories.

With the system of determining the minimum support prices, the cost of production estimated by the Commission for Agricultural Costs and Prices plays an important role. The cost of cultivation is estimated based on scientifically selected clusters of about nine thousand farms in the country. The data are collected on a daily basis through a cost-accounting method by regularly appointed supervisors. The data collection and its summary tables are made by the state universities, and is not influenced in any way. These summary tables are passed on to the Economic and Statistical Advisor in the Ministry of Agriculture of the Government of India. The Economic and Statistical Advisor works out the average cost of cultivation for all the crops covered under the MSP and forwards these to the Commission for Agricultural Costs and Prices, which uses these estimates after projecting the costs to the current crops. Such an extensive procedure cannot be ignored in determination of MSP, though other factors like stocks in hand, international (border) prices, consumer price index, etc. do enter into the calculations of the commission.

The important point to note is that the system of cost calculations considers only the actual costs incurred by the farmer himself. No social costs incurred by the government, and suffered by society, is taken into account in the calculations of the commission. The once-a-week power cuts for industry and businesses as well as the domestic sector not only cost the government in terms of diverting the power from paying sectors to the non- paying sector, it adversely affects secondary and tertiary sector production and government tax income from the production and business volumes foregone.

Domestic sector suffering also has huge social costs. These costs calculated in monetary terms cannot be less than INR 50 billion. Adding this 50 billion to the INR 60 billion subsidy would make a total of INR 110 billion worth of direct social costs to the people of Punjab, which would never enter into the calculations of the commission. This subsidy and other related social costs are in fact the very heavy opportunity cost of producing the rice crop in the state. Since these costs are not considered by the commission in the calculation of the cost of production of agricultural commodities, to the extent that the cost of production is considered in the determination of minimum support prices, these prices remain lower than the actual economic cost of production of these commodities. As a consequence of lower procurement prices, the issue prices remain lower. Thus the real benefit of these subsidies given in the form of social costs goes to the consumers of these commodities. The major part of these costs in Punjab are incurred on the growing of rice and also wheat, which are sold at more than 90% of MSP to the government procurement agencies, and these grains are consumed outside of Punjab; the government of Punjab is therefore subsidizing the consumers of the other states, and not their own farmers.

Farmers of Punjab remain under an illusion that they are receiving subsidy in the form of a free electricity supply to the tube-wells. They are in fact serving as a conduit for the power subsidy flowing to the consumers of the other states. Thus, Punjab state is subsidizing the consumers of food grains in other states, not the Punjab farmers, despite the huge social costs incurred every year. There is a need for the farmers to realize that they are living under an illusion, and also for the state government to realize that they are wasting precious resouces, which is placing the farmers and other sectors of the society at a tremendous disadvantage. A consequence is the state's mounting debt, below normal functioning of the secondary and tertiary sectors of the economy, domestic sector suffering, and, above all, a drain on resources for additional production of power and its inefficient transmission to the consumers of power. Farmers' costs also increase, because the inadequate-quality supply of electric power forces them to use diesel pumps, which are a very high cost proposition. Over-exploitation of ground water and its inefficient use because of free power supply has adversely affected the water balance of the state, and the water table in central Punjab is receding by more than 2 feet every year. As a result, the farmers are resorting to submersible pumps, which imposes quite a heavy cost in itself. More importantly, when a submersible pump is installed, the water supply in the adjoining shallow pumps is adversely affected. Thus the small farmers, who may not have sufficient resources for installing the submersible pumps, suffer a great deal. The situation thus becomes unsustainable.

3. POLICY PRESCRIPTIONS

Subsidies play an important role in a welfare state in the context of income transfer to the disadvantaged sections of the society. Yet in order to achieve this objective, the following points need to be adhered to.

- Subsidies must be targeted to reach the deserving beneficiaries only.
- Undifferentiated across-the-board subsidies do not serve the purpose and amount to diverting the flow of scarce resources to unintended recipients.
- The subsidies should be temporary in nature so that vested interests are not created which would try to continue the subsidies beyond their necessity.
- Subsidies should be partial and never absolutely free, which will restrain the user from overuse and misuse of the subsidized items.
- As far as possible, the subsidies should be of an investment type that generate and improve the capacity of the beneficiaries to be independent.
- Subsidies should constitute the bare minimum and should continue only as long as is absolutely necessary.
- Consumer subsidies should be given directly to the intended beneficiaries in the form of food stamps, to avoid pilferage of rationed goods and to provide the consumer with a choice of the food items he/she would like to purchase.
- Farm subsidies should preferably be investment subsidies that improve productivity, production, and efficiency of resource use such as land development, laying of underground water channels, purchase of machinery and appropriate implements, installation of tube-wells, etc.
- Input subsidies that promote over-exploitation of scarce resources, such as the underground water table, and cause environmental degradation such as poisoning of soils and pollution of air and water should be avoided and, if considered essential, should be restricted to the minimum necessary. This category includes supply of electric power, fertilizers, pesticides, etc.
- If the subsidies are necessary for promotional purposes, the subsidies should be partial. No subsidies should be supplied absolutely free. For instance, if electric power for the farm sector is to be subsidized, it should be partial and on metered supply of power. A flat-rate system is as harmful as a free supply, because, after paying flat-rate charges, the user applies no restraint and over-exploits and even misuses the scarce resource.

- Subsidies should be targeted to the small and marginal farmers only. Across-the-board undifferentiated subsidies carry no economic logic. Such a system of subsidies tends to gravitate towards the undeserving large farms.

- It needs to be realized that, for the commodities that are procured by the national government (mainly food grains) at minimum support prices and to the extent that the cost of cultivation of these commodities enters into the calculations for determining these prices, the system does not allow the subsidies to flow into the pockets of producers. Since cost estimates account for only the actual costs incurred by the producer, the cost of production estimates include only the subsidized costs. For subsidies such as power supply, the cost to the farmers is counted as zero. Accordingly the support prices remain low and benefit flows to the consumers in the form of lower issue prices, not to the farmers.

- Input subsidies can also be given to the targeted beneficiaries directly in the form of *security printed input stamps*, and it should be left to the farmer to decide which combination of inputs he would like to purchase with these stamps. In this case market prices should prevail for the inputs used by the farmers. This will target the beneficiaries, improve their input management capacity, and would provide a check on excessive use and misuse of scarce resources.

It needs to be realized that, unfortunately, any benefit to the farm sector in the form of subsidies or higher prices translates mainly into the escalation of land rent, and the actual cultivators, especially the tenants, suffer in the process; hence there is a need to effectively implement the rent control provisions of the law.

Chapter 14

Public-Private Partnership and Policy Reforms for Effective Agricultural Research, Development, and Training: A Viewpoint

Sant S. Virmani

Ex-Principal Scientist, Plant Breeding, Genetics and Biotechnology Division, International Rice Research Institute, Manila, Philippines

1. THE ISSUE

Agriculture is a core subject that requires serious attention from most of the national governments of the world to meet their food, feed, and fiber needs. In developing countries of Asia and Latin America tremendous progress has been made during the past five decades in meeting their agriculture-based needs. This was achieved through the development and dissemination of modern technologies in various crops as well as livestock and deploying certain policy reforms with the support of mainly public institutions. During the past two decades private and NGO sector institutions have also contributed significantly in some countries in this herculean task. Recently, this subject has also been receiving attention in African countries. The impact of these efforts for developing and disseminating various agricultural technologies in different countries has depended on the extent of support and efficiency of their public, private, and NGO institutions, the extent of their mutual collaboration, and policy support provided by the national governments. Past experience has shown that none of these three sectors was adequate and proficient enough individually to meet the gigantic requirements of a country on a sustainable basis. Each sector has its inherent strengths and weaknesses. Countries where maximum success has been achieved in developing agriculture are those where public, private, and NGO sectors have been deployed adequately, collaboratively, and proficiently in such a way that their efforts have been complementary in

Agricultural Sustainability. DOI: http://dx.doi.org/10.1016/B978-0-12-404560-6.00014-9
263

meeting their needs on a sustainable basis. This note is intended to emphasize to national and state governments the significance of public-private and NGO partnership and the associated policy reforms that would help them to develop more effective agricultural research, training, and development systems to meet their food, feed, and fiber needs on a sustainable basis.

2. STRENGTHS AND WEAKNESSES OF PUBLIC, PRIVATE, AND NGO SECTORS

Different countries in the world have different capabilities to develop and disseminate the required agricultural technologies to meet their national objectives by utilizing their available resources in public, private, and NGO sectors. The net results of these efforts depend upon how adequately and skillfully these resources are developed and deployed, recognizing their strengths and weaknesses. Generally speaking, in developing countries the public sector is stronger than private and NGO sectors in mobilizing the required resources for generating agricultural technologies, although this sector is weak in the optimization of dissemination and utilization. On the other hand, private and NGO sectors are generally more efficient in dissemination of agricultural technologies than is the public sector, on account of their less bureaucratic procedures for mobilizing and deploying resources to meet their objectives. Between private and NGO sectors, the former is more capable in mobilizing the required resources for technology generation and wide scale dissemination. The latter, operating at the grass root level, has the advantage in evaluating and disseminating a new technology compared with both public and private sectors. Policy support is the mandate of the public sector and is the key to mobilization and deployment of resources of all the sectors in a country for development and dissemination of agricultural technologies. Considering these realities, it is very important that national governments should mobilize and deploy all three sectors in such a way that their strengths are maximized and weaknesses are minimized for developing and disseminating the required agricultural technologies which would meet their goal of producing sufficient food, feed, and fiber on a sustainable basis.

Development and dissemination of appropriate agricultural technologies requires suitably trained and adequate human resources to do various research and extension tasks. Traditionally, the public sector has been involved in basic and strategic research and has been playing a key role in meeting the training needs in most countries. Private and NGO institutions depend mostly on public sector institutions to meet their human resource needs. However, most of the fresh graduates have little idea about the working environment of private and NGO institutions involved in applied research; hence they are required to spend some time in orientation to the new working environment of these sectors and their work culture before starting actual assignment on their job. With active participation of private and NGO sectors in the development and dissemination of agricultural technologies, it would be useful to establish their collaboration with the public sector so that these sectors would get the required human resources appropriately trained for developing and disseminating suitable agricultural technologies in a country. In such a collaborative arrangement, public institutions can also organize training courses as per the needs of the private and NGO sector institutions at the latter's cost. In return, private and NGO sectors can acquaint the public sector students with their unique working environment and its specific needs so that they can be immediately put on the job after finishing their training.

3.   INTERNATIONAL COLLABORATION

During the past five decades international collaboration under the leadership of the international agricultural research and development organizations (CGIAR centers, FAO and other UN agencies) have played a pivotal role in the development and dissemination of agricultural technologies around the world, especially in developing countries. Bilateral collaboration between advanced countries and developing countries has also been more effective in the development of agricultural research, training, and extension activities in several countries with technical support of CGIAR Centers and policy support of national governments. With future challenges of increasing demand for agricultural products on a sustainable and environment-friendly basis with decreasing agricultural land and water resources, it is important to recognize that intricacies of agricultural research, training, and extension are also increasing. There is a great need to equip young researchers with cutting-edge technologies. In order to meet this challenge, international collaboration can also be utilized as much as possible in strengthening public-private sector partnership for the required effective agricultural research, training, extension, and policy support systems. Some countries, notably China, have taken it seriously and deployed policy tools to provide fully funded scholarships to talented youth to obtain the necessary research experience abroad in advanced countries and CGIAR Centers and also ensuring that they return and contribute towards nation building. Such programs have a great potential in countries like India, which possess a big resource of ambitious youth. The government of India has initiated some brain-gain programs aimed at attracting non-resident Indian scientists already trained in advanced countries. However, there is still a great scope to assist young scientists in attaining appropriate training abroad to further strengthen the science and technology sectors.

4.   MECHANISM(S) FOR ESTABLISHING PUBLIC, PRIVATE, AND NGO PARTNERSHIP

A major challenge for establishing public-private partnership (PPP) lies in identifying a suitable mechanism. Each country or its units may need to develop and/or design its own suitable mechanism for establishing such a partnership depending on their political, administrative and social structure. The author proposes the following guidelines for establishing such mechanisms based on his observations and experience obtained while working with different countries.

- The participating partners should enjoy equal respect and authority to present their viewpoints. Many attempts to establish PPP do not make the desired progress because public sector representatives tend to treat the representatives of other sectors with bias and suspicion.
- In order to develop model(s) for PPP, several small group meetings should be held at national and sub-national levels involving key decision makers of each sector at these levels. Each meeting should be moderated by a third party having no vested interest in the technology for which the partnership is being established and it should aim to come up with concrete decisions that would contribute towards developing the desired model of partnership. Many such meetings in the past have resulted in highlighting the needs of such a partnership rather than specifying a mechanism for attaining them.
- The third party can also establish collaboration with appropriate agencies, including international agricultural centers, to learn about the models of public-private partnership operating successfully in other countries and share these with the target countries with constructive modifications, if necessary, to suit local conditions. It should also attempt to identify and/ or validate issues constraining PPP in the country of operation and monitor the progress of the established PPP.
- The international agricultural research centers and selected advanced countries should be utilized under the PPP to develop human resources of a country, on a sustainable basis, in the required advanced agricultural technologies to meet the current and future challenges of agricultural research and development.
- The ultimate goal of the partnership should be specified, keeping in view the interests of farmers, consumers, private sector, and public sector. Roles of each sector in developing, evaluating, and disseminating a technology should be clearly defined and agreed upon in the model(s) developed and agreed upon by the participating sectors.
- If necessary, an interim model of partnership should be developed and tried at the targeted level with a mutual understanding that it can be modified and improved over a period of time based on the experience gained during its implementation.
- The model should be one in which all the partners would be willing to invest their resources to the maximum to meet their interest as well as national interest.

5. POLICY SUPPORT

For the success of public-private partnerships in serving the interest of farmers, consumers, and the public-private sector partners, it is very important that the national and/or state governments should develop and provide appropriate and unbiased policy support that would motivate participating partners to do their best to develop and disseminate a technology. The national policy support should also help to strengthen linkages with international centers and selected advanced countries for the purpose of training the required human resources in the advanced agricultural technologies. If necessary, the private sector should also help financially in this endeavor. The jointly developed model of partnership should be a win-win situation for all the participating partners, stakeholders, and the country at large. Development and supply of such a policy support is the responsibility of the concerned government operating at a national or regional level and is the key to the success of public-private partnership.

Chapter 15

Contract-Farming for Production and Procurement of Mint—Lessons from Personal Experience

Tarlok S. Sahota

Thunder Bay Agriculture Research Station, Thunder Bay, Ontario, Canada

1. INTRODUCTION

Contract-farming is a production and marketing/procurement system wherein producers agree to grow a crop (produce) at a pre-agreed market price for procurement by another party, usually a public or private company/corporation. Both the producers and the buyers are bound by a written and signed contract agreement that specifies the terms and conditions of the relationship between the two parties, including, but not limited to, the procurement prices. Contract farmers are sometimes also called "out-growers" when they are linked with a large farm or processing plant which supports production planning, input supply, extension advice, and transport. The contract-farming approach is used for a wide variety of agricultural products (FAO, 2008). Such farming/procurement systems are also referred to as "corporate farming", especially when the corporate houses have their own centralized crop production units and they need to supplement their production/ requirements through out-growers. However, corporate farming may or may not necessarily include out-growers. This note summarizes my own experience in organizing and managing contract-farming programs within and outside India.

2. EFFECTIVE COMMUNICATION—FOUNDATION FOR A GOOD START

In general, farmers are often cautious, if not suspicious, about the sincerity of private companies, especially with respect to their commitment and honoring of contracts, more so at the initial stages of a program. It is important to explain

Agricultural Sustainability. DOI: http://dx.doi.org/10.1016/B978-0-12-404560-6.00015-0

what the company and its program is about, what it could provide in terms of benefits to farmers, and what would be expected of farmers, through farmers' meetings and media. It is certainly a good idea to employ the services of a qualified, experienced, and knowledgeable person who has roots in the farming community of the area of operation. Such a person would be trusted more than an outsider. During initial meetings, farmers (particularly those in developing countries) will test the knowledge of the development staff by asking them questions to which they already know the answers. Not being able to answer their questions will mostly be perceived as lack of knowledge, no matter how insignificant or irrelevant these questions may sound. For example, at a district level farmers' meeting, jointly organized by the Department of Agriculture, Punjab, a private company, and the Punjab Agro Industries Corporation (PAIC) of Chandigarh, India, to promote a contract-farming program of the PAIC, one of the farmers asked a young (relatively inexperienced) company manager how much grains on a single cob of corn would weigh. He was unable to answer the question. I was also at the meeting representing the PAIC as a consultant to promote their contract-farming program, aimed at diversifying the cropping systems in the Punjab State of India. I was able to answer the farmer's question satisfactorily by saying that on average the grains per cob would weigh about 100 grams. It is obvious that grain weight on a cob could vary from cob to cob, but even then there would be an average cob weight. My answer was based on a quick calculation of the local corn grain yield divided by the plant population. I knew that, under local conditions, corn plants would have on average 1.1 cobs per plant. Apart from the scientific knowledge, presence of mind and common sense proves a significant asset in such meetings.

Paying attention to the local customs and traditions is another important point, which must be considered before approaching farmers or organizing farmers' meetings, particularly while working in a cross-cultural context. For example, before organizing a contractual cotton production and procurement program in the Gangola State, Nigeria, I was aware that it was a custom to meet traditional area chiefs and seek their permission/blessings before starting a program in their area, even if the program had state and/or Federal Government's approval. This applies all the more so when one is a foreigner. I went to the area chief with gifts (as per the custom) and mentioned that I came from a culture that teaches respect for elders and hence came to pay my respects and seek his blessings. Finding a common thread with the new/ foreign area's community helps a lot. Since I am from a progressive farming community myself, that common thread has always been there.

3.   START ON A LOW SCALE AND BUILD CONFIDENCE IN THE COMMUNITY BEFORE EXPANDING

In late 1996, I took on responsibility as Director Mint Development to initiate and develop a contract mint farming and procurement program in the State of Punjab (India) for a leading global mint company. I inherited five acres of nursery scattered at five places across the state, and two staff who were not from Punjab. It took me two days in a car to visit those five acres, but within only five years we were able to expand the program to over 10,000 acres with some 2,200 farmers growing mint and nearly two dozen field, office, and warehouse staff. In 1997, I started at two locations—Tohra (District Patiala) and Dyalpura Mirza (District Bathinda)—with around 100 acres each at both the locations. These locations weren't necessarily ideal, but the company strategically opted to be somewhat removed from the traditional mint-growing areas. I started with two mint species—spearmint and peppermint (which were relatively alien to Punjab)—at both the locations in order to ensure that, even if one species were to fail, there would not be a total program failure. Farmers in Tohra liked spearmint more than the peppermint, and the reverse was true for the other location. At both the locations, we installed two small tank steam distillation units to instill confidence in the farmers (that the company was there to stay) and also to convince the management of the company to provide greater investment in large-scale distillation units, which was eventually done, based on the success in a small area.

4.   CLUSTER APPROACH FOR EXPANSION

I adopted a cluster approach for promoting the program in a firm and sustainable manner. I still recall how I then explained my approach of expansion to my manager, Scott M. Bolton (who is a gem of a person and was more of a friend than a boss): as a small pebble, when dropped in a pond of water, makes a little dip mark and initiates the waves that spread towards the shore, in a similar manner I said, I would pick up locations and spread around those locations. This approach worked very well for us and by 2001–2002 we had nearly a dozen such clusters, though Jagraon in District Ludhiana remained our biggest cluster.

5.   OVERCOMING OUR OWN INHIBITIONS FOR EXPANSION

Peppermint and spearmint planting stock was imported from the USA, and multiplied by tissue culture at the biotechnology laboratory of Punjab Agricultural University, Ludhiana, India. The company was very possessive of the imported quality stock and was reluctant to pass it on (especially to distillers associated with our competitors). I was able to overcome this inhibition, which was getting in the way of expanding the program on a commercial scale. Being from the area, I was well versed and could identify distillers that could be relied upon. Inclusion of such distillers was also a necessity to minimize capital investment by the company. At the same time, company investment was also required to keep distillers under control; in establishing new distillation units, distillers were given a subtle message that the company could erect mint distillation units in their areas and create competition for them if they dishonoured contracts. Thus the competition was turned into cooperation by involving outside distillers who got normal distillation charges from the company and the assured procurement and procurement prices.

6. EXPANSION WAS REQUIRED TO CONQUER OPPOSITION FROM WITHIN THE COMPANY

The company for which I worked was the top international mint company of the time, known for supplying quality mint oils and blends to its customers (mainly big cosmetic and household product manufacturers such as Colgate-Palmolive) who compelled it to move the production and procurement to India. Because of the existing malpractices in the trade, Colgate, for example, would not like to buy oil from local traders. However, direct procurement from the market, based on quality tests, could be cheaper than procurement through contract-farming, especially during the years when the production of mint oils was high and prices were low, and more so at the initial stages of the program when overhead costs were relatively high. Thus the procurement section from within the company would usually oppose the contract-farming program; this was despite the fact that direct procurement from the market could not ensure stable supplies over the years, mainly because of year-to-year fluctuation in the acreage under mint farming. I was able to win over the opposition from within the company by expanding the program quickly; the increased oil production and procurement helped in lowering the overhead cost per unit of oil. *Mentha arvensis* (Japanese mint), which gave 1.5–2.0 times higher oil yield than peppermint or spearmint, was added to the program to increase the procurement oil

TABLE 15.1 Contract Mint Oil Farm Gate Prices (2000–2001)

Mint Species	Time/Type of Planting	Oil Price	
		(USD/kg)[a]	(INR/kg)
Peppermint	October–December	13.57	570
	January	12.62	530
	February–March	11.43	480
	Intercropped in other crops	11.90	500
Spearmint/scotchmint	Up to mid February	11.90	500
	After mid February	11.43	480
Japanese mint	Up to mid February	7.14	300
	After mid February	5.95	250

Source: Sahota (2000).
[a] *At INR 42 = 1 USD. At start of the program in 1997, farm gate mint oil prices were consistent for all planting times: INR 570, 500, and 300/kg for peppermint, spearmint/scotchmint, and Japanese mint, respectively. Distillation charges @ USD 1.67 were cost free to farmers.*

volumes. However, because of the relatively low oil yield from peppermint and spearmint, procurement prices for their oils were kept competitive with those of Japanese mint (see Table 15.1), to provide incentive to the farmers.

7. BALANCING COMPANY'S AND FARMERS' INTERESTS

One of the biggest challenges in contract-farming programs is to strike a balance between the company's and the farmers' interests. The company's customers would like to buy oils at low prices because of cut-throat competition from their competitors. On the contrary, because of yearly increase in input costs, farmers would like the procurement prices to be increased every year. Four years after our initiation of the contract mint program, I was asked to lower the procurement price from the agreed INR 570/kg peppermint oil to INR 550/kg by the new Managing Director of the company. I couldn't afford to lose the trust of my farmers by going back on my commitment; I therefore refused to do that. However, I did find ways to bring down the overall peppermint oil procurement price by:

- Lowering the price for the late-planted crop, for which oil quality was lower, by convincing farmers and taking them into confidence.
- Lowering the procurement price for peppermint intercropped in green peas, a practice innovated and promoted by us. Lower oil prices were compensated for by increased mint oil yield as a result of the synergistic effects of intercropping, and the green peas—harvested before mint picked up growth—brought in additional income to farmers.
- As mentioned before, contract production/procurement volumes were increased to lower overhead costs per kg oil.

By 2000, the company was able to make substantial savings by procuring contract mint oils as compared with direct procurement from the market in other parts of India.

8. RESEARCH AND DEVELOPMENT SUPPORT FOR QUICK EXPANSION

My background and experience in agricultural research helped me in conducting development-oriented research, both on-station (company owned/operated research plots) and on-farm. If the findings from a particular technique/practice were the same at the company-operated research plots and also at farmers' fields, that was immediately passed on to farmers for adoption. In a few years, we revolutionized the way farmers cultivate mint crops in Punjab. The innovative ways of growing mint not only increased oil and planting stock yield, but also improved oil quality in addition to making it possible to grow mint in otherwise unsuitable areas. For example,

TABLE 15.2 Effect of Methods of Planting on First Cut Peppermint Oil Yield at Bareilly, UP, India

Method of Planting	Oil Yield (kg/acre)
Broad bed and furrow 45–90–45 cm	28.3
Ridges 60 cm apart	25.4
Ridges 60 cm apart, covered with manure	26.8
Flat 45 cm rows	22.6

Source: Sahota (2000).

the Indomint method (my own innovation) of planting rootstock at the base of broad ridges 45 cm thick and 90 cm apart, rather than the existing practice of planting mint on 60 cm × 60 cm ridges, enabled us to successfully introduce mint to sandy/salty soils (Table 15.2). With evaporating water, the soil salts would go to the top of the ridge, leaving the roots free of excessive salts; the roots would proliferate inside the ridges where top fertile soil had accumulated and would increase root and oil yield. Another innovation, the "Dr. Sahota method" of planting mint tops, one in each of the four corners of small square beds of 90 cm^2, surrounded by furrows, made by running a potato ridger (alternate tine removed) in two opposite directions, helped to increase l-carvone (key constituent in spearmint) from 55% (normal) to 70%. Such high carvone oil could be blended with low carvone oil, mainly from late-planted or intercropped spearmint, to maintain the overall quality at a superior level. Mint plants proliferate best under semi-aquatic conditions, which were provided by this and the "broad bed and furrow" method. Irrigation in the furrow supplied moisture, and the middle part of the bed provided dry loose soil for growth of root/suckers and, being broad, ample space was available on the bed for intercrops such as peas and onions. The practice also economized on the precious and scarce water source. Initially developed only for peppermint, this method of planting on "broad bed and furrow" was soon applied to other species of mint as well. The success of this method could be estimated from its quick adoption by the farmers, i.e. from an acre under this method in 1997 to about 6,700 acres by 2001–2002 (two thirds of the total contract mint acreage).

Some other important innovations developed under my supervision are as follows.

- Using 10 cm tops of mint plants for planting in the "Dr. Sahota method" and "broad bed and furrow" method in moist soil after furrow irrigation. Before this, mint had always been planted by rootstock that would not be

TABLE 15.3 Effect of Methods of Planting on Oil Yield of Peppermint and Japanese Mint (Total from Two Cuts at Khasi Kalan District Ludhina, Punjab, India)

Method of Planting	Oil Yield (kg/acre)	
	Peppermint	Japanese Mint
Broad bed and furrow	25	45
Dr. Tarlok Singh Sahota method	28	54

Source: Sahota (2000).

ready for planting till January–February. The practice of planting mint by "tops" (i) enabled planting of peppermint early (October–December), which increased crop length and improved both oil yield and quality (Table 15.3), (ii) helped in spreading the planting period, thus making effective use of available labor and resources, and (iii) made it possible to divert some areas from wheat to peppermint, which was quite a challenging task. Previously, attempts had been made to prolong the crop duration by delaying harvesting, but it didn't work, because the crop would be caught in rains with the resultant reduction in both oil yield and quality.

- Preparing nursery plantlets of Japanese mint by planting rootstock in small poly bags in January helped to utilize the area vacated by late-harvested potatoes for planting in March. The yield of the crop thus raised from nursery was comparable in every sense to the crop planted from rootstock at optimum planting time in January.

- New intercropping systems such as: (i) spearmint in first and second year sugarcane, which would be harvested before sugarcane picked up growth. Frequent irrigations to spearmint improved growth of both spearmint and sugarcane, and the effective ground cover provided by spearmint helped to suppress weed growth at initial stages of sugarcane; (ii) intercropping peppermint in October-planted sugarcane, taking advantage of slow growth of sugarcane in winter months, and in early spring/and summer; (iii) intercultivation of onions, garlic, and green peas that would be harvested before mint picked up growth. Mint growth during early (winter) months is mostly lateral and/or underground. These intercropping strategies provided additional income to growers and saved on area required for intercrops, if they were to be cultivated as sole crops (Table 15.4).

- Farm trials on farmers' fields helped to demonstrate higher profitability from the mint crops and/or mint-based cropping systems (Tables 15.5 and 15.6).

TABLE 15.4 Returns per Acre from Intercrops in Fall-Planted Sugarcane in a Trial on a Farmer's Field in District Faridkot, Punjab, India (1999–2000)

Intercrop	Duration	Yield (kg/acre)	Gross Income (INR)	Cost of Cultivation (INR)	Net Income (INR)	Net Income (USD)
Onion	Nov 12 to Apr 30	3,200	9,600	8,600	1,000	23.8
Gobhi sarson/ mustard	Nov 12 to Apr 12	700	8,400	1,880	6,520	155.2
Wheat	Nov 12 to Apr 15	2,000	11,600	2,497	9,103	216.7
Peppermint	Nov 12 to Jun 15	28	15,960	4,165	11,795	280.8

Source: Sahota (2000).

TABLE 15.5 Returns Per Acre from Rabi (Winter) Crops at Simbli, District Hoshiarpur, India (1999–2000)

Crop	Duration	Yield (kg/acre)	Gross Income (INR)	Cost of Cultivation (INR)	Net Income (INR)	Net Income (USD)
Pea	Oct 10 to Jan 10	2,100	12,600	7,000	5,600	133.3
Potato	Sep 25 to Dec 12	8,000	12,000	10,500	1,500	35.7
Sunflower	Feb 20 to May 30	900	9,000	4,500	4,500	107.1
Wheat	Nov 15 to Apr 27	1,900	12,600	5,200	7,400	176.2
Peppermint[a]	Jan 3 to Jun 14	2,500	14,250	6,450	7,800	185.7
Peppermint[b]	Jun 15 to Dec 31	6,400	24,000	4,000	20,000	476.2

Source: Sahota (2000).
[a] For oil production.
[b] For rootstock production

TABLE 15.6 Net Return Per Acre from Crops Following Potato or Peas at Simbli, District Hoshiarpur, India (1999–2000)

Crops	Duration (two crops)	Net Return per Acre	
		INR	USD
Potato–sunflower	Sep 25 to May 30	6,000	142.9
Potato–wheat	Sep 4 to April 27	8,900	211.9
Potato–peppermint	Sep 25 to Jun 14	9,300	221.4
Peas–sunflower	Oct 10 to May 30	10,100	240.5
Peas–peppermint	Oct 10 to Jun 14	13,400	319.0

9.   FARMERS FIRST

Being primary producers, farmers are at the mercy of procurement agents/traders in many marketing systems, particularly in developing parts of the world. A contract-farming system largely removes many of the uncertainties associated with marketing of the produce by the farmers. However, agricultural production being largely weather dependent, and particularly the quality parameters being highly variable, there always remains scope for disappointment of farmers. A transparent and honest procurement and payment system serves as a morale booster for the farmers and earns their loyalty. I adopted a farmer friendly approach under which payments to farmers were made on a priority basis and, at times, trusted growers and distillers in need would also be paid in advance. Farmers with record-breaking mint oil yields were awarded/honored in our annual farmers' meetings in which most, if not all, contract growers/distillers would participate. This raised the self-esteem of such farmers and enabled others to learn from them. Recognition and deserved appreciation at appropriate time(s) could be valued more than money. The farmers' loyalty resulting from these efforts was evident in a year with very high mint oil prices. Despite the price in the open market being above the contract prices, only 4 out of 2,200 farmers dishonored the contracts.

10.   MEETING QUALITY GOALS

Due to climatic differences, the quality of the mint oils produced in Punjab was not comparable to that produced in the relatively cooler Willamette Valley, Oregon, which was the original production hub of the company in the USA. In the search for superior quality mint oil, we conducted mint production and distillation trials in cooler areas outside Punjab. Although we could identify some areas producing mint oil of quality comparable to that from the Willamette

TABLE 15.7 Effect of Time of Planting on First Cut Oil Yield (kg/acre) of Peppermint and Spearmint at Bareilly, UP, India

Time of Planting	Peppermint	Spearmint
October 1	35.3	–
November 1	35.7	–
December 1	33.0	37.2
January 1	30.2	32.0
February 1	23.0	27.0
March 1	19.8	20.8

Source: Sahota (2000).

TABLE 15.8 Effect of Delay in Irrigation on First Cut Peppermint Oil Yield at Bareilly, UP, India

Irrigation	Oil Yield (kg/acre)
Normal; no delay	24.1
3–5 days delay in 3rd irrigation	22.3
3–5 days delay in 3rd & 4th irrigation	21.4
3–5 days delay in 3rd, 4th, & 5th irrigation	19.6

Source: Sahota (2000).

Valley, the yield was too low to support a commercially viable contract mint-farming program. There are also some areas within Punjab where relatively better quality mint oils could be produced, but smaller land holdings limited our expansion in these areas. An alternative was to upgrade the quality of oil by blending mint oils. While adding active constituents, e.g., "Menthol", to peppermint or Japanese mint oils to raise otherwise low "Menthol" content in these oils would be considered as an adulteration and thus not permitted/acceptable, mixing the same species oil from one location to another is not considered an adulteration. We could balance the active constituents in the mint oils by blending oils of the same species from different locations. This type of blending was done in the USA as well by blending oils from India and the USA. We also focused on agronomic management to improve mint oil quality by adopting

TABLE 15.9 Effect of Nitrogen on Peppermint Oil Yield at Bathinda, Punjab, India

N (kg/acre)	Oil Yield (kg/acre)
69	19.1
92	30.0
115	22.5
138	23.1

Source: Sahota (2000).

TABLE 15.10 Effect of Phosphorus on Peppermint Oil Yield at Bathinda, Punjab, India

P_2O_5 (kg/acre)	Oil Yield (kg/acre)
46	19.1
69	25.3
92	35.8

Source: Sahota (2000).

TABLE 15.11 Effect of Zinc on Peppermint and Spearmint Oil Yield (Total from Two Cuts at Bareilly, UP, India)

Zn (kg/acre)	Oil Yield (kg/acre)	
	Peppermint	Japanese Mint
0	30.7	35.1
2	30.8	37.4
4	34.3	40.6

Source: Sahota (2000).

practices that promoted growth and minimized stresses (early planting, new methods of planting, cropping systems, fertilizer nutrients, timely irrigation, insect-pests control, time of harvesting, etc.; Tables 15.7 to 15.12). The stressed crop would produce undesirable oil constituents impacting on quality and odor. We applied strict checks to critically analyze the oil for quality components,

TABLE 15.12 Optimum Time for Harvesting January-Planted Peppermint at Bareilly, UP, India

Harvesting at	Oil Yield (kg/acre)	
	Peppermint	Japanese Mint
24–25 nodes	28.7	34.6
Cumulative daily temperature 2300°C	32.6	35.1
Cumulative daily temperature 2450°C	35.0	39.9
Cumulative daily temperature 2600°C	33.7	34.5

Source: Sahota (2000).

and appropriate corrective measures were taken as and when required. For example, replacing the old-style rusty iron condensers and separators with aluminum ones at our distillation units and at some of our contract distillers' units helped to keep the mint oils clean and retain their original natural color.

11. ADVISORY AND EXTENSION SERVICES

Pocket size brochures containing necessary information on practices for every stage of mint production, from pre-planting to harvesting and distillation, were provided to the farmers. Besides, popular Punjabi newspapers and farm magazines willingly published my articles, which helped to spread technical need-based and pertinent messages to all in good time (Sahota, 2000). Regular field visits and staff meetings were conducted on the fields on a rotational basis, which helped in the development of human resources as well as in maintaining close contact with the farmers.

12. SYNERGY WITH OTHER ORGANIZATIONS

While organizations working in the same market often have competing interests, there might still be some scope to collaborate or to find synergies. Another major player, Hindustan Lever Ltd, was involved in contract cultivation and procurement of *Basmati* rice in and around Jagraon, District Ludhiana, Punjab. As *Basmati* matures and vacates fields later than coarse grain paddy rice, sowing of a subsequent wheat crop is delayed, resulting in yield loss. However, this fitted well with the plantation of peppermint early in the fall. *Basmati* rice is one of the crops that respond well to soil organic matter; crown and rootstock of peppermint provided easily decomposable organic matter, thereby improving the overall productivity from the *Basmati*–peppermint cropping system.

The staff of Hindustan Lever Ltd. willingly provided us with a list of their contract Basmati rice growers who could be approached and/or persuaded to plant mint after Basmati rice. Working in association has not only provided positive synergies to both the companies but has also benefited farmers.

13.   CONCLUDING REMARKS

Contract-farming may prove a useful approach in promoting a particular crop and/or cropping system, as the ensured procurement encourages the farmers by overcoming marketing and price-related uncertainties. It also facilitates the provision of focussed training and guidance to the farmers for adoption of innovation(s). However, the alternative crop/technology promoted in contract-farming programs has to be economically more rewarding than the conventional crops/practices; otherwise such programs will not be viable and sustainable. For example, a huge contract-farming program was launched by Punjab Agro Foodgrains Corporation (PAFC) to diversify the cropping pattern in Punjab by replacing an eco-damaging paddy rice-wheat system with alternative crops (such as maize, Basmati rice, durum wheat, barley, mungbean, guar, canola, and sunflower) that could be less water-intensive, more eco-friendly, and economically more rewarding. After running a successful contract-farming program with mint, my services were sought to plan and support this contract-farming program. After overcoming the initial "hiccups", the program proved successful and spread from around 9,030 hectares in 2002 to over 102,400 hectares towards the end of the last decade. However, withdrawal of concessions in September 2010, coupled with considerable increase in the Federal Government's minimum support price for paddy and wheat, had an adverse effect on further spread of the contract-farming area under the alternative crops (http://www.punjabagro.gov.in/pafc-c-farming.html). No wonder PAFC's contract-farming area dropped to ~34,000 hectares in 2010–11 and to as low as ~4,800 hectares in 2011–12. It is evident that farmers are willing to try to adopt sustainable technologies. However, lack of policy support is the biggest constraint to agricultural change, especially in the programs led by the government/public sector organizations. In the case of private organizations, focus on short-term profits sometimes promotes malpractices such as inconsistency in policies and fake commitments to the farmers, which could prove lethal for the program. I had to leave the mint company in early 2002; thereafter in the hands of new management the company didn't honor the contract prices/commitments and died its own death.

Farmers all over the world are hard-working and generous. They will be willing to cover an extra mile for someone who is respectful and sincere to them. I salute farmers and God who created farmers to feed the world!

ACKNOWLEDGEMENTS

The author is grateful to Jim Todd, the then Vice President of A. M. Todd Company USA, Scott Bolton, the then Director of International Field

Operations, A. M. Todd Company USA, Bhagirath Biyani, MD Indomint Agriproducts Ltd till 2000, field and support staff, farmers and distillers for their untiring efforts and wholehearted support to the contract mint-farming program. And, above all to the Lord Almighty for giving me strength and abilities to successfully establish contract-farming programs.

REFERENCES

FAO, 2008. Contract Farming. Contract Farming Resource Centre. FAO, Rome, 2008.

Sahota, T.S. 2000. Mint cultivation. Indomint Agriproducts Limited, SCO 2475–76, Sector 22-C, Chandigarh, India. [The booklet is a compilation of the author's articles on mint in Punjabi dailies and farm magazines].

Index

Printed in the United States
By Bookmasters